# Lecture Notes in Physics

## Volume 1039

W0225637

The series Lecture Notes in Physics (LNP), founded in 1969, reports new developments in physics research and teaching - quickly and informally, but with a high quality and the explicit aim to summarize and communicate current knowledge in an accessible way. Books published in this series are conceived as bridging material between advanced graduate textbooks and the forefront of research and to serve three purposes:

- to be a compact and modern up-to-date source of reference on a well-defined topic;
- to serve as an accessible introduction to the field to postgraduate students and non-specialist researchers from related areas;
- to be a source of advanced teaching material for specialized seminars, courses and schools.

Both monographs and multi-author volumes will be considered for publication. Edited volumes should however consist of a very limited number of contributions only. Proceedings will not be considered for LNP.

Volumes published in LNP are disseminated both in print and in electronic formats, the electronic archive being available at springerlink.com. The series content is indexed, abstracted and referenced by many abstracting and information services, bibliographic networks, subscription agencies, library networks, and consortia.

Proposals should be sent to a member of the Editorial Board, or directly to the responsible editor at Springer:

Dr Lisa Scalone
lisa.scalone@springernature.com

Jean-Pierre Minier • Martin Ferrand •
Christophe Henry

# Understanding Turbulent Systems

## Progress in Particle Dynamics Modeling

 Springer

Jean-Pierre Minier
CEREA, EDF R&D, ENPC
Institut Polytechnique de Paris
Chatou, France

Martin Ferrand
CEREA, EDF R&D, ENPC
Institut Polytechnique de Paris
Chatou, France

Christophe Henry
Inria, CNRS, Calisto team
Université Côte d'Azur
Sophia Antipolis, France

ISSN 0075-8450       ISSN 1616-6361   (electronic)
Lecture Notes in Physics
ISBN 978-3-031-84465-2      ISBN 978-3-031-84466-9   (eBook)
https://doi.org/10.1007/978-3-031-84466-9

This work was supported by Électricité de France.

This Springer imprint is published by the registered company Springer Nature Switzerland AG
The registered company address is: Gewerbestrasse 11, 6330 Cham, Switzerland

If disposing of this product, please recycle the paper.

# Preface

The purpose of this book is to present the statistical description of turbulent poly-dispersed two-phase flows based on the probability density function (PDF) approach. Adopting the point of view of the physicist, the presentation focuses on the analysis of the physical content of stochastic formulations used to model the dynamics of discrete particles in non-fully resolved turbulent flows. We follow a step-by-step approach to introduce this multi-scale and multi-physics topic, and bring out current challenges to emphasize not only how but why PDF models are developed. By investigating state-of-the-art models, the book also invites researchers to address the open issues presented and, to that effect, new ideas are proposed.

Starting with examples from daily situations and covering the basic equations as well as going through the reasons calling for a statistical treatment, this book can serve as an introduction for readers not yet familiar with the topic. Since we provide detailed accounts of the specific challenges we are faced with when considering statistical descriptions of discrete particle dynamics in random media with non-zero time and space correlations, the book is also intended for practitioners in the field. Finally, new ideas and cutting-edge formulations are developed in the hope to overcome present limitations and embolden specialists to pursue their own views.

The book is structured accordingly. Chapters 1–3 are dedicated to setting the stage, introducing examples as well as fundamental equations, while Chap. 4 is a reminder of the mathematical background for readers less acquainted with the probabilistic framework. Chapters 5–7 describe the rationale behind the elaboration of present PDF models together with details on their formulations and numerical applications. New models are then developed for the velocity of the fluid seen in Chaps. 8 and 11. After indicating connections with soft matter in Chap. 9, new results for the study of model reduction are given in Chap. 10. Specific issues about particle collisions are addressed in Chap. 12 before proposing a conclusion.

Chatou, France
Chatou, France
Sophia Antipolis, France
December 2024

Jean-Pierre Minier
Martin Ferrand
Christophe Henry

**Competing Interests** The authors have no competing interests to declare that are relevant to the content of this manuscript.

# Contents

# Acronyms

## List of Abbrevations

| | |
|---|---|
| BBO | Basset-Boussinesq-Oseen |
| BBGKY | Bogoliubov–Born–Green–Kirkwood–Yvon |
| BG | BackGround (Model Without Structures in a Flow) |
| CFD | Computational Fluid Dynamics |
| CTE | Crossing-Trajectory Effect |
| DEM | Discrete-Elements Models |
| DNS | Direct Numerical Simulation |
| DPD | Dissipative Particle Dynamics |
| DSMC | Direct Simulation Monte Carlo |
| FDF | Filtered Density Function |
| FENE | Finitely-Extendable-Nonlinear-Elastic (model) |
| FND | Furutsu-Novikov-Donsker |
| FP | Fokker-Planck |
| GENERIC | General Equation for Non-Equilibrium Reversible-Irreversible Coupling (approach) |
| GLM | Generalized Langevin Model |
| HIT | Homogeneous Isotropic Turbulence |
| IP | Isotropization of Production |
| K41 | Kolmogorov 1941 (Theory) |
| K62 | Kolmogorov 1962 (Theory) |
| LES | Large-Eddy Simulation |
| LMDF | Lagrangian Mass Density Function |
| LFDF | Lagrangian Filtered Density Function |
| LRR | Launder, Reece and Rodi |
| LRR-IP | Launder, Reece and Rodi Model Based on Isotropization of Production |
| MCAC | Monte Carlo Aggregation Code |
| MD | Molecular Dynamics |
| MDF | Mass Density Function |
| ms | Mean-square |
| NS | Navier-Stokes |

| ODE | Ordinary Differential Equation |
| OU | Ornstein-Uhlenbeck |
| PDE | Partial Differential Equation |
| PDF | Probability Density Function |
| RANS | Reynolds-Averaged Navier-Stokes |
| SDE | Stochastic Differential Equation |
| SDPD | Smoothed Dissipative Particle Dynamics |
| SGS | Sub-Grid Scale |
| SLM | Simplified Langevin Model |
| SPH | Smoothed Particle Hydrodynamics |
| VT | Vortex Tube (Model with Structures in a Flow) |

## List of Symbols

The following convention is used throughout the book: bold symbols indicate vectors (e.g., $\mathbf{A}$) or tensors (e.g., $\mathbf{B}$). The components within a vector/tensor are designated with subscripts (like $\mathbf{A}_i$, or $\mathbf{B}_{ij}$). In addition, we rely on Einstein's summation convention to imply summation over all the values of the index (for instance to mean that $A_{ij}X_j = \sum_j A_{ij}X_j$). To avoid confusion whether an indice refers to a summation, we use the notation $[\cdot]$ to designate an indice where no summation is implied (in subscripts or superscripts). We also use capital letters to designate variables written in physical space (e.g., $Z$) and small letters to designate the same variables written in sample space (e.g., $z$).

## Roman Letters

| $A$ | Drift coefficient in SDEs | (Any) |
| $B$ | Diffusion coefficient in SDEs | (Any) |
| $b_{\parallel}$ | Csanady factor in the direction parallel to $\langle \mathbf{U}_r \rangle$ | $(\varnothing)$ |
| $b_{\perp}$ | Csanady factor in the direction transverse to $\langle \mathbf{U}_r \rangle$ | $(\varnothing)$ |
| $C_0$ | Kolmogorov constant | $(\varnothing)$ |
| $C_A$ | Added-Mass coefficient | $(\varnothing)$ |
| $C_D$ | Drag coefficient | $(\varnothing)$ |
| $C_T$ | Ratio of Lagrangian to Eulerian timescales | $(\varnothing)$ |
| $\mathcal{D}$ | Diffusion coefficient | (Any) |
| $\mathbf{D}$ | Diffusion matrix (tensor) | (Any) |
| $dt$ | Increment of time | (s) |
| $d_p$ | Particle diameter | (m) |
| $\mathcal{E}$ | Observation frame | $(\varnothing)$ |
| $f$ | Distribution function | $(\varnothing)$ |
| $\mathcal{f}$ | Generic function | (Any) |
| $\mathbf{F}$ | Force (vector) | (N) |

| | | |
|---|---|---|
| $\mathcal{F}_p^E$ | Eulerian mass density function for particles | $(\varnothing)$ |
| $\mathcal{F}_p^L$ | Lagrangian mass density function for particles | $(\varnothing)$ |
| $\mathbf{g}$ | Gravitational acceleration | $(\text{m.s}^{-2})$ |
| $\mathscr{G}$ | Generic function | (Any) |
| $\mathbf{G}$ | Matrix for return-to-equilibrium terms | (Any) |
| $\mathbf{H}$ | Matrix for crossing-trajectory effect | $(\varnothing)$ |
| $I_p$ | Particle moment of inertia | $(\text{kg.m}^2)$ |
| $k_B$ | Boltzmann constant | $(\text{J.K}^{-1})$ |
| $k_f$ | Fluid turbulent kinetic energy | $(\text{m}^2.\text{s}^{-2})$ |
| $K$ | Diffusion coefficient | (Any) |
| $l$ | Length scale | (m) |
| $\mathbf{L}$ | Diffusion matrix | $(\text{m}^2.\text{s}^{-3})$ |
| $L_E$ | Characteristic Eulerian length scale | (m) |
| $L_f$ | Characteristic fluid length scale | (m) |
| $\mathbf{M}$ | Torque (vector) | (N.m) |
| $m_p$ | Particle mass | (kg) |
| $M_p$ | Total mass of particles the domain | (kg) |
| $N$ | Number | $(\varnothing)$ |
| $\mathcal{N}(0, 1)$ | Gaussian distribution with zero mean and unit standard deviation | $\varnothing$ |
| $N_p$ | Number of particles | $(\varnothing)$ |
| $p$ | Probability | $(\varnothing)$ |
| $\mathbb{P}$ | Probability | $(\varnothing)$ |
| $P_f$ | Fluid pressure | (Pa) |
| $\mathcal{P}_{k_f}$ | Turbulent kinetic energy production | $(\text{m}^2.\text{s}^{-3})$ |
| $p^L$ | Unconditional Lagrangian PDF | $(\varnothing)$ |
| $\mathbf{r}$ | Spatial separation vector | (m) |
| $\mathcal{R}$ | Auto-correlation function | $(\varnothing)$ |
| Re | Fluid Reynolds number based on large-scales | $(\varnothing)$ |
| $\text{Re}_\lambda$ | Fluid Reynolds number based on Taylor scale | $(\varnothing)$ |
| $\text{Re}_p$ | Particle Reynolds number | $(\varnothing)$ |
| $\mathbf{R}_f$ | Fluid Reynolds tensor | $(\text{m}^2.\text{s}^{-2})$ |
| $R_{f,ij}$ | Component of the fluid Reynolds stress | $(\text{kg.m}^{-1}.\text{s}^{-2})$ |
| $St$ | Stokes number | $(\varnothing)$ |
| $St_{\eta_K}$ | Stokes number based on Kolmogorov timescale | $(\varnothing)$ |
| $S_p$ | Particle surface | $(\text{m}^2)$ |
| $t$ | Time | (s) |
| $T$ | Timescale | (s) |
| $T_E$ | Fluid Eulerian timescale | (s) |
| $T_L$ | Fluid Lagrangian timescale | (s) |
| $u_K$ | Kolmogorov velocity scale | $(\text{m.s}^{-1})$ |
| $\mathbf{U}_d$ | Drift velocity (vector) | $(\text{m.s}^{-1})$ |
| $\mathbf{U}_f$ | Fluid velocity (vector) | $(\text{m.s}^{-1})$ |
| $\mathbf{U}_p$ | Particle translational velocity (vector) | $(\text{m.s}^{-1})$ |
| $\mathbf{U}_r$ | Relative velocity (vector) | $(\text{m.s}^{-1})$ |

| | | |
|---|---|---|
| $\mathbf{U_s}$ | Fluid velocity seen by particles (vector) | $(\mathrm{m.s^{-1}})$ |
| $\mathbf{u}$ | Fluctuating part of the velocity (vector) | $(\mathrm{m.s^{-1}})$ |
| $u_*$ | Friction velocity at the wall | $(\mathrm{m.s^{-1}})$ |
| $\mathcal{V}$ | Volume | $(\mathrm{m^3})$ |
| $\mathcal{V}_p$ | Particle volume | $(\mathrm{m^3})$ |
| $\mathbf{V_p}$ | Particle translational velocity (in sample space) | $(\mathrm{m.s^{-1}})$ |
| $\mathbf{V_s}$ | Fluid velocity seen by particles (in sample space) | $(\mathrm{m.s^{-1}})$ |
| $W$ | Wiener process | $(\varnothing)$ |
| $\mathbf{X_f}$ | Position of a fluid particle (vector) | $(\mathrm{m})$ |
| $\mathbf{X_p}$ | Position of the particle center of mass (vector) | $(\mathrm{m})$ |
| $\mathbf{y_p}$ | Particle position (in sample space) | $(\mathrm{m})$ |
| $\mathbf{Z}$ | State vector | (Composed) |
| $\mathbf{z}$ | State vector (in sample space) | (Composed) |
| $\mathbf{Z_c}$ | Complementary part of the state vector | (Composed) |
| $\mathbf{Z_p}$ | Particle state vector | (Composed) |

## Greek Letters

| | | |
|---|---|---|
| $\alpha_f$ | Fluid volume fraction | $(\varnothing)$ |
| $\alpha_p$ | Particle volume fraction | $(\varnothing)$ |
| $\delta(\cdot)$ | Dirac function | $(\varnothing)$ |
| $\delta_{ij}$ | Kronecker's symbol | $(\varnothing)$ |
| $\Delta\cdot$ | Increment of a variable | $(\varnothing)$ |
| $\eta_K$ | Kolmogorov length scale | $(\mathrm{m})$ |
| $\epsilon_f$ | Dissipation rate of the turbulent kinetic energy | $(\mathrm{m^2.s^{-3}})$ |
| $\mu_f$ | Fluid dynamic viscosity | $(\mathrm{Pa.s})$ |
| $\nu_f$ | Fluid kinematic viscosity | $(\mathrm{m^2.s^{-1}})$ |
| $\omega_f$ | Turbulent frequency | $(\mathrm{s^{-1}})$ |
| $\boldsymbol{\omega_p}$ | Particle angular velocity (vector) | $(\mathrm{s^{-1}})$ |
| $\phi$ | Scalar quantity (e.g. mass fraction, fluid temperature) | (Any) |
| $\boldsymbol{\Phi}$ | Field quantity (e.g. velocity) | (Any) |
| $\Psi$ | Potential energy | $(\mathrm{J})$ |
| $\rho_f$ | Fluid density | $(\mathrm{kg.m^{-3}})$ |
| $\rho_p$ | Particle density | $(\mathrm{kg.m^{-3}})$ |
| $\sigma$ | Diffusion coefficient in an OU process | (Any) |
| $\tau$ | Timescale | $(\mathrm{s})$ |
| $\tau_K$ | Kolmogorov time scale | $(\mathrm{s})$ |
| $\tau_p$ | Particle relaxation time | $(\mathrm{s})$ |
| $\tau_{wall}$ | Wall shear stress | $(\mathrm{Pa})$ |
| $\Theta$ | Temperature | $(\mathrm{K})$ |
| $\xi$ | Noise term | $(\varnothing)$ |
| $\zeta(t)$ | Gaussian random variable | $(\varnothing)$ |

## Subscripts

| | | |
|---|---|---|
| 0 | Related to the initial state | (Any) |
| buoy | Related to buoyancy | (Any) |
| c,p | Related to the collisional partner | (Any) |
| E | Related to Eulerian quantities | (Any) |
| ext | External to the system considered | (Any) |
| f | Fluid quantity | (Any) |
| grav | Related to gravity | (Any) |
| int | Internal to the system considered | (Any) |
| K | Related to Kolmogorov theory | (Any) |
| L | Related to Lagrangian quantities | (Any) |
| ls | Related to spatial filter | (Any) |
| p | Particle quantity | (Any) |
| s | Fluid quantity sampled along particle trajectories | (Any) |
| s-p | Related to a single-phase situation | (Any) |
| t-p | Related to a two-phase situation | (Any) |
| $\parallel$ | Component aligned with a given direction | (Any) |
| $\perp$ | Component(s) transverse a given direction | (Any) |

## Superscripts

| | | |
|---|---|---|
| $ac$ | Quantity after a collision | (Any) |
| $bc$ | Quantity before a collision | (Any) |
| $i$ | Quantity related to particle labeled $i$ | (Any) |
| $r$ | Reduced description / formulation | (Any) |
| $+$ | Dimensionless quantity | ($\varnothing$) |

## Special Notations

| | | |
|---|---|---|
| $\{\cdot\}$ | Set of variables | (Any) |
| $\langle\cdot\rangle$ | Statistical average (ensemble) | (Any) |
| $\langle\cdot\rangle_{ls}$ | Statistical average (spatial filter) | (Any) |
| $\lvert\cdot\rvert$ | Norm of a vector | (Any) |
| $\partial$ | Partial derivative | ($\varnothing$) |
| $D\cdot/Dt$ | Material derivative $\partial\cdot/\partial t + U_i\partial\cdot/\partial x_i$ | ($s^{-1}$) |
| $\widehat{\cdot}$ or $\widetilde{\cdot}$ | Rescaled variables | (Any) |

# Setting the Stage

1

**Abstract**

When encountering investigations in a scientific field we are not familiar with, we tend to ask the following typical questions:

(i) What is the subject about?
(ii) What are the main points and the typical issues?
(iii) What are the properties of current model formulations?
(iv) What are the remaining challenges for which new ideas are called for?

The purpose of the present book is to provide answers to these questions for the scientific domain constituted by *statistical descriptions of discrete particle dynamics in turbulent flows*. In that sense, this book is a multi-fold invitation whose nature is outlined in this chapter which serves as an introduction and set the stage.

**Chapter Content** After an overview of the challenges that we are faced with and the chosen standpoint to address them in Sect. 1.1, we precise what this book is, or is not, about in Sect. 1.2 before detailing the structure in Sect. 1.3.

## 1.1 Introducing Turbulent Dispersed Two-Phase Flows

### 1.1.1 A Multi-Scale and Multi-Physics Problem

This book is, first, an invitation to consider as a subject worthy of attention the wide range of situations where small discrete elements are embedded in turbulent flows. These situations occur often at a human scale and in our daily environments, with the small discrete elements being either bubbles, droplets or solid particles.

© The Author(s) 2025
J.-P. Minier et al., *Understanding Turbulent Systems*,
Lecture Notes in Physics 1039, https://doi.org/10.1007/978-3-031-84466-9_1

1

These turbulent dispersed two-phase flows display complex behavior due to the interplay of two fundamental interactions, the fluid-particle and particle-particle interactions, compounded by the turbulence of the carrier flow. Referring to such discrete elements as 'particles' (be they droplets, bubbles or actual solid particles), such flows are therefore also called particle-laden flows. They make up an intriguing subject with connections with soft matter, involving similarities and differences between Brownian and turbulence 'noises', as well as with macroscopic physics, involving the search for reduced constitutive relations based on scale separations and the use of fast-variable elimination techniques.

## 1.1.2  Challenges in Statistical Physics

Turbulent particle-laden flows is not a subject where the basic laws are unknown but where the huge number of interacting degrees of freedom call for reduced, or coarse-grained, statistical descriptions to be developed. Since we are basically considering transport and collision phenomena, it could be believed that these processes are well described by kinetic theories. In the general case of non-fully resolved turbulent flows, we are however dealing with particles influenced by random media with non-zero time and space correlations. The second invitation is therefore to recognize the limitations of kinetic-based descriptions and to address the challenges leading us to extend the classical framework, for fluid-particle as well as particle-particle interactions. This is done by revisiting the definition of the particle state vector which gathers the variables retained to describe particle dynamics based on the key notion of slow and fast variables. In recent years, significant progress has been made following this probabilistic density function (PDF) approach to discrete particle dynamics in turbulent flows. Yet, there are remaining challenges in statistical physics that call for new ideas and new developments.

## 1.1.3  A Physicist's Approach

To address the various issues related to statistical descriptions of turbulent dispersed two-phase flows, a wide range of standpoints are possible and it is important to be aware of which one is retained in any work. To that effect, it proves useful to consider two typical viewpoints, referred to as the physicist and the engineer viewpoints, respectively. In short, the physicist point of view is mostly centered on the analysis of the physical nature of different models (What do they represent? What do they contain? What is missing?) whereas the engineering point of view is mostly focused on the assessment of the predictive capacities of certain model formulations (How are theoretical expressions implemented? How precise are we? What is expected from future developments?).

These two standpoints are not in competition and it is even better to regard them as two components playing different roles in the life cycle of a scientific model, as illustrated in Fig. 1.1.

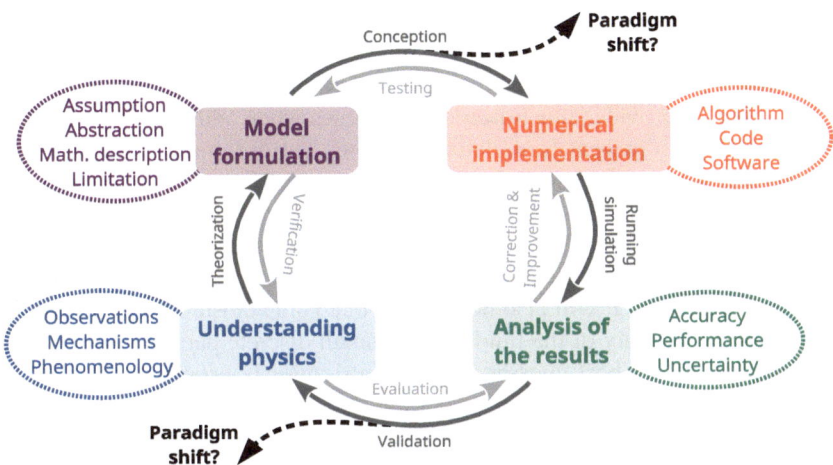

**Fig. 1.1** Illustration of how the two standpoints (the physicist point of view on the left and the engineering point of view on the right) are intertwined in the life cycle of scientific models. This sketch highlights how models constantly evolve as they are tested in new configurations/applications and how a sudden shift in the paradigm can lead to the development of a whole new field/category of models

There is, also, no hierarchy between the physicist and engineering viewpoints but there is, perhaps, a chronological order to be respected since it is best to start with the analysis of the physical content of a model before addressing its practical assessment. At this stage, the situation is different depending on which subject is considered and whether there is a well-accepted common knowledge on model contents. For example, this is (more or less) the current status in single-phase turbulence where the background pertaining to classical models is rather well established and shared within this community, as manifested in the classical textbook by Pope [1] with the relevant exception of the PDF approach (see the specific chapter in [1, chapter 12]). However, this is not the case for dispersed two-phase flows models which are more intricate since we are dealing with two different phases having different natures, namely a continuous phase and a set of discrete particles, compounded by the use of stochastic models which are often less familiar for researchers used to continuum mechanics methods. There is often confusion as to what is attempted by these stochastic models, their true meaning, or how they should be manipulated (see accounts in [2, 3]). Moreover, in turbulent dispersed two-phase flows modeling, we are not in the situation where the macroscopic level of description, represented by a closed set of equations for averaged quantities, has been established beforehand and can serve as a safeguard. On the opposite, PDF models are precisely used to derive those very mean-field equations. This is a typical example of a bottom-up approach and, consequently, clarifying the physical content of stochastic formulations is an essential step not to be overlooked. A third invitation is therefore to investigate the physics-based properties of model formulations.

### 1.1.4    Charted and Uncharted Territories

A bottom-up approach in which models are developed from first principles is attractive since it provides much-needed guidelines when additional physics has to be accounted for in present formulations or when current model closures must be revisited. Indeed, it will be seen that a satisfactory mathematical framework based on PDFs, thus loosely called 'the PDF approach', has already been set up, giving us a clear view on which classes of stochastic models are best suited to capture the time-evolution of the state variables selected to describe turbulent particle-laden flows. However, it will also be recalled that some closure proposals (e.g. the timescale of the velocity of the fluid seen by small discrete particles) appear more, at the moment, as our best guesses rather than rigorously-established expressions. To overcome the limitations of present formulations, assess the range of validity of some hypothesis introduced without justification (such as an eddy or turbulent viscosity), and unify what is often addressed separately (such as transport and collective effects), new ideas and physically-based principles are called for. This represents a fourth invitation and, to exemplify this approach, some attempts are developed in this work.

## 1.2    What This Book Is About and What It Is Not About?

Drawing on the above distinction between the two different standpoints, the developments presented in this book are essentially discussed from the physicist point of view. This means that we are mostly concerned with the *modeling approach* more than elaborating on variants among the same class of models. This also means that the emphasis is put on analyzing the *physical content* of the different models which are introduced more than obtaining quantitative estimates of the predictive capacities of specific formulations. More precisely, once the statistical issue is brought out, we wish to clarify: Why do we consider some specific stochastic models? What is their physical content (or the lack of it) and what are the phenomena that are either captured, approximated, or left out? What improvements can be expected by further developments within the same class of models and what requires leaps into altogether new modeling constructions? In a nutshell, it can therefore be said that, taking the standpoint provided by the modern formulation of stochastic processes and focusing on the description of the particle phase, this book proposes a step-by-step pedagogical presentation of current models while pointing out new directions and remaining unexplored territories.

## 1.3 Description of the Book Contents

This book is organized as follows. Chapter 2 provides insights into the physics of particle-laden turbulent flows through a selection of practical situations. Chapter 3 describes the governing particle equations and introduces the central issue in statistical physics. Chapter 4 presents a unified formalism leading from particle stochastic models to macroscopic equations so as to concentrate on key stochastic processes, and explains the double hierarchy in terms of one- or N-PDF methods and of particle variables which need to be retained. Chapter 5 concentrates on discrete particle transport by turbulent flows without particle collisions. The key point is to demonstrate the failure of kinetic-based descriptions in non-fully-resolved turbulent flows, which leads to introduce an extended set of particle-attached variables, in particular the velocity of the fluid seen. Chapter 6 describes the physics of particle dispersion and revisits the construction of present stochastic models for the velocity of the fluid seen, based on Langevin type of models developed in single-phase turbulent flows. A selection of numerical simulations is discussed in Chap. 7 to illustrate a number of physically-relevant points through practical applications. Walking off the map of current formulations, new ideas and formulations are developed in Chap. 8 to encourage further research. Since several issues appear as the counterparts of what is addressed in soft matter, similarities and differences are discussed in Chap. 9 to build bridges but also point to special issues. Recent and even new results are detailed in Chap. 10 to detail rigorous and physically-meaningful methods to eliminate fast variables in stochastic models. This is directly in relation with the derivation of closure laws as local or non-local source terms and has direct impact on the formulation of constitutive relations in macroscopic models. In Chap. 11, the source terms representing particle back effects on turbulent flows are analyzed and discussed through the presentation of a numerical application. Chapter 12 concentrates on particle collisions without particle transport and revisits the classical Boltzmann picture along with the central assumption of molecular chaos. It is shown that this classical account is too limited for particles in turbulent flows and that extended formulations are needed. Finally, conclusions and perspectives are discussed in Chap. 13.

## References

1. S. Pope, *Turbulent Flows* (Cambridge University Press, 2000). https://doi.org/10.1017/CBO9780511840531
2. J.P. Minier, Phys. Rep. **665**, 1 (2016). https://doi.org/10.1016/j.physrep.2016.10.007
3. J.P. Minier, E. Peirano, Phys. Rep. **352**(1–3), 1 (2001). https://doi.org/10.1016/S0370-1573(01)00011-4

# The Physics of Dispersed Two-Phase Flows

## 2

**Abstract**

What is a dispersed two-phase flow and what are its main characteristics? To provide answers to these questions, this chapter presents a selection of examples taken from a wide range of natural and industrial applications. Each example sheds light on some of the key challenges faced in this complex multidisciplinary field and on the range of questions that have to be addressed. They help us to bring out the nature and main characteristics of our topic and to decide which modeling road we wish to follow to set our venture into motion.

**Chapter Content** We begin our journey with the description of the selected examples in Sect. 2.1. What can be drawn about these examples is proposed in Sect. 2.2, before choosing in Sect. 2.3 the modeling road to travel on.

## 2.1 A Rich Tapestry of Applications

Let us start with some definitions of the terminology used throughout this work. In the field of fluid dynamics, a multiphase flow corresponds to the simultaneous flow of at least two non-miscible thermodynamical phases. A phase is labeled 'continuous' if it occupies a continually connected region of space, while a phase is named 'disperse' if it occupies disconnected regions of space.

This book focuses on dispersed two-phase flows, which constitute a subclass within multiphase flows. As it follows from their name, dispersed two-phase flows involve a dispersed phase, present as a set of discrete elements (e.g. bubbles, droplets, solid particles), embedded in a continuous one (also referred to as the carrier phase), which can be a gas or a liquid flow.

© The Author(s) 2025
J.-P. Minier et al., *Understanding Turbulent Systems*,
Lecture Notes in Physics 1039, https://doi.org/10.1007/978-3-031-84466-9_2

Dispersed two-phase flows are encountered in large-scale settings in the universe and in the atmosphere as well as occurring in many environmental and industrial situations in our daily activities. In fact, it is even difficult to do justice to the richness and variety of applications involving dispersed two-phase flows. In the following, we discuss a few examples which illustrate the wide variety of situations encountered and the corresponding challenges to be addressed.

### 2.1.1   Interstellar and Molecular Clouds

*What Is an Interstellar/Molecular Cloud?*  An interstellar cloud corresponds to a region in the space between star systems with a higher-than-average amount of matter compared to the background interstellar medium. When thermodynamical conditions and the density and size of the matter present in the cloud are such that they allow the formation of molecules (most commonly molecular hydrogen $H_2$), it is called a molecular cloud. These clouds are usually formed by a mixture of molecular gases (like hydrogen, helium) and solid particles (like dust) suspended in space, making them a typical example of dispersed flows [1].

*Where Do Interstellar/Molecular Clouds Occur?*  Interstellar and molecular clouds are commonly encountered in the space between star systems but also sometimes in the space between galaxies. This includes the two following objects:

- Dark nebulae refer to molecular clouds dense enough to absorb visible light.
  As other molecular clouds, dark nebulae contain molecular gases (like hydrogen, helium) but these are transparent to the visible light. What makes such nebulae appear dark is the presence of a high density of sub-micrometer dust particles, often coated with frozen carbon monoxide and nitrogen, which block the passage of light. The shape of dark nebulae can be very irregular: as illustrated in Fig. 2.1a, dark nebulae display sometimes convoluted serpentine shapes which mimic spectacular figures.
- Stellar nurseries refer to a type of molecular cloud where stars form inside them.
  Within the densest regions of some molecular clouds, the amount of dust and gas can be so high that gravitational forces trigger a collapse of dust and gas. This gravitational collapse is believed to be the beginning of star formation. Since the materials collapsing towards the (future) young star possess some angular momentum, it usually leads to the formation of a protoplanetary disk in the latter stages of the process. As displayed in Fig. 2.1b, this corresponds to a pancake or ring-shaped accretion disk of matter (composed of gas, dust, stones, planetesimals or asteroids) orbiting around the young star.

*What Are the Modeling Issues in Molecular Clouds?*  The above description only skims the surface of the complexity at play in the context of molecular clouds. Therefore, depending on the context of each study, the objectives of the models that need to be developed can vary significantly, leading to models with different

(a)                                                                              (b)

**Fig. 2.1** Illustrations of some features occurring in molecular clouds in space. (**a**) Dark nebula IC 434, known as the Horsehead due to its shape. The nebula is about 1500 light years away from Earth within the Orion constellation, with a size of roughly 3.5 light years where light is absorbed by the nebula. The Flame Nebula is also visible, on the left. Cropped image from Wikimedia Commons by Gianni Lacroce, used under CC BY-SA 4.0. (**b**) Artist view of planetesimals, stones and planets within the protoplanetary disk around a young star. Reproduced from Wikimedia Commons, by Merikanto, used under CC BY-SA 4.0

levels of complexity. For instance, models that mimic the absorption of light in dark nebulae require two pieces of information: (i) the nature and amount of dust within each region of the nebula and (ii) a physical model for light absorption by micrometer-sized particles (e.g. using the Rayleigh or Mie scattering theories). On the other side, models for the evolution of a protoplanetary disk requires to describe the dynamics of gas and dust with a range of densities (from rarefied gases to dense granular media) while including gravitational effects (for planet formation), magneto-hydrodynamic aspects (since many planetesimals have electric and/or magnetic properties) as well as scenarios for collisions between such objects (e.g. accretion of small dust by large planetesimals or catastrophic impact between two planetoids). This has led to the development of innovative two-phase flow approaches [2]. Molecular clouds also illustrate how tiny microscopic objects can nevertheless impact phenomena acting at astrophysical scales (here over several light years).

## 2.1.2   Sediment Transport

*What Is Sediment Transport?* Sediments correspond to naturally-occurring solid materials coming from the deterioration of rocks, soils and/or minerals (through the processes of erosion or weathering). Once formed, these sediments are subsequently transported by the action of a flow and by the force of gravity. Gravity ultimately leads to the settling of sediments which accumulate in deposits at the bottom bed: this process is called 'sedimentation'.

(a)                                                          (b)

**Fig. 2.2** Illustrations of how sediment transport occurs in glaciers. (**a**) Medial moraine formed by the accumulation of debris between two joining glaciers in the Nuussuaq Peninsula in Greenland. Reproduced from Nuussuaq peninsula moraines by Algkalv, used under CC BY-SA 3.0. (**b**) Tiny sediment suspended in meltwater entering Lake Louise in Canada, turning the lake's color grey or milky white (hence the name 'glacial milk'). Cropped image from Wikimedia Commons, by WikiPedant, used under CC BY-SA 4.0

*Where Does Sediment Transport Occur?* In natural systems, sediments can be transported by various types of fluids in a range of environments. Among them, this includes the following categories:

- Glacial transport refers to the motion of sediments by ice.

    As glaciers move down the slope of a mountain, they can carry sediment materials of various sizes (ranging from meter-sized rocks down to micrometer-sized clay or silts). These sediments can deposit on the edges of a glacier to form an accumulation of unconsolidated debris: such a formation is called a 'moraine'. Figure 2.2a displays a moraine formed at the junction between two glaciers, leading here to a ridge of debris in-between.

    As glaciers move down a slope, they can also pulverize rocks by mechanical grinding at the interface between ice and ground. This eventually leads to the formation of very fine-grained materials, with a size typically between a few micrometers up to a few tenths of micrometers. These tiny materials are called rock flour (or glacial flour). In fact, as the ice melts, rock flour becomes suspended in the meltwater and, as this water reaches a lake, it can sometimes make the water appear cloudy/milky (as shown in Fig. 2.2b).

- Fluvial transport corresponds to the displacement of sediments by water in rivers.

    As water flows downstream rivers, tributaries or streams, it can carry various amounts of sediments. The size of these sediments typically ranges from a few nanometers (e.g. silt) up to a few millimeters (e.g. pebbles), or even a few meters (e.g. rocks or tree trunks during flash floods). The concentration and the nature of the sediments transported by a river vary due to subtle balance between the erosion of the riverbanks and the deposition of suspended sediments on riverbeds. As displayed in Fig. 2.3a, this results not only in complex river geometries

 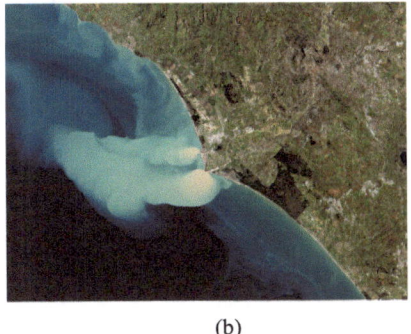

(a)                                                                (b)

**Fig. 2.3** Illustrations of how sediment transport occurs in rivers. (**a**) Surkhob river in Tadjikistan forming at the confluence between the Muksu river (rather turquoise water on the left) and the Kyzyl-Suu river (brownish water on the right). Swirls are visible at the junction, where mixing occurs. Reproduced from Surkhob by Irangeologist, used under CC BY-SA 3.0. (**b**) Sediment carried by the Tiber River after heavy rainfall on 5 February 2019 pouring into the Thyrrhenian Sea to form a sediment plume stretching over 28 km from the Italian coast. Cropped image from Sediment plume at sea, by ESA (contains modified Copernicus Sentinel data from 2019), used under CC BY-SA 3.0 IGO

(e.g. with meanders) but sometimes also in fascinating blends of colors at the confluence between two rivers (with swirls and surface ripples).

At a river mouth, freshwater from the river flows into a larger body of water (e.g. in lakes, seas or oceans) [3]. Sediments suspended in such freshwaters tend to settle down due to the slower motion in the larger body of water (which can even be stagnant as in lakes) and due to the change in the water composition (e.g. higher salinity in seawaters than in freshwaters). As shown in Fig. 2.3b, this leads to the formation of plumes of sediments that play a key role in the dynamics of river deltas. These river deltas have always been of great importance for mankind, among other reasons for agricultural purposes (due to the fertility of the ground coming from the deposition of mineral and organic materials, as in the Ganges Delta in India) as well as for construction purposes (with sand/gravel quarries in older deltas).

- Coastal transport occurs in near-shore environments when sediments are set in motion by waves and currents.

  Sediments are naturally suspended in seawaters. As indicated, some of them come from the influx of sediments provided by freshwaters in river deltas (see Fig. 2.3b) [3]. Another source of sediments in seawater is the settling of heavy sediments that are transported by the atmosphere due to wind effects (see Fig. 2.5b) or following volcanic eruptions (since plumes contain ashes and fragmented volcanic materials). Coastal erosion also plays a role in the influx of sediment in seawaters: this is particularly intense during storms that tend to move larger sediments due to the stronger currents and waves, which can quickly modify the near-shore landscape (as illustrated in Fig. 2.4a).

(a)                                              (b)

**Fig. 2.4** Illustrations of how sediment transport occurs in coastal regions. (**a**) Coastal erosion at Point Peron in Western Australia, showing the aftermath of a storm on a near-shore road in August 2020. Reproduced from Wikimedia Commons, by Calistemon, used under CC BY-SA 4.0. (**b**) Ripples in the sand formed by underwater motion that are emerging at a low tide on a beach near Carn Towan in England. Reproduced from Ripples in the sand at low tide by Rod Allday, used under CC BY-SA 2.0

   The dynamics of sediments in coastal environments is mostly driven by wind-generated waves, as well as by tides, storm surges and currents. In fact, there is a subtle equilibrium between the sediment sizes and the intensity of the waves/currents/tides. This plays a role not only in the formation of a variety of beaches (e.g. beaches made of sand, gravels or pebbles), but also in the shape of the coastline (dynamics driven by erosion and deposition) as well as in the formation of centimeter-sized ripples on the seafloor near the shore (see Fig. 2.4b).

- Aeolian transport refers to the movement of sediments in the atmosphere.

   Sediments below a few millimeters (like sand, dust or ash) can easily become airborne under the action of the wind. Among the numerous intriguing examples in natural environments, Fig. 2.5c illustrates how dust devils are formed when tiny sediments are lifted up due to a swirling flow (more catastrophic events happen during tornadoes, where larger objects can be lifted up). Similarly, tiny sand particles forming dunes in deserts can be blown away by the wind, as shown in Fig. 2.5a. This contributes to the remarkable dynamics of sand dunes, whose shape and location constantly evolve under the action of the wind and especially gusts [4]. In fact, the subtle balance between the settling of sediments and their motion due to the wind is responsible for the formation of sand ripples visible in Fig. 2.5a (note that they are similar to underwater ripples seen in Fig. 2.4b).

   While the transport of heavy sand grains usually occurs over short distances (typically a few centimeters for sand ripples), the tiniest and lightest sediments can be transported in the atmosphere over much larger distances. For instance, this occurs in dust storms such as the one displayed in Fig. 2.5d, where tiny and light dust grains can be transported in the air over several kilometers. Another well-known illustration is the inter-continental transport of sand, where very fine sand grains reach the upper atmosphere and move across seas or oceans (like the Atlantic Ocean as visible in Fig. 2.5b).

(a)                                           (b)

(c)                                 (d)

**Fig. 2.5** Illustrations of how sediment transport occurs in the air. (**a**) Sand being blowing off dunes by the wind in the Sahara desert, near the town of Mahmid El Ghezlane in Morocco. Reproduced from Wikimedia Commons by Anderson Sady, used under CC BY-SA 3.0. (**b**) Dust and sand being carried by the wind over the Atlantic Ocean towards the islands of Cabo Verde (570 km off the west coast of Senegal and Mauritania). The sand is coming from the Sahara and Sahel region. Reproduced from Cabo Verde, by ESA (contains modified Copernicus Sentinel data from 2018), used under CC BY-SA 3.0 IGO. (**c**) Dust devil. Cropped image from Dust devil, by Peretz Partensky, used under CC BY-SA 2.0. (**d**) Dust storm headed towards the city of Phoenix in Arizona (USA). Reproduced from Phoenix dust storm, by Alan Stark, used under CC BY-SA 2.0

*What Are the Modeling Issues in Sediment Transport?*  As it transpires from these selected examples, a wide range of issues are encountered in sediment transport. To name a few, sediment transport affects human activities in various ways including: agriculture (soil fertility can be enhanced by the frequent influx of sediments rich in microorganisms and minerals, as in river deltas [3]), construction (landslides, mudflow or fast floods might have dramatic consequences on ground stability), transportation (e.g. ashes contained in volcanic plumes affect aircraft engines) as well as human health (e.g. with the effect of air pollution on respiratory diseases [5], like those induced by smoke fogs or dust storms). For these reasons, a number of models have been designed either to make predictions (e.g. forecast warnings on dust storms or dispersion of pollutants following industrial incidents), or to optimize some processes (like the frequency at which preemptive avalanches should be triggered) or even to design mitigation measures (such as to reduce the impact of sediments and debris on dams and their safety [6]).

### 2.1.3   Flows with Bubbles and Droplets

*What Are Bubble/Droplet Flows?*  A bubble flow refers to a type of dispersed two-phase flow, where discrete gas bubbles are suspended in a liquid flow (e.g. air or nitrogen bubbles in water). Alternatively, a droplet flow designates flows where liquid droplets are dispersed in a fluid (e.g. water droplets in air or oil droplets in water).

*Where Do Bubble/Droplet Flows Occur?*  Bubble and droplet flows occur frequently in our daily life. Among numerous examples, one can cite:

- Water droplets in the atmosphere.

  Water is found on Earth in a variety of forms (liquid water, solid ice and water vapor) and its presence in the atmosphere plays a crucial role in both climate and weather. For instance, warm clouds (see Fig. 2.6a) and fogs (see Fig. 2.6c) are made of microscopic droplets of liquid water, which can stem from the condensation of water vapor on so-called condensation nuclei (like dust, smoke or other tiny particles in suspension in the air). In fogs, these droplets evaporate due to solar radiation and/or to humidity changes. In clouds, continuous condensation can lead to the growth of droplets until they reach sizes large enough to start falling down. These larger falling droplets can then collide with smaller ones and coalesce, i.e. combine to form even larger drops. When this downward velocity is high enough, droplets start to fall as precipitation: it is raining (see Fig. 2.6a).

  As displayed in Fig. 2.6b, rain drops impacting a layer of water can produce the so-called splashing phenomenon, with a rising ring of water forming at the edge of the surface where the water drop spreads and with tiny secondary droplets detaching from this ring. Such splashing phenomena are responsible for the fascinating sounds associated with rain.

  Alternatively, as shown in Fig. 2.6d, tiny droplets can be emitted near chimneys (possibly mixed with tiny solid particles). The transport of these droplets by the underlying airflow leads to the dispersion of a smoke plume. These plumes can either disappear (due to the evaporation of droplets and/or to the dilution of droplets through mixing with a dryer air) or merge with other plumes/clouds to form thick clouds (which might appear dark if tiny soot particles are also suspended).
- Sparkling beverages.

  Bubble flows are occurring in all sparkling beverages. This comes from the presence of a high quantity of dissolved gas (especially carbon dioxide), which can be naturally present in the water or artificially added to obtain

(a)

(b)

(c)

(d)

**Fig. 2.6** Illustrations of droplet flows in nature. (**a**) Rain cloud over an island. Reproduced from Wikimedia Commons by AhmadAlQasim, used under CC BY-SA 4.0. (**b**) Water droplet impacting a bucket of water, leading to a splashing phenomenon (secondary droplets are seen detaching from the rising ring of water). Reproduced from Splash, by Aceebee, used under CC BY-SA 2.0. (**c**) The Golden Gate Bridge in San Francisco covered by a dense fog that severely hampers visibility. Reproduced from The bridge (August 2013), by Frank Schulenburg, used under CC BY-SA 3.0. (**d**) A smoke plume dispersing in the air after its emission from a chimney. Reproduced from A smoke plume from the Dunbar Cement Works chimney, by Walter Baxter, used under CC BY-SA 2.0

sparkling drinks (like champagne or beer). When poured in a glass, bubbles form spontaneously around nucleation spots (typically impurities like dirt or tiny air bubbles trapped in defects of the glass). As illustrated in Fig. 2.7a, these bubbles rise in the glass since they are lighter than the surrounding fluid.

- Boiling water.

    Air bubbles can also be quickly formed when heating water close to the boiling point. As visible in Fig. 2.7b, small bubbles form at discrete point at the bottom of a pan that is heated above the boiling temperature. As soon as a bubble is large enough, it detaches from the bottom surface and quickly rises towards the top.

- Soap bubbles.

    Soap bubbles are formed when a small amount of air becomes enclosed in a thin film of soap, as displayed in Fig. 2.7c. These bubbles usually last for short periods of time (typically a few seconds) before bursting. In the absence of wind, these bubbles quickly take the shape of a sphere. Soap bubbles are often used during recreational activities for children but can also be assembled together in industrial facilities to create a foam.

**Fig. 2.7** Illustrations of bubble flows in our daily life. (**a**) Rising bubbles in a flute glass containing champagne. Reproduced from Champagne glass by Quinn Dombrowski, used under CC BY-SA 2.0. (**b**) Boiling water in a transparent cooking pan. Reproduced from Kochendes Wasser, by Markus Schweiss, used under CC BY-SA 3.0. (**c**) Soap bubble floating in the air (with reflection). Cropped image from Wikimedia Commons, by Mýdlová Bublina, used under CC BY-SA 4.0

*What Are the Modeling Issues in Bubble/Droplet Flows?* Many of the challenges associated to these flows are related to a phase transition of water (between vapor, liquid and solid states). For instance, many industrial processes rely on water to transfer heat (e.g. heat exchangers in boilers, thermal and nuclear reactors [7]). This implies that an amount of water is heated up until it evaporates. Hence, while there is solely liquid water at the bottom of such heat exchangers, the amount of vapor bubbles gradually increases as we reach the upper stages, going through a variety of flow regimes: discrete bubbles are formed in the liquid water near the bottom (leading to a bubble flow), and start to coalesce as they rise to form large pockets that can be as large as the pipe diameter (leading to a slug flow) while the region near the top is composed of discrete droplets suspended in an airflow (leading to a droplet flow or sometimes to an annular flow when there is a thin liquid layer on the surface of the pipe). Understanding these phenomena not only requires to master the physics of phase transitions but also to model properly how bubbles/droplets grow through coalescence and how they deform as they are carried by the flow.

## 2.1.4  Flows with Living Organisms

*What Are Flows with Living Organisms?* Living organisms are omnipresent in nature. Among them, one usually distinguishes between prokaryotic organisms and eukaryotic organisms (depending on whether the nucleus is bound by a membrane). The majority of prokaryotes correspond to unicellular organisms (e.g. bacteria), with sizes typically within the micrometer-scale. Most eukaryotes are multicellular (e.g. algae, fungi, plants or animals). As living organisms move in a flow, either passively by being convected by the flow (e.g. algae) or actively with self-propulsion techniques (e.g. fishes), they can alter the surrounding flow. These organisms are referred to as active matter, by contrast with passive objects (like solid inorganic particles).

*Where Do Such Flows Occur?* Various examples of flows with living organisms exist in nature, including:

- Marine plankton corresponds to organisms that drift in water but are unable to actively propel themselves against currents.

  Marine plankton includes a wide variety of species (like bacteria, alga, fungi, fish eggs and even jellyfish). They are often classified in sub-categories like phytoplankton or zooplankton. Phytoplankton corresponds to the type of plankton that are autotroph, meaning that they obtain energy through photosynthesis (like diatoms). These micro-organisms hence typically live in the upper layer of the ocean where they receive enough sunlight. Their activity depends not only on the amount of light received, but also on the temperature, pH and amount of nutrients dissolved in water (such as nitrate or phosphate). When optimal conditions are met, this can lead to a so-called 'bloom', where their number grows significantly in a short period of time (see Fig. 2.8a for an algae bloom in a river). Some of these phytoplankton are even bioluminescent (like dinoflagellates), meaning that they emit light when subject to mechanical stimulus: this can lead to exciting and colorful observations of waves crashing on a beach, as displayed in Fig. 2.8b.
- Animals can move and affect their environment.

  Compared to plankton, larger animals (like humans, fishes or birds) can propel themselves to move either with or against the flow. As displayed in Fig. 2.9a, fishes are able to swim underwater. By doing so, they affect the fluid around them, which induces here some motion of the nearby kelp (a large marine algae that grows from the seafloor vertically into large brown leaves).

(a)                                                                (b)

**Fig. 2.8** Illustrations of plankton in marine systems. (**a**) Algae bloom in the Spree river: leaves are visible on the river's surface while variations in algae concentration reveal complex features (like swirls). Reproduced from Heatwave causing algae bloom in the Spree by Lars Plougmann, used under CC BY-SA 2.0. (**b**) Phytoplankton glowing in the night due to bioluminescence triggered by the agitation of waves crashing on Chabahar beaches (Iran). Reproduced from Wikimedia Commons, by Safa Daneshvar, used under CC BY-SA 4.0

(a)                                                         (b)

**Fig. 2.9** Illustrations of animal motion in their environment. (**a**) Fishes evolving underwater in the Monterey Bay Aquarium near a kelp forest (a large marine brown algae). Reproduced from Wikimedia Commons, by Amadscientist, used under CC BY-SA 3.0. (**b**) A group of starlings (birds) over fields near Gretna (England), forming a large flock with elaborate shapes and patterns in the evening sky. Reproduced from A wedge of starlings, by Walter Baxter, used under CC BY-SA 2.0

Another striking features when observing a large group of animals is that they can form flocks (or swarms) that move together like a fluid. This is illustrated in Fig. 2.9b for birds flying over a field: they form elaborate patterns that are constantly evolving in time.

*What Are the Modeling Issues in Such Flows?* Models for the motion of living organisms address two important aspects: the motility of each individual and the emergence of patterns in swarms/flocks. First, when dealing with microscopic or nanoscopic swimmers (like plankton), the motility of each individual is often described using an intrinsic velocity (whose orientation can be aligned with the direction of the swimmer or different from it). This can lead to complex behavior, where microswimmers can navigate in turbulent flows or get trapped in specific regions (like swirls) [8]. Second, studies of swarms/flocks have revealed the complex ordered or disordered systems that arise [9]. Since the motion of flocks resembles those of a fluid, continuum theories for the dynamics of swarms at large scales are also being developed.

## 2.2  What Can Be Learned from These Examples?

### 2.2.1  An Inter-Disciplinary Topic

First and foremost, these selected examples highlight the inter-disciplinary nature of dispersed two-phase flows. This translates that this subject involves a wide variety of scientific disciplines. In fact, it appears that the topic of dispersed two-phase flow is related to various branches of sciences, especially in life science and physical science. This topic even involves a number of scientific disciplines, which can be viewed as subdivisions of these scientific branches, such as:

- physics (e.g. fluid mechanics for flows, classical mechanics for particle dynamics, thermodynamics of heated/cooled fluids, astrophysics for planetary systems);
- chemistry (e.g. surface and interface sciences, thermochemistry of reactive flows);
- engineering (e.g. hydraulics, material sciences for deformable objects);
- geology (e.g. sedimentology for deltas, volcanology for atmospheric eruptions);
- biology (e.g. anatomy and physiology of cells/plants and its impact on motility);
- ecology (e.g. evolution of a population of cells or plants);
- mathematics (e.g. computational mathematics for algorithms, mathematical analysis for differential equations, probability and statistics for random processes).

This large variety of scientific disciplines partly explains the different terminology used depending on the context of each study. For instance, this has led to classifications of two-phase flows in terms of the type of fluids involved (e.g. liquid/liquid, liquid/gas, liquid/solid, or gas/solid flows) or according to the state of the dispersed phase (e.g. with bubble flows or granular flows). In this book, we do not engage in a debate as to which classification is more appropriate. Instead, we shed light on how models are designed based on the phenomena that should be accounted for and based on the objectives of a study.

Due to its inter-disciplinary nature, coming up with adequate descriptions of two-phase flows implies to combine knowledge gained from several scientific disciplines. The difficulty with such multi-physics simulations is that it is not sufficient to have a good understanding of each individual aspect of the process (e.g. a description of the phenomenology and governing equations for fluid mechanics only). One also has to take into account all the interactions between the multiple aspects of the system. One illustration of such issues is given in Chap. 11 for the modification of the fluid flow due to the presence of particles in suspension.

## 2.2.2  A Multiscale Problem

Another important challenge of dispersed two-phase flows is that they involve a wide range of spatial and temporal scales. For instance, the sediments transported in rivers can display sizes going from a few micrometers (e.g. silt and clay) up to a few meters (e.g. rocks during flash floods). Similarly, the size of aerosols transported in atmospheric flows goes from a few nanometers (e.g. soot particles) up to a few millimeters (e.g. large dust or sand). This usually implies that models should account for polydispersity in particle sizes. The only exception is when studies focus specifically on a single size of particles, a case commonly referred to as monodispersed particle sizes. Furthermore, these suspended particles are transported by a flow over regions going from a few centimeters or less (e.g. ripples in sand, whose shapes change within a few minutes) up to kilometers (e.g. sand storms that can last for several days).

These scale separations have tremendous consequences on modeling approaches, leading us to choose the level of information to be contained in a given simulation right from the outset. For instance, it is possible to have microscopic information on the flow structures and to explicitly track the motion of each individual particles when dealing with the dynamics of microparticles in the portion of a pipe of a few millimeters. Yet, such a task is currently out of reach even with presently-available computational facilities for the case of microscopic aerosols dispersed in the air by winds across Europe. Hence, large-scale models require a-priori knowledge on what should be explicitly solved and on what can be approximated.

### 2.2.3   An Intricate Range of Phenomena

Dispersed two-phase flows are complex systems that involve a diversity of phenomena, processes and mechanisms. To name a few, natural and industrial fluid flows are often turbulent, i.e. characterized by apparently-random changes in the pressure and velocity fields (with swirls or other intermittent features). Meanwhile, particle dynamics involves phenomena like particle transport, particle deposition, particle resuspension or particle agglomeration. Particle transport or dispersion usually occurs due to a subtle balance between (i) the hydrodynamic/aerodynamic forces exerted by the underlying fluid on the particles and (ii) the settling of solid particles under gravity. This explains why the tiniest and lightest particles can remain in suspension for much longer times and thus be transported over longer distances than larger and heavier ones. In fact, specific studies are devoted to colloidal particles, which are micrometer-sized particles that are small enough so that Brownian motion counteracts settling, meaning that they do not sediment [10–12]. This picture can be much more complex, as in river deltas, where the settling rate of sediments is increased at the river mouth due to a combination of effects related to hydrodynamic transport (e.g. slower fluid motion in the seawater) and to particle agglomeration (i.e. formation of larger clusters as tiny sediments stick to each other more rapidly in seawater than in freshwater due to differences in the composition of water) [3]. Particle resuspension designates the process by which objects lying on a surface are detached from it under the action of a flow [13]. This process, together with particle deposition (which is the opposite process), plays a key role in wind-blown sand [4] or in the evolution of riverbeds [3]. Additional effects can also occur when particles are not hard solid objects. This is the case with bubbles or droplets, which can grow and deform depending on the flow outside/inside them and on surface tension at the air/water interface [14]. Similarly, elongated and flexible fibers can deform and modify their shape according to the local velocity field and to their mechanical properties (like cellulose in the papermaking industry [15]). In the remainder of this book, we consider that such objects are treated as single entities and we are essentially concerned with their dynamics in turbulent flows. In that sense, apart from their different densities or similar properties, we do not distinguish between solid particles, droplets or bubbles. We simply refer to them as 'discrete particles' or simply 'particles'. We also focus our attention on the case of passive particles (in contrast with active particles), i.e. particles without self-propulsion capabilities.

In spite of this complexity, it is possible to identify some basic principles. This is particularly the case when analyzing the various forces involved in the dynamics of particles in a flow and three fundamental interactions can be distinguished: fluid-particle, particle-surface and particle-particle interactions. The fluid-particle interaction is at play in particle transport and dispersion (as can be seen from aerosols in the atmosphere or sediments in rivers) and includes therefore the hydro-dynamic/aerodynamic forces acting between a fluid and a particle in suspension within it. The particle-surface interaction is involved in the deposition of particles on boundary surfaces (e.g. with the build-up of limescale in the interior of pipes). The particle-particle interaction is present in the formation (or breakup) of aggregates or when saltating particles hit layers of deposited sand particles. Both particle-surface and particle-particle interactions imply short-range interactions related to the physico-chemistry of surfaces. Hence, compared to hydrodynamical scales, these scales are usually much smaller and such interactions can be regarded as near-contact ones.

Another remark is that, for dispersed two-phase flows, we are essentially dealing with a two-step mechanism. The first one is the transport step involved in particle dispersion but also responsible for bringing discrete particles nearby. The second step is the collision step (whether it is actual collisions, agglomeration, etc.) which can only take place when particles are in the immediate vicinity of one another, as the result of the transport step. This successive mechanism is the reason why issues related to the transport step are given more emphasis in this book (see Chaps. 5–11) due their intrinsic relevance but also to pave the way for specific analysis of the open questions related to particle-particle interactions in flows (see Chap. 12).

## 2.3    On Which Modeling Road Do We Travel?

### 2.3.1    Various Modeling Standpoints

The previous examples also reveal that dispersed two-phase flows involve a variety of issues regarding our current understanding of the phenomena at play. For instance, this includes questions like: Where do the suspended particles come from? What is their nature? How are they transported/dispersed by the underlying flow? What is the fate of these objects? Where do they accumulate in natural or industrial environments? Do they have an impact on human health? Depending on the question(s) addressed in each case, the type of model retained varies (especially in terms of the phenomena that are chosen to be explicitly solved or not).

Another important feature that arises when considering models is the inherent motivation underlying such developments. One usually distinguishes three main categories of tools:

- predictive models, which are tools that provide the most accurate forecast possible within a given time (e.g. forecast of the concentration of pollen or pollutants in the air to alert populations at risk);
- optimization techniques, which explore a range of alternative situations/designs to select the best element with respect to some criteria (e.g. optimal design of a water turbine in a dam to minimize the damage on blades due to the impact of suspended sediments);
- control problems, which are related to models whose purpose is to drive a system to a desired state, while minimizing some criteria (delay or error) and ensuring a given level of stability (e.g. controlling the amount of particles suspended in an industrial flow).

In the remainder of this book, we focus specifically on predictive models. Should readers be interested in adding additional aspects related to optimization techniques, they are therefore referred to existing textbooks on optimization [16] or control problems [17].

### 2.3.2  Milestones for the Predictive Modeling Approach

When developing predictive models, it is useful to distinguish two typical viewpoints: the physicist and the engineering points of view (see also Fig. 1.1). In essence, the first viewpoint is focused on the physical content of a model while the second one is related to its numerical implementation in a software.

In this book, we adopt essentially the physicist point of view. This implies that we focus our attention on the physical models developed according to the physical phenomena and mechanisms identified through observations and/or experiments. Such physical models are theoretical formulations that reproduce (some of) these observations using a mathematical description. Hence, they are usually based on governing equations (like the Navier-Stokes equation for fluid mechanics or Newton's law of mechanics for particle dynamics that are detailed in Chap. 3). It follows that a great deal of effort is spend here to recall the physical and/or mathematical assumptions used to derive these governing equations. This is particularly relevant due to the multi-disciplinary nature of two-phase flows, which requires to hold together and combine equations derived in different contexts or even scientific disciplines. As a result, exacting attention should also be devoted to assessing that a high level of coherency and compatibility between these interacting formulations is achieved. This careful analysis should be carried out through a multi-faceted prism that covers aspects related to the information contained in each formulation (e.g. the velocity or pressure fields), the level of description (e.g. instantaneous quantities or statistical values like an average), the underlying assumptions (which phenomena are accounted for or neglected?) and the range of applicability.

### 2.3.3  Starting to Address Complexity in Models

Through these issues, we are touching upon the general question of 'how to address complexity?' This is, of course, far too vast a subject to be discussed in detail in the following pages. Nevertheless, this question lies in the background and is connected to some points discussed in later developments. For instance, it will be seen that the selection of the variables retained to characterize particles as mechanical systems is related to our choice to describe as 'noise' or 'disorder' some external effects. One of the main themes of this book is to bring forth that this is a key point if we are to devise complete and well-based descriptions. These aspects should therefore be carefully weighted. To put it in more philosophical terms, we should not be too quick in qualifying as noise some effects, since what we may call 'chaos' is perhaps an order whose reading we have not yet learned. A second example concerns the modeling approach to follow when trying to model one variable which is displaying 'complex behavior'. Should we retain the idea of devising a single model, which is then likely to become complex? Or should we consider that complexity is perhaps best captured by considering random alternation between several models, each of which being simple?

## References

1. M. Chevance, J.D. Kruijssen, E. Vazquez-Semadeni, F. Nakamura, R. Klessen, J. Ballesteros-Paredes, S.i. Inutsuka, A. Adamo, P. Hennebelle, Space Sci. Rev. **216**, 1 (2020). https://doi.org/10.1007/s11214-020-00674-x
2. V. Springel, Annu. Rev. Astron. Astrophys. **48**(1), 391 (2010). https://doi.org/10.1146/annurev-astro-081309-130914
3. P. Allen, *Sediment Routing Systems: The Fate of Sediment from Source to Sink* (Cambridge University Press, 2017). https://doi.org/10.1017/9781316135754
4. J.F. Kok, E.J. Parteli, T.I. Michaels, D.B. Karam, Rep. Prog. Phys. **75**(10), 106901 (2012). https://doi.org/10.1088/0034-4885/75/10/106901
5. F.H. Dominski, J.H.L. Branco, G. Buonanno, L. Stabile, M.G. da Silva, A. Andrade, Environ. Res. **201**, 111487 (2021). https://doi.org/10.1016/j.envres.2021.111487
6. N. Adamo, N. Al-Ansari, V. Sissakian, J. Laue, S. Knutsson, J. Earth Sci. Geotech. Eng. **11**(1), 27 (2021). https://doi.org/10.47260/jesge/1112
7. R. Shah, D. Sekulic, *Handbook of Heat Transfer*, vol. 3 (1998). https://www.academia.edu/1958878/Handbook_of_heat_transfer
8. C. Bechinger, R. Di Leonardo, H. Löwen, C. Reichhardt, G. Volpe, G. Volpe, Rev. Mod. Phys. **88**(4), 045006 (2016). https://doi.org/10.1103/RevModPhys.88.045006
9. M.C. Marchetti, J.F. Joanny, S. Ramaswamy, T.B. Liverpool, J. Prost, M. Rao, R.A. Simha, Rev. Mod. Phys. **85**(3), 1143 (2013). https://doi.org/10.1103/RevModPhys.85.1143
10. M. Elimelech, J. Gregory, X. Jia, *Particle Deposition and Aggregation: Measurement, Modelling and Simulation* (Butterworth-Heinemann, 2013). https://doi.org/10.1016/C2013-0-04548-3
11. R. Hunter, *Foundations of Colloid Science* (Oxford University Press, 2001). https://global.oup.com/academic/product/foundations-of-colloid-science-9780198505020
12. J.N. Israelachvili, *Intermolecular and Surface Forces* (Academic Press, 2011). https://doi.org/10.1016/C2011-0-05119-0

13. C. Henry, J.P. Minier, S. Brambilla, Phys. Rep. **1007**, 1 (2023). https://doi.org/10.1016/j.physrep.2022.12.005
14. S.L. Anna, Annu. Rev. Fluid Mech. **48**(1), 285 (2016). https://doi.org/10.1146/annurev-fluid-122414-034425
15. F. Lundell, L.D. Söderberg, P.H. Alfredsson, Annu. Rev. Fluid Mech. **43**, 195 (2011). https://doi.org/10.1146/annurev-fluid-122109-160700
16. S.S. Rao, *Engineering Optimization: Theory and Practice* (John Wiley & Sons, 2019). https://doi.org/10.1002/9781119454816
17. J.C. Doyle, B.A. Francis, A.R. Tannenbaum, *Feedback Control Theory* (Dover Publications, 2009)

# The Governing Equations of Dispersed Two-phase Flows

**3**

**Abstract**

Adopting the physicist point of view and pursuing a predictive modeling approach raise immediately the following questions: How do we describe a disperse two-phase flow? And what is the structure of the governing equations retained to capture the characteristics of the disperse two-phase flows we wish to simulate? To answer these questions, this chapter provides an account of the governing equations that are either used or referred to in this book. It is thus written as an introduction for readers unfamiliar with the main equations, like the Navier-Stokes equations for the dynamics of fluids and the particle equations of motion.

**Chapter Content** After recalling in Sect. 3.1 the formulations used for the fluid and particle phases, the Navier-Stokes equations for fluid dynamics are presented in Sect. 3.2, before considering the particle equations of motion in Sect. 3.3. Once the fundamental equations are formulated, the key statistical challenge induced by the turbulent nature of these flows is brought out in Sect. 3.4.

## 3.1 Two Different Formulations

A dispersed two-phase flow is a composite system involving a continuous phase and a dispersed phase (as illustrated by the examples displayed in Chap. 2). Given the field/particle nature of these composite flows, it appears natural to describe the continuous phase using a field-based approach (or Eulerian description) and the dispersed phase with a trajectory-based approach (or Lagrangian description), as represented in Fig. 3.1.

© The Author(s) 2025
J.-P. Minier et al., *Understanding Turbulent Systems*,
Lecture Notes in Physics 1039, https://doi.org/10.1007/978-3-031-84466-9_3

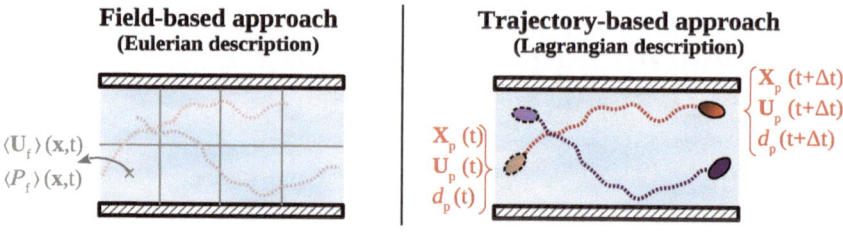

**Fig. 3.1** Illustration of the two types of approaches used to model dispersed two-phase flows: Field-based approaches (with information on the velocity $\mathbf{U}_f$ and pressure $P_f$ fields) and trajectory-based approaches (with the tracking of the particle position $\mathbf{X}_p$, velocity $\mathbf{U}_p$ or diameter $d_p$)

- Field-based approaches consist in describing a flow with a set of fields (e.g. velocity, pressure, temperature) whose evolution in space and time is expressed by transport equations. These governing equations are recalled in Sect. 3.2.
- Trajectory-based approaches consist in tracking in time the evolution of a set of variables attached to each object (e.g. position, velocity) which define the particle state vector. Since this particle state vector always includes the particle location, we are then following discrete elements both in space and time, with evolution equations represented by the traditional ordinary differential equations (ODE) of classical mechanics. These governing equations are detailed in Sect. 3.3.

In the following, more emphasis is placed on the particle phase since their statistical treatment is the main theme of the present book, as indicated in Sect. 3.4.2.

## 3.2 The Navier-Stokes Equations for Fluid Flows

The dynamics of a fluid is best described at the continuum level using the Navier-Stokes (NS) equations which gather three balance equations for the fluid mass, momentum and energy densities. Each balance equation is derived from conservation principles to express the rate of change of these field variables over a small control volume (see a classical account in [1] and a much-recommended new presentation in [2]).

In their most generic form, the NS equations correspond to the transport equations:

$$\frac{\partial \rho_f}{\partial t} + \frac{\partial \left( \rho_f \, U_{f,i} \right)}{\partial x_i} = 0 \,, \tag{3.1a}$$

$$\frac{\partial \left( \rho_f \, U_{f,i} \right)}{\partial t} + \frac{\partial \left( \rho_f \, U_{f,k} \, U_{f,i} \right)}{\partial x_k} = \frac{\partial \mathfrak{S}_{int,ik}}{\partial x_k} + f_{ext,i} \,, \tag{3.1b}$$

$$\frac{\partial \left( \rho_f \, \mathfrak{E}_f \right)}{\partial t} + \frac{\partial \left( \rho_f \, U_{f,k} \, \mathfrak{E}_f \right)}{\partial x_k} = -\frac{\partial j_{\mathfrak{E}f,i}}{\partial x_i} - \mathfrak{S}_{int,ij} \frac{\partial U_{f,j}}{\partial x_i} \,, \tag{3.1c}$$

where $\mathbf{U}_f(t, \mathbf{x})$ is the fluid velocity, $\rho_f$ the fluid density and $\mathcal{E}_f$ the fluid specific internal energy (the fluid internal energy per unit mass). The right-hand side of the momentum equation, Eq. (3.1b), includes two contributions: the stress tensor due to internal forces $\mathfrak{S}_{int}$ (e.g. accounting for pressure or viscous stresses) and the external forces acting on the fluid $\mathbf{f}_{ext}$ (e.g. gravity or back-effect of immersed particles on the flow), while the right-hand side of the energy equation, Eq. (3.1c), involves $\mathbf{j}_{\mathcal{E}_f}$ the diffusive flux of internal energy. The energy equation can also be written in terms of the fluid entropy or an equivalent thermodynamic potential, such as its free energy, enthalpy or Gibbs free energy. In practice, handling a thermodynamic potential is not convenient since these potentials are not easy to measure or to control directly and the energy equation is often expressed in terms of an intensive variable which is typically the fluid temperature. The energy equation is needed to have a close set of equations in Eq. (3.1) as it intervenes in the equation of state relating, for instance, the fluid density to its pressure and temperature (the canonical example of such an equation of state is the ideal gas law).

The NS equations given above are for a single-component fluid system. When considering a reactive flow involving a number $N_{f,\,species}$ of species, which are different molecular components taking part in $N_{f,\,reaction}$ chemical reactions, the NS equations need to be extended. Note that these species are treated individually in the different chemical reactions but are nevertheless considered as mixed at the molecular level so that we are not in the situation of a multi-phase system but, more precisely, of a single-phase but multi-component fluid system. Then, Eq. (3.1) are still valid for the fluid mixture but must be supplemented with equations for the mass fractions $\Phi_f^{[l]}$ of the different species (with $l = 1, \ldots, N_{f,\,species}$), which write

$$\frac{\partial}{\partial t}\left(\rho_f \, \Phi_f^{[l]}\right) + \frac{\partial}{\partial x_k}\left(\rho_f \, U_{f,k} \, \Phi_f^{[l]}\right) = -\frac{\partial}{\partial x_i}\left(j_{\Phi_f^{[l]},i}\right) + \rho_f \, S_f^{[l]}. \qquad (3.2)$$

On the right-hand side of Eq. (3.2), $\mathbf{j}_{\Phi_f^{[l]}}$ is the diffusive flux of the species $[l]$ resulting from molecular interactions. This flux represents the difference between the actual flux of the species $[l]$ and the flux of $[l]$ from the average velocity $\mathbf{U}_f$, making it similar to a drift velocity (note that, due to its molecular origin, $\mathbf{j}_{\Phi_f^{[l]}}$ appears indeed as a diffusive flux at the continuum level of description). The second term on the right-hand side of Eq. (3.2) stands for the reactive source term corresponding to the formation or destruction of the species $[l]$ by the chemical reactions and can be written as $\rho_f \, S_f^{[l]} = \nu_{\Phi_f^{[l]}} \, \Gamma_f$, with $\Gamma_f$ the mass rate of reaction per unit volume and $\nu_{\Phi_f^{[l]}}$ is related to the stoichiometric coefficient of the species $[l]$ (subject to a conservation of the mass of the system, so that $\sum_{l=1}^{N_{f,\,species}} \nu_{\Phi_f^{[l]}} = 0$), if we refer, for simplicity, to the case of a single reaction. In the usual case of endothermic or exothermic chemical reactions, a local source term $S_{\mathcal{E}_f}$ accounting for the release (or consumption) of heat must also be added to the right-hand side of the energy equation Eq. (3.1c). In the present work, we are not interested in reactive flows and we do not need to specify the expressions of reactive source terms. The important point is, however, that these terms are obtained from the instantaneous set of scalar variables $\Phi_f$ and its temperature (or another energy variable) through usually non-linear but known relations. This is relevant when modeling turbulent reactive flows since one of the key advantages of the PDF approach is precisely its ability to treat such source terms without approximation (see comprehensive discussions in [3, 4]).

In the remainder of this book, we focus specifically on the case of incompressible flows, in which the density $\rho_f$ remains constant, and of Newtonian fluids, in which the viscous stress evolves linearly with the local strain rate

$$\mathfrak{S}_{\text{int},ik} = -P_f \, \delta_{ik} + \rho_f \, \nu_f \left( \frac{\partial U_{f,i}}{\partial x_k} + \frac{\partial U_{f,k}}{\partial x_i} \right) , \tag{3.3}$$

with $P_f$ the fluid pressure and where the kinematic viscosity $\nu_f$ is constant. The terminology of Newtonian fluids is traditionally used to refer to fluids where not only the Newton's law, Eq. (3.3), is valid but where the Fick's and Fourier's laws also apply. The Fick's law states that the diffusive flux of the scalar [$l$] is given by

$$j_{\Phi_f^{[l]},i} = -\rho_f \, \Gamma_f^{[l]} \, \frac{\partial \Phi_f^{[l]}}{\partial x_i} , \tag{3.4}$$

where $\Gamma_f^{[l]}$ is the diffusivity of the scalar. In a similar manner, the Fourier's law for the diffusive flux entering the transport equation for the fluid temperature noted $\Theta_f$ states that

$$j_{\Theta_f,i} = -\lambda_f \, \frac{\partial \Theta_f}{\partial x_i} , \tag{3.5}$$

with $\lambda_f$ the thermal diffusivity. This thermal diffusivity is given by $\lambda_f = K_{\Theta_f}/(\rho_f \, C_{f,\text{P}})$ where $K_{\Theta_f}$ is the fluid thermal conductivity and $C_{f,\text{P}}$ the heat capacity at constant pressure. Note that the diffusive flux of the internal energy in Eq. (3.1c) is given by $\mathbf{j}_{\mathcal{E}_f} = \rho_f \, C_{f,\text{P}} \, \mathbf{j}_{\Theta_f}$ and that the flux $\mathbf{j}_{\Theta_f}$ appears once the equation for the internal energy is transformed into an equation for the fluid temperature using the heat capacity at constant pressure [2].

With the Newton, Fourier and Fick closure laws given in Eqs. (3.3)–(3.5), the NS equations for incompressible flows take the form

$$\frac{\partial U_{f,k}}{\partial x_k} = 0 , \tag{3.6a}$$

$$\frac{\partial U_{f,i}}{\partial t} + U_{f,k} \frac{\partial U_{f,i}}{\partial x_k} = -\frac{1}{\rho_f} \frac{\partial P_f}{\partial x_i} + \frac{\partial}{\partial x_k} \left[ \nu_f \left( \frac{\partial U_{f,i}}{\partial x_k} + \frac{\partial U_{f,k}}{\partial x_i} \right) \right] + g_i , \tag{3.6b}$$

$$\frac{\partial \Phi_f^{[l]}}{\partial t} + U_{f,k} \frac{\partial \Phi_f^{[l]}}{\partial x_k} = \frac{\partial}{\partial x_k} \left[ \Gamma_f^{[l]} \frac{\partial \Phi_f^{[l]}}{\partial x_k} \right] + S_f^{[l]} . \tag{3.6c}$$

In these equations, we have considered that $\mathbf{f}_{\text{ext}} = \rho_f \mathbf{g}$ with $\mathbf{g}$ the gravitational acceleration and we have conserved the molecular transport coefficients, $\nu_f$ and $\Gamma_f^{[l]}$, inside the derivatives for situations when they depend on the value of a transported scalar, for example when they are functions of the fluid temperature $\Theta_f$ which needs therefore to be included as an active scalar since it has an effect on the fluid dynamics. In the rest of this work, we consider essentially constant-property Newtonian fluids and the NS equations become [5]:

$$\frac{\partial U_{f,k}}{\partial x_k} = 0 \,, \tag{3.7a}$$

$$\frac{\partial U_{f,i}}{\partial t} + U_{f,k} \frac{\partial U_{f,i}}{\partial x_k} = -\frac{1}{\rho_f} \frac{\partial P_f}{\partial x_i} + \nu_f \frac{\partial^2 U_{f,i}}{\partial x_k \partial x_k} + g_i \,, \tag{3.7b}$$

$$\frac{\partial \Phi_f^{[l]}}{\partial t} + U_{f,k} \frac{\partial \Phi_f^{[l]}}{\partial x_k} = \Gamma_f^{[l]} \frac{\partial^2 \Phi_f^{[l]}}{\partial x_k \partial x_k} + S_f^{[l]} \,. \tag{3.7c}$$

Given the similarity between the Fick and Fourier laws in Eqs. (3.4) and (3.5), we have gathered in Eq. (3.7c) the transport equations for the species mass fractions (if any) with the one for the temperature, or the chosen thermodynamic potential, in one extended set of scalars, still noted $\Phi_f$ for the sake of simplicity (thus, for $l = N_{f,\text{species}} + 1$, we have $\Phi_f^{[l]} = \Theta_f$ and $\Gamma_f^{[l]} = \lambda_f$, as well as $S_f^{[l]} = S_{\mathcal{E}_f}$). For such constant-property fluids, note that the dynamics of the flow is accounted for by the first two equations, Eqs. (3.7a) and (3.7b), while Eq. (3.7c) represents the transport equation for passive scalars (they have no influence on the flow dynamics). This equation for scalars is therefore written for the sake of completeness since it is referred to (without reactive source terms) when discussing the diffusive limit for passive scalars in Sect. 10.3.1.

In the case of a dispersed two-phase flow system, the NS equations in Eq. (3.7) are to be understood as being written at locations $\mathbf{x}$ where no particles are present at time $t$. For a complete description of disperse two-phase flows, they need to be complemented with the formulation of the particle equations of motion.

## 3.3 The Equations of Motion for Particles

To describe the particle phase, it is natural to adopt a trajectory-based approach (also called a Lagrangian point of view). This approach consists in tracking explicitly each particle by solving the evolution equations for the state vector $\mathbf{Z}_p = (\mathbf{X}_p, \mathbf{U}_p, \mathbf{\Omega}_p)$, where $\mathbf{X}_p$ is the location of the particle center of mass while $\mathbf{U}_p$ and $\mathbf{\Omega}_p$ are the translational and rotational velocities, respectively (more details on how to select the relevant variables entering the state vector are provided in Chap. 5).

Applying the fundamental laws of classical mechanics, the time evolution of the
particle state vector is given by the following ODEs:

$$\frac{d\mathbf{X_p}}{dt} = \mathbf{U_p} \, , \tag{3.8a}$$

$$m_p \frac{d\mathbf{U_p}}{dt} = \mathbf{F}_{f \to p} + \mathbf{F}_{p \to p} + \mathbf{F}_{ext} \, , \tag{3.8b}$$

$$I_p \frac{d\boldsymbol{\Omega_p}}{dt} = \mathbf{M}_{f \to p} + \mathbf{M}_{p \to p} \, , \tag{3.8c}$$

where $m_p$ is the particle mass and $I_p$ its moment of inertia. In Eqs. (3.8b)–(3.8c),
$\mathbf{F}_{f \to p}$ and $\mathbf{M}_{f \to p}$ represent the forces and torques exerted by the fluid on each
particle, $\mathbf{F}_{p \to p}$ and $\mathbf{M}_{p \to p}$ the forces and torques due to particle-particle interactions
and $\mathbf{F}_{ext}$ the forces due to external fields (such as gravity). Solving these equations
of motion requires to obtain closed expressions for the various forces and torques
entering Eq. (3.8). In the following, we briefly recall the key forces and their
expressions.

### 3.3.1  Interactions with External Fields

One of the most common interaction between particles and external fields is the role
of gravity (other contributions like electro-magnetic fields are left out of the present
book). Gravity forces arise due to the weight of the particle and are given by:

$$\mathbf{F}_{grav \to p} = \mathcal{V}_p \, \rho_p \, \mathbf{g} \, , \tag{3.9}$$

where $\rho_p$ is the particle density, $\mathcal{V}_p$ its volume and $\mathbf{g}$ the gravity acceleration.

### 3.3.2  Interactions with the Flow Field

As depicted in Fig. 3.2, particle-fluid interactions have two origins: a contribution
related to the fluid pressure gradient and viscous effects and a contribution induced
by the relative velocity between the fluid and the particle. The different origins of
these forces stem from the decomposition of the fluid velocity field into two parts [6,
7]: an undisturbed velocity field (as if the discrete particle were not present) and a
disturbance velocity field created by the presence and motion of the discrete particle
which is considered in the analytical treatment.

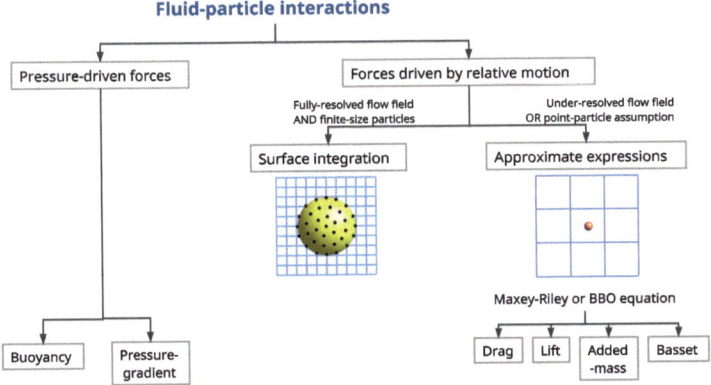

**Fig. 3.2** Illustration of the various ways to express the hydrodynamic force exerted on an object immersed in a fluid. Reprinted with permission from [8]. © 2023 Elsevier B.V. All rights reserved

First, pressure-driven forces are actually decomposed in two contributions:

- The buoyancy $\mathbf{F}_{\text{buoy}}$, which corresponds to the weight of the displaced fluid (following Archimedes' principle). Combining buoyancy with gravity, we can express directly the apparent weight of a body immersed in a fluid:

$$\mathbf{F}_{\text{grav}} + \mathbf{F}_{\text{buoy}} = \mathcal{V}_{\text{p}} \left( \rho_{\text{p}} - \rho_{\text{f}} \right) \mathbf{g} . \tag{3.10}$$

- The pressure-gradient force, which is the force exerted by the velocity field undisturbed by the presence of the particle, expressed as [9]:

$$\mathbf{F}_{\text{Press−Grad}} = \mathcal{V}_{\text{p}} \, \rho_{\text{f}} \, \frac{\text{D}\mathbf{U}_{\text{f}}}{\text{D}t} , \tag{3.11}$$

where $\text{D}/\text{D}t = \partial/\partial t + \mathbf{U}_{\text{f}} \cdot \nabla$ is the material derivative along a fluid element.

Second, the fluid forces related to velocity differences between a particle and the fluid can be evaluated from two approaches: either by using a fully-resolved flow field where the force is estimated by integrating the fluid stress tensor along the entire surface of the particle, as illustrated in the center of Fig. 3.2, or by using directly approximate expressions, often referred to as making a point-particle approximation (for reasons detailed below). Provided that the flow can be explicitly solved around each individual object, the former approach is well adapted to large particles but requires a fine-grained solution of the fluid velocity field (much smaller than the particle scale) as well as to keep track of the location of each particle surface within the flow. The latter approach is more relevant when dealing with particles smaller or comparable to the smallest turbulent eddies or when the flow field is not explicitly solved at the particle scale, which is far less demanding but requires simplifications to be made. In the case of the exact integration of the fluid stresses over particle surfaces, note that we obtain the forces as a result. On the other hand, resorting to approximate expressions means that the forces acting on particles constitute an input in the description of particle dynamics in fluid flows. These approximate expressions are indeed traditionally derived under the assumption of point-like particles, but this terminology is not strictly true (since variations at the particle scale can be accounted for, as with the Faxen terms to be considered below). What is actually assumed is that fluid variations at particle scale can be accounted for through universal formulas (as with the Faxen terms), if need be. This is in line with the basic tenets of the Kolmogorov theory, as long as particle sizes remain smaller or comparable to the small scales in a turbulent flow (see Sect. 5.3). In the present work, we are basically interested in developing a statistical treatment of particle dynamics of turbulent flows and, for that reason, we need to start with explicit fundamental governing equations. From this point onwards, we therefore evolve in the framework of given approximate expressions for the hydrodynamical forces.

The derivation of hydrodynamic forces acting on a particle embedded in a fluid flow has a long history, going back to the nineteenth century with the Basset-Boussinesq-Oseen (BBO) expression. However, in spite of numerous studies, it cannot be said that this issue has been completely solved, though a state-of-the-art formulation has emerged. Present formulations retain the drag, lift, added-mass and Basset (or history) forces [6, 7, 9–11] (see also Fig. 3.2). When added to the pressure-driven contributions, this gives the following approximation:

$$
\mathbf{F}_{f \to p} = \underbrace{\mathbf{F}_{f \to p}^{Press-Grad} + \mathbf{F}_{f \to p}^{Buoy}}_{\mathbf{F}_{pressure-driven}} + \underbrace{\mathbf{F}_{f \to p}^{Drag} + \mathbf{F}_{f \to p}^{Lift} + \mathbf{F}_{f \to p}^{Added-Mass} + \mathbf{F}_{f \to p}^{Basset}}_{\mathbf{F}_{relative-motion}}.
$$

$$(3.12)$$

Yet, the expressions of the fluid forces due to the relative motion vary according to different authors. The lift force is a special case since several expressions are proposed but each one obtained in a very specific or asymptotic situation. The result is a catalog of widely different formulations [10, 12] so that there is, at the moment, no well-established consensus about a unique expression (the 'optimal lift force' in McLaughlin's works [13, 14] is perhaps the most general expression available). Approximate expressions for the Basset force are also still debated and, consequently, present models for these two forces cannot be regarded as being reliable enough. In the present work, we are more interested in a general and well-accepted form to develop probabilistic approaches and, for these reasons, the lift and Basset forces are left out. When lift forces are not considered, the particle rotation $\mathbf{\Omega}_p$ does not play an explicit role in the dynamics of spherical particles and can also be left out of the particle state vector which is then reduced to $\mathbf{Z}_p = (\mathbf{X}_p, \mathbf{U}_p)$. Similarly, we do not consider electro- or thermo-phoresis phenomena which, if present, are accounted for by the addition of corresponding formulations in the particle momentum equation (see discussion in [15, section 2.3]). Nevertheless, it is important to realize that the framework described below is an open one and can easily be extended to account for these effects. Hence, present restrictions should not be seen as limitations but rather as an invitation to widen the range of applications by introducing future developments related to these additional physical phenomena.

For small spherical particles with diameters $d_p \sim \eta_K$, with $\eta_K$ the Kolmogorov length scale that represents the smallest length scale for fluid motion in a turbulent flow (to be defined in Sect. 5.3), the reference formulation of the particle momentum equation is derived in [6] as

$$
m_p \frac{dU_p}{dt} = \frac{\pi d_p^3}{6} \rho_f \frac{DU_s^{\mathcal{V}_p}}{Dt} + \frac{\pi d_p^3}{6} (\rho_p - \rho_f) \mathbf{g}
$$
$$
+ \frac{1}{2} \frac{\pi d_p^2}{4} \rho_f C_D |\mathbf{U}_s^{\mathcal{S}_p} - \mathbf{U}_p| (\mathbf{U}_s^{\mathcal{S}_p} - \mathbf{U}_p) + \frac{\pi d_p^3}{6} \rho_f C_A \left( \frac{DU_s^{\mathcal{V}_p}}{Dt} - \frac{dU_p}{dt} \right).
$$

$$(3.13)$$

In Eq. (3.13), the first two terms on the right-hand side are the fluid acceleration and the buoyancy force, the third term is the general form of the drag force written with the drag coefficient $C_D$ while the fourth term is the added-mass force written with a coefficient $C_A$ (usually taken as $C_A = 1/2$). The added-mass force depends on the difference between the acceleration of the fluid seen (along its own trajectory) and the particle acceleration while the drag force is expressed in terms of the velocity slip $\mathbf{U}_s^{\mathcal{S}_p} - \mathbf{U}_p$. These expressions involve $\mathbf{U}_s^{\mathcal{S}_p}$ and $\mathbf{U}_s^{\mathcal{V}_p}$, which are the fluid velocities averaged over the surface $\mathcal{S}_p$ and the volume $\mathcal{V}_p$ of the particle, respectively, i.e.

$$
\mathbf{U}_s^{\mathcal{S}_p} = \frac{1}{\mathcal{S}_p} \int_{\mathcal{S}_p} \mathbf{U}_f(t, \mathbf{r}) \, d\mathbf{r} \,,
$$

$$(3.14)$$

$$
\mathbf{U}_s^{\mathcal{V}_p} = \frac{1}{\mathcal{V}_p} \int_{\mathcal{V}_p} \mathbf{U}_f(t, \mathbf{r}) \, d\mathbf{r} \,.
$$

$$(3.15)$$

These velocities are expressed by a series expansion around the 'velocity' at the particle center, which introduces the notion of the velocity of the fluid seen, $\widehat{\mathbf{U}}_s(t) = \mathbf{U}_f^{ud}(t, \mathbf{X}_p(t))$

$$
\mathbf{U}_s^{\mathcal{S}_p} \simeq \widehat{\mathbf{U}}_s + \frac{d_p^2}{24} \left( \nabla^2 \mathbf{U}_f^{ud} \right)(t, \mathbf{X}_p(t)) \,,
$$

$$(3.16)$$

$$
\mathbf{U}_s^{\mathcal{V}_p} \simeq \widehat{\mathbf{U}}_s + \frac{d_p^2}{40} \left( \nabla^2 \mathbf{U}_f^{ud} \right)(t, \mathbf{X}_p(t)) \,.
$$

$$(3.17)$$

The notation $\mathbf{U}_f^{ud}(t, \mathbf{X}_p(t))$ designates the 'undisturbed velocity field' that would exist if the particle were not present at the location $\mathbf{x} = \mathbf{X}_p(t)$ at time $t$ [6, 7] and the last terms added to $\widehat{\mathbf{U}}_s(t)$ on the right-hand side of Eqs. (3.16) and (3.17) are the Faxen terms [6, 7, 10].

For small particles (say $d_p \leq \eta_K$), the Faxen terms are small enough to be neglected which means that, for the expression of these hydrodynamical forces, particles are considered as mere points. This corresponds to the point-wise particle approximation mentioned above and a classical form of the particle momentum equation is

$$
m_p \frac{dU_p}{dt} = \frac{\pi d_p^3}{6} \rho_f \frac{DU_s}{Dt} + \frac{\pi d_p^3}{6} \left( \rho_p - \rho_f \right) g
$$
$$
+ \frac{1}{2} \frac{\pi d_p^2}{4} \rho_f \, C_D | U_s - U_p | (U_s - U_p) + \frac{\pi d_p^3}{6} \, C_A \, \rho_f \left( \frac{DU_s}{Dt} - \frac{dU_p}{dt} \right) . \tag{3.18}
$$

The velocity of the fluid seen is now given as the local instantaneous value of the fluid velocity at the same time and at the particle position, i.e. $U_s(t) = U_f(t, X_p(t))$, where $U_f(t, x)$ represents the fluid velocity field without having to distinguish undisturbed and disturbed fluid velocity fields anymore. For particles heavier than the fluid $\rho_p \gg \rho_f$, it can be shown that the drag force is the dominant one [6, 7, 16] and the particle momentum equation is then further reduced to

$$
\frac{dU_p}{dt} = \frac{U_s - U_p}{\tau_p} + g \tag{3.19}
$$

where the drag force is written so as to bring out the particle relaxation timescale $\tau_p$

$$
\tau_p = \frac{\rho_p}{\rho_f} \frac{4 \, d_p}{3 \, C_D | U_s - U_p |}. \tag{3.20}
$$

The form given in Eq. (3.19) is referred to as the reduced particle momentum equation. The particle relaxation timescale $\tau_p$, which is a measure of particle inertia, is the key notion for particle transport. More precisely, $\tau_p$ represents the timescale needed for particle velocities to adjust to the local fluid velocity seen. In the Stokes regime, which is valid when $Re_p \leq 1$ (with the particle Reynolds number $Re_p = |U_s - U_p| d_p / \nu_f$), the drag coefficient is $C_D = 24/Re_p$. In that case, we retrieve the classical form $F_{f \rightarrow p}^{\text{Drag}} = 3 \pi \rho_f \, \nu_f \, d_p \left( U_s - U_p \right)$ and the particle relaxation timescale is given by the well-known expression

$$
\tau_p = \frac{\rho_p}{\rho_f} \frac{d_p^2}{18 \nu_f}. \tag{3.21}
$$

For general values of $Re_p$, the drag coefficient is obtained through empirical correlations and an often-retained formula is [10, 17]

$$
C_D = \begin{cases} \dfrac{24}{Re_p} \left[ 1 + 0.15 \, Re_p^{0.687} \right] & \text{if } Re_p \leq 1000, \\ 0.44 & \text{if } Re_p \geq 1000. \end{cases} \tag{3.22}
$$

This correlation is for isolated particles or when particle concentration is not high enough for collective effects, representing hydrodynamical influences of each particle on its neighbors (the 'wake effect'), to be significant. If needed, particle wake effects can be accounted for with modified correlations based, for example, on the local particle volumetric fraction [10, 18].

In most cases involving particles or droplets in a turbulent flow, the preceding expressions provide a satisfactory picture and are sufficient to describe the influence of fluid flows on particle dynamics. For diameters larger than the Kolmogorov length scale, it is seen from the general form of the particle momentum equation, cf. Eq. (3.13), and the expressions given in Eqs. (3.14)–(3.15) that a non-negligible particle size induces a filtering effect and that fluid velocity fluctuations with length scales smaller than $d_p$ tend to be smoothed out and act as an underlying noise. Specific developments can then be considered (see for example interesting proposals in [19, 20] among other ideas), but since we concentrate essentially on turbulent dispersion and collisional effects, the particle point-wise approximation is retained in the present work from now on (see further comments in Sect. 3.4.2).

Even when added-mass forces are neglected, the acceleration of the fluid seen (i.e., $DU_s/Dt$) is sometimes retained in the particle velocity equation which is then written as

$$\frac{dU_p}{dt} = -\frac{1}{\rho_p}\nabla P_f + \frac{U_s - U_p}{\tau_p} + g, \tag{3.23}$$

where we have used the Euler form of the NS equations by neglecting viscosity to relate the fluid pressure gradient and acceleration as

$$\frac{DU_s}{Dt} = -\frac{1}{\rho_f}\nabla P_f + g. \tag{3.24}$$

This leads to an alternative form of Eq. (3.23), which is

$$\frac{dU_p}{dt} = \frac{\rho_f}{\rho_p}\frac{DU_s}{Dt} + \frac{U_s - U_p}{\tau_p} + \left(1 - \frac{\rho_f}{\rho_p}\right)g. \tag{3.25}$$

Equations (3.23) or (3.25) are referred to as the extended particle momentum equation.

**To Summarize** two forms of the particle momentum equation were derived and are used in the rest of this work. The first one is the reduced particle momentum equation, Eq. (3.19), which involves only the drag force written with the particle relaxation timescale $\tau_p$. The second form is the extended particle momentum equation, Eqs. (3.23) or (3.25), which contains the same drag force but also the fluid pressure-gradient (or the fluid acceleration term).

### 3.3.2.1 A Note on Brownian Effects

So far, we have only considered forces arising from the continuum description of fluid flows. However, small enough particles are sensitive to Brownian effects due to the molecular nature of the fluid. This vague notion of 'small enough' is quantified by introducing the criterion that, in the absence of a fluid flow, the particle settling velocity is counterbalanced by the random velocities imparted by collisions with the molecules of the fluid. This implies that such particles do not sediment, as the colloids discussed in Sect. 2.2.3. Using a one-dimensional formulation in the direction aligned with gravity and the equipartition theorem of statistical physics, the diameter $d_p$ of these particles can be estimated by

$$\sqrt{\frac{k_B \, \Theta_f}{m_p}} \sim \tau_p \, g \implies d_p \simeq \left( \frac{\rho_f^2 v_f^2}{\rho_p^3 \, g^2} k_B \, \Theta_f \right)^{1/7}, \tag{3.26}$$

where $k_B$ is the Boltzmann constant. This defines more precisely colloidal particles which have therefore a diameter of the order of a few microns ($d_p \leq 1 - 2 \, \mu m$) if we consider air at room temperature while, for $d_p \geq 5 - 10 \, \mu m$, particles are called inertial and Brownian effects become more and more negligible as the particle size increases. In practice, it is best to include Brownian motion in the particle momentum equation whatever the particle diameter so that its effects diminish continuously when considering increasing particle diameters without having to introduce an artificial cut-off between colloidal and inertial particles. Although implementing Brownian effects is straightforward, it is connected to interesting questions and is addressed in more details in Chap. 9. Moreover, Brownian motions are typically expressed by Wiener processes. Since these fundamental stochastic processes are only introduced in Sect. 4, it is best to address the issues related to Brownian effects later on.

### 3.3.2.2 Remarks on Bubbly Flows

Compared to solid particles and droplets in a gas, the case of small bubbles in a liquid raises specific concerns and a few comments are in order. First, bubbles tend to have larger sizes (due to surface tension and the resulting inner pressure) which implies that the volume fraction occupied by the bubble phase becomes quickly appreciable. In practice, this volumetric back effect from bubbles to the liquid phase is important and the liquid-phase equations need to be modified accordingly (through some interface tracking methods). Second, even if the point-wise approximation is less valid for bubbles, we can still follow their center-of-mass motion. Yet, since bubble density is much smaller than the liquid one ($\rho_p \ll \rho_f$), we cannot neglect the added-mass force anymore. This leads to the bubble momentum equation

$$\frac{d\mathbf{U}_p}{dt} = -\frac{1}{\rho_p} \left( \frac{1 + C_A}{1 + C_A \rho_f / \rho_p} \right) \nabla P_f + \frac{\mathbf{U}_s - \mathbf{U}_p}{\tau_p} + \mathbf{g}. \tag{3.27}$$

Assuming that we are in the Stokes regime, the relaxation timescale in Eq. (3.27) is transformed to become

$$\tau_p = \left(\frac{\rho_p + C_A \rho_f}{\rho_f}\right) \frac{d_p^2}{18\nu_f}. \tag{3.28}$$

As a consequence, we must know not only $U_s$ but also $\nabla P_f$ or $DU_s/Dt$. Apart from the different constants appearing in front of the pressure-gradient term and in the definition of $\tau_p$, it is seen that Eq. (3.27) is quite similar to the extended particle equation of motion, Eq. (3.23), and does not involve additional terms. This points to the interest of being able to treat the extended form of the particle equation of motion for small solid particles since we can include bubbly flows with minor changes.

### 3.3.3  Interactions Between Particles

For discrete particles embedded in a fluid flow, the previous interactions always apply. In several situations, they need to be supplemented by particle-particle interactions, which can take various forms depending on the context of the study (a more detailed account is given in [8, Section 4]). For instance, some of these forces arise even when particles are not in contact (e.g. long-range electrostatic interactions between charged objects, capillary forces between two particles connected by a liquid bridge, short-range van der Waals interactions between two bodies) while other forces occur solely when particles are in contact (such as adhesion or friction forces). Particle-particle interactions are therefore of a more general nature than particle collisions which correspond to a limit case when the hard-sphere model is retained (no interaction at distance between two particles and an infinite repulsive potential at some 'contact distance' taken as the sum of the two particle radius). If we introduce a general notation, $F_{p \to p}^{[i]-[j]}$ for such an interaction between two particles labeled as $[i]$ and $[j]$ respectively, an important feature is that this interaction depends (a minima) upon the two particle positions, which we indicate by writing $F_{p \to p}^{[i]-[j]}(X_p^{[i]}, X_p^{[j]})$. This dependency is for two-particle interactions involving forces between pairs of discrete particles and, in some circumstances, it must be extended to include a subset of $N_p$ particles (with $N_p \geq 3$) which need therefore to be tracked simultaneously. This remains valid for collisions between discrete particles even in the point-particle approximation which is the limit case of this general situation. The important outcome is that we must, at least, have access to the correlated dynamics of pairs of particles, in contrast to the interactions with external fields and with the fluid flow.

Another noteworthy feature of particle-particle interactions is that their relative significance with respect to the hydrodynamical forces is essentially a function of the particle concentration in the flow. When the particle concentration is 'low enough', we are dealing with dilute particle-laden flows and we can neglect particle-particle interactions altogether. In other words, a discrete particle does not 'see' or 'meet' other discrete particles and we can proceed by considering each discrete particle as being isolated in the flow (we simply add up one-particle effects over the number of particles considered). From a physical standpoint, this means that we are focusing on modeling particle transport by fluid flows. This is a situation of utmost practical relevance and is taken as the basic physical situation in which additional phenomena can be introduced once the modeling framework is properly set up. For these reasons, we start by addressing dilute particle suspensions from Chaps. 5 to 10. For moderately dense particle flows, we then discuss particle back effects on the fluid flow in Chap. 11 (where we revisit the different regimes and the somewhat fuzzy limits between dilute and dense suspensions). Whereas the elementary collision event between two discrete particles is simple to formulate, the statistical treatment of particle-particle interactions in fluid flows raises intricate issues concerning how we should handle transport and collision in the same description, some of which remain open, and is given special attention in Chap. 12.

## 3.4    Adopting a Statistical Approach

The NS equations are generally obtained at the hydrodynamical level of description by writing balance equations for fluid mass, momentum and energy densities and by applying the Newton, Fourier and Fick laws which are, at that level of description, phenomenological or empirical closures. These equations can also be derived from the kinetic theory (see [21, chapter 7], among several references), and discussing the similarities and differences with respect to this reference theory provides the background for turbulent dispersed two-phase flow modeling. In the following, we introduce the main ideas behind the statistical formulation developed from Chap. 4 by trying to answer general questions such as 'what are we looking for?' and 'what is the interest of a PDF approach?'.

### 3.4.1   The Roads to the Macroscopic Level of Description

*What Are the Arrival and Departure Halls?*  If the hydrodynamical equations mark the end of the road for the application of the kinetic theory, they constitute the starting point when we turn our attention towards the all-important phenomena occurring in the vast majority of flows: turbulence. This is sketched in Fig. 3.3. It is first to be noted that, contrary to density or molecular viscosity which are properties of a fluid, turbulence is a characteristic of a flow. Studying single-phase turbulence per se is not the central objective of the present book and we refer essentially to existing textbooks [22, 23] and, in particular, to the reference book by Pope on turbulent flows [5]. Nevertheless, since many of the developments introduced for dispersed two-phase flows in the following chapters rely heavily on turbulence theory (to be recalled in Sect. 5.3) and turbulence models (to be discussed in Sect. 6.2 and later on in Sect. 10.3), it is useful to clarify the guidelines of the enterprise in which we are about to embark.

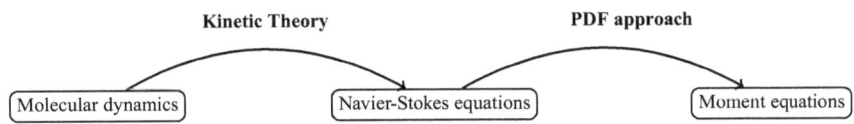

**Fig. 3.3** Sketch of the levels where the kinetic theory and the PDF approach operate: the kinetic theory starts from molecular dynamics and leads to the NS equations while the PDF approach starts from the NS equations and leads to the moments of interest in a turbulent flow. This representation is given for the fluid phase for which the governing equations are the NS equations but is relevant for a more complex system with the intermediate level being the governing equations at the hydrodynamical level of description as done in Sect. 3.1

*Why Do We Need a Statistical Approach?* At such an early stage in the developments, we can limit ourselves to a crudely-cut picture and describe turbulence as a competition between convective effects (essentially destabilizing) and viscous ones (essentially stabilizing). When considering turbulent flows, there are no new fundamental equations to work out and the governing equations are still the NS equations. Yet, in a turbulent flow, the velocity field exhibits variations, or fluctuations, which cover a wide range of time and space scales. Given initial and boundary conditions, it is possible to solve the NS equations to obtain what is then a fully-resolved fluid flow. This is the purpose of direct numerical simulation (DNS), which is however tractable only in simple geometries and for not-too-intense turbulence (thus, justifying its nickname of 'brute-force calculation'). To pursue the parallel with kinetic theory, this is similar to molecular dynamics (MD) where all the molecules present in a domain are explicitly tracked at the same time. When such a direct method is not applicable, we are faced with a system having a huge number of degrees of freedom. By regarding the velocity and pressure fields of a turbulent flow as random fields, we can adopt a statistical viewpoint and begin the search for reduced statistical descriptions to derive the 'macroscopic equations'. This name of macroscopic equations is retained to continue the analogy with classical statistical physics but refers now to the first few moments (or averages) of the velocity field, as well as the pressure or temperature fields if need be. This encompasses formulations based on different statistical operators, such as the classical Reynolds-averaged Navier-Stokes (RANS) approach where true probabilistic averages are considered as well as the more recent large-eddy simulation (LES) method where locally spatial averaging is applied, since these operators can be set in the same framework. The terminology of microscopic and macroscopic equations remains relevant, provided that they are regarded as referring to the level of description more than to different volume sizes (see a detailed discussion in [24, section 3]). It is therefore retained when describing the mathematical theoretical background to be PDF formulation in Sect. 4.1.

*What Is the Leading Idea of the PDF Approach?* In turbulence, the traditional modeling approach consists in applying the statistical operator to the NS equations to derive directly the transport equations satisfied by the few moments of interest.

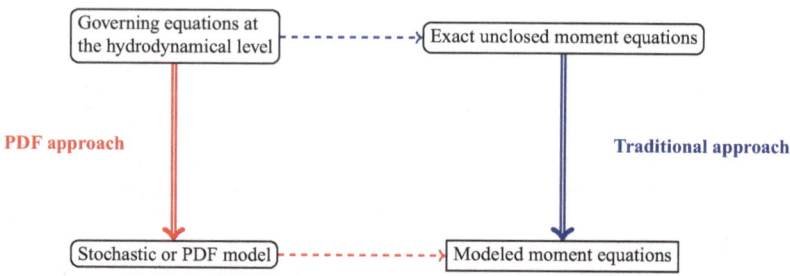

**Fig. 3.4** Illustration of the different ways to obtain the moments of interest. The traditional approach derives the exact unclosed moment equations and then model them (cf. blue lines) while the PDF approach replaces the governing equations at the hydrodynamical level of description by a stochastic model from which the modeled moment equations are extracted (cf. red lines). The double arrow indicates a modeling step and the dashed line an exact operation

These moment equations are exact but unclosed, due to the non-linearity of the NS equations. They are then closed by resorting to phenomenological relations often based on a gradient hypothesis such as Newton's law. As recalled at the beginning of this section, this traditional or moment approach mirrors the one used in continuum mechanics to obtain the NS equations themselves. In the PDF approach, modeling is also achieved through a two-step process but with the two steps applied in the reverse order. This means that the NS equations are first replaced by a stochastic process mimicking the key features of turbulent flows and, once this is done, moments are extracted from the stochastic model or its distributions (hence the name of the method). It follows that the PDF approach is, at least in spirit, similar to kinetic theory, where MD is replaced by the Boltzmann equation from which the hydrodynamical equations are derived. As illustrated in Fig. 3.4, we can say that, in the traditional approach, the first step is exact while the second one is modeled, whereas it is the other way around in the PDF approach. This presentation was made essentially with reference to the fluid phase since the different levels of description are easily understood (they are well exemplified by known formulations). The leading idea of the PDF approach is however of a general nature and can be carried out to more complex systems where we start from governing equations expressed at the hydrodynamical level of description from which we aim to extract a reduced statistical description due to the underlying effects of turbulence.

*What Is the Main Interest of the PDF Approach?* Its familiar aspect should not hide the fact that the traditional modeling approach rests on two assumptions: first, we can close the system of equations directly at the macroscopic level with physically-meaningful relations; second, the information available is sufficient to address more complex physics than the dynamics of a turbulent flow. It turns out that both assumptions run quickly into trouble. Difficulties with the second assumption are readily illustrated by turbulent reactive flows and ... turbulent dispersed two-phase flows. Indeed, if we go back to the transport equation for the mass fractions of molecular species involved in chemical reactions given in Eq. (3.7c) and apply an averaging operator, it is immediately seen that obtaining the mean values of the source terms $S_f^{[l]}(\mathbf{\Phi})$ for the species $[l]$ becomes an open issue: we are faced with the usual, but often inextricable, problem of having to express the mean value $\langle S_f^{[l]}(\mathbf{\Phi}) \rangle$ knowing only the first- and second-order moments of $\mathbf{\Phi}$. The same issue appears in the two-phase flow situation since we have to express the mean value of the particle relaxation timescale $\tau_p$ which, as indicated by Eqs. (3.20) and (3.22), is a non-linear function of the particle variables and, especially, its diameter which can vary over orders of magnitude in poly-dispersed two-phase flows. On the contrary, the PDF approach generates the distribution of these variables and can therefore treat without approximation such terms as well as any source term, however complex, as long as it is a function of the variables entering the PDF description. With respect to the first assumption mentioned above, it may be fair to say that the difficulties encountered are shared by all modeling approaches faced with the formidable challenge of turbulence. Due in great part to their particle-based formulations which are well-suited to handle transport phenomena, PDF approaches have nevertheless brought in significant insights, especially to clarify the intrinsic diffusive or convective nature of many terms. A theoretically-relevant and practically-important example is the case of point-source dispersion where PDF methods capture, with minimum input, the continuous switch from the near-field convective regime to the far-field diffusive spread (this is the first numerical example presented in Chap. 7).

In short, the attractiveness of PDF methods come from their ability to treat without approximation transport and local source terms, as well as building a flexible framework that can accommodate more complex physical phenomena. One such application being turbulent dispersed two-phase flows.

*Is It Really Similar to Kinetic Theory?* The principles sound alike but the PDF approach differs noticeably from kinetic theory in terms of the levels of description where it operates (the departure and arrival levels) and of the physical systems to which it is applied. In the classical kinetic theory picture, the starting point is made up by a large number of elementary particles undergoing free flights (or evolving in a known potential field) and collisions and the end point is a set of fields related through the NS equations. We have thus moved from particles to fields, each representing a different physical object operating at a different level of description. In single-phase turbulence, we start with fields, then introduce stochastic particles (or their distributions) from which new fields, which are the moments of interest, are derived. In turbulent dispersed two-phase flows, this is further compounded since the starting point is now a combination of fields and particles. Not only are the mathematical objects of fields and particles playing different roles compared to kinetic theory but the basic physical phenomena are not the same. And when a turbulence model is involved, we are not dealing anymore with particles traveling in void or in fully resolved fields between collisions but with discrete particles being transported and agitated by non-fully resolved velocity fields. Taking a slightly more general perspective but still focusing on the dispersed phase, this means that we are considering particles influenced by media with non-zero time and space correlations, which is an intermediate situation between pure order and pure disorder. This constitutes a brand new challenge.

*What Do We Mean by a Moment/PDF Model?* Since we consider a composite field-and-particle system, we would expect the PDF approach to include both the continuous and dispersed phases in its description. This is indeed possible (see [16, section 8]) and even desirable if we wish to set up a comprehensive framework. In this work, we choose however to commit the modeling efforts to address more specifically the dispersed or particle phase. This implies that the macroscopic description of the fluid phase is considered as available beforehand. For example, when devising statistical models for discrete particle dynamics, the fluid mean velocity field (or the velocity second-order moments) is considered as known. This hybrid formulation is therefore referred to as a moment/PDF model.

*Why Going for a Moment/PDF Model?* We do so to concentrate on the statistical description of the dispersed phase which, at the moment, is the less explored and most challenging part in an overall attempt at a full PDF description. Another incentive is that this allows to develop stochastic models for discrete particles as outgrowths of the ones used for the description of the carrier turbulent flow that contains these particles. For instance, when using the mean fluid velocity field in a particle model, it is important to note that the governing (modeled) equation of that very field is itself derived from a PDF method for single-phase turbulence. This constitutes an essential characteristic of the PDF approaches developed in this work and this consistency issue (the fact that PDF models for discrete inertial particles revert to known PDF models for fluid particles when particle inertia vanishes) plays a central role. In that sense, focusing on the statistical models for the dispersed phase does not mean that we are just going halfway in the overall fluid-particle two-phase flow description. We are in fact going up to a point where the goal of reaching such a complete formulation becomes quite attainable.

### 3.4.2 The Issue to Address in Statistical Physics

With these clarifications, we first consider discrete particle dynamics expressed by the system

$$\frac{d\mathbf{X}_p}{dt} = \mathbf{U}_p \,, \tag{3.29a}$$

$$\frac{d\mathbf{U}_p}{dt} = \frac{\mathbf{U}_s - \mathbf{U}_p}{\tau_p} + \mathbf{g} \,, \tag{3.29b}$$

where $\tau_p$ is given by Eqs. (3.20) and (3.22). As indicated above, this corresponds to the reduced particle equation of motion. However, by doing so, we have stripped

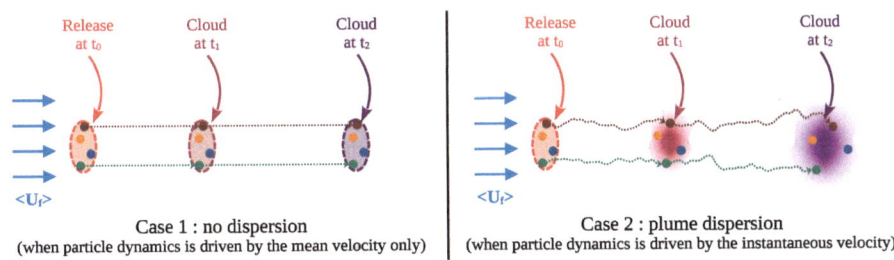

**Fig. 3.5** Sketch of the outcomes of simulations of a point-source dispersion problem. When the mean fluid velocity is used in the particle momentum equation, particles released with the same initial conditions follow the same trajectory and we fail to capture the phenomenon of plume dispersion (left-hand side). When a model for the instantaneous fluid velocity encountered by particles is introduced in the particle momentum equation, the simulations reproduce the plume spread due to the different fluctuating fluid velocities (right-hand side)

down the problem to its bare statistical essential which is the treatment of the velocity of the fluid seen $\mathbf{U}_s(t)$. Indeed, the advantage of Eq. (3.29b) is to bring out that the real issue is how to model the 'driving force', $\mathbf{U}_s(t)$, more than the various forms taken by the particle momentum equation depending on which forces are retained. This can be rephrased using an analogy where the particle momentum equation is regarded as a (non-linear) filter from which we derive statistics of interest which are outputs of the filter. However, the real issue is to model the input (or the driving force) of this filter, which is the instantaneous velocity of the fluid seen $\mathbf{U}_s(t)$.

We cannot reduce the problem anymore and, in Eq. (3.29), $\mathbf{U}_s(t)$ must represent the instantaneous fluid velocity seen by discrete particles. This is revealed by considering the case of point-source dispersion, as illustrated in Fig. 3.5. Let us imagine that we release a number of particles from a given location and with a given initial velocity in a turbulent flow where only the mean velocity field is known. If we limit ourselves to the available information and plug in the mean value $\langle \mathbf{U}_s(t) \rangle$ in Eq. (3.29), we have a purely deterministic system and all the particles follow the same trajectory since they start with the same initial conditions. Clearly, we would not only fail to capture the correct spread of the plume but we would even fail to predict that there is a plume. Note that this fluid mean velocity field could be exact, for that matter, without changing the problem. There is no avoiding that we must handle the instantaneous fluid velocity seen in the discrete particle equation of motion.

In the rare situations where the fluid flow is fully resolved and where we have access to the instantaneous velocity field $\mathbf{U}_f(t, \mathbf{x})$, $\mathbf{U}_s(t)$ is known and we are dealing with a fully-characterized deterministic time signal as the input of the particle-momentum filter, or a series of such time functions if we are considering a number of discrete particles. In that case, we can proceed with the particle kinetic variables $(\mathbf{X}_p, \mathbf{U}_p)$ as in classical mechanics. However, for the vast majority of turbulent flows the number of degrees of freedom is so high (cf. Sect. 5.3) that we have only access to a limited information, usually in the form of the first few moments of the fluid velocity field. The loss of information implied by such coarse-grained descriptions is reflected by the representation of $\mathbf{U}_s(t)$ as a stochastic process.

At first sight, a statistical treatment of the extended particle equation of motion is more difficult. Indeed, if we consider, for instance, the form given in Eq. (3.25), it appears that the issue of modeling $\mathbf{U}_s(t)$ is now compounded by the fact that we must simulate also the velocity increments along a fluid particle trajectory $D\mathbf{U}_s/Dt$: we must now generate the pair of correlated stochastic processes $(\mathbf{U}_s, D\mathbf{U}_s/Dt)$. However, it turns out that new insights into the models for the velocity of the fluid seen, described in Chap. 8 (in particular in Sect. 8.2 and the recent proposal based on a two-step formulation in Sect. 8.3.2), contain such a correlated sub-model for $D\mathbf{U}_s/Dt$. Therefore, considering first the reduced particle equation of motion does not imply that we end up in deadlock when we move to the extended particle equation of motion. It is nevertheless best to do so in a step-by-step manner and clarify first in the following chapters the issues raised by the simplified formulation in Eq. (3.29).

We can now enter the world of probabilistic formulations.

## References

1. R.B. Bird, C. Curtiss, R. Armstrong, O. Hassager, *Dynamics of Polymeric Liquids*, 2nd edn. (John Wiley and Sons, Singapore, 2002). https://www.wiley.com/en-us/Dynamics+of+Polymeric+Liquids%2C+2+Volume+Set%2C+2nd+Edition-p-9780471518440
2. D.C. Venerus, H.C. Öttinger, *A Modern Course in Transport Phenomena* (Cambridge University Press, Cambridge, 2018). https://www.cambridge.org/us/universitypress/subjects/engineering/chemical-engineering/modern-course-transport-phenomena
3. S.B. Pope, Progress Energy Combustion Sci. **11**(2), 119 (1985). https://doi.org/10.1016/0360-1285(85)90002-4
4. D.C. Haworth, Progress Energy Combustion Sci. **36**(2), 168 (2010). https://doi.org/10.1016/j.pecs.2009.09.003
5. S. Pope, *Turbulent Flows* (Cambridge University Press, Cambridge, 2000). https://doi.org/10.1017/CBO9780511840531
6. R. Gatignol, J. Mécanique Théorique et Appl. **2**(2), 143 (1983)
7. M.R. Maxey, J.J. Riley, Phys. Fluids **26**(4), 883 (1983). https://doi.org/10.1063/1.864230
8. C. Henry, J.P. Minier, S. Brambilla, Phys. Rep. **1007**, 1 (2023). https://doi.org/10.1016/j.physrep.2022.12.005
9. J.G.M. Kuerten, Flow Turbulence Combustion **97**(3), 689 (2016). https://doi.org/10.1007/s10494-016-9765-y
10. R. Clift, J.R. Grace, M. Weber, *Bubbles, Drops and Particles* (Academic Press, Cambridge, 1978). https://store.doverpublications.com/products/9780486317748
11. L. Brandt, F. Coletti, Ann. Rev. Fluid Mech. **54**, 159 (2022). https://doi.org/10.1146/annurev-fluid-030121-021103
12. C. Henry, J.P. Minier, Progress Energy Combustion Sci. **45**, 1 (2014). https://doi.org/10.1016/j.pecs.2014.06.001
13. J.B. McLaughlin, J. Fluid Mech. **224**, 261 (1991). https://doi.org/10.1017/S0022112091001751
14. Q. Wang, K. Squires, M. Chen, J. McLaughlin, Int. J. Multiphase Flow **23**(4), 749 (1997). https://doi.org/10.1016/S0301-9322(97)00014-1

15. J.P. Minier, Progress Energy Combustion Sci. **50**, 1 (2015). https://doi.org/10.1016/j.pecs.2015. 02.003

16. J.P. Minier, E. Peirano, Phys. Rep. **352**(1–3), 1 (2001). https://doi.org/10.1016/S0370-1573(01)00011-4

17. C.E. Brennen, *Fundamentals of Multiphase Flow* (Cambridge University Press, Cambridge, 2005). https://doi.org/10.1017/CBO9780511807169

18. Z. Cheng, A. Wachs, Int. J. Multiphase Flow **167**, 104524 (2023). https://doi.org/10.1016/j. ijmultiphaseflow.2023.104524

19. T. Elperin, N. Kleeorin, V.S. L'vov, I. Rogachevskii, D. Sokoloff, Phys. Rev. E **66**(3), 036302 (2002). https://doi.org/10.1103/PhysRevE.66.036302

20. T. Elperin, N. Kleeorin, M.A. Liberman, V.S. L'vov, I. Rogachevskii, Environ. Fluid Mech. **7**, 173 (2007). https://doi.org/10.1007/s10652-007-9019-6

21. H.C. Öttinger, *Beyond Equilibrium Thermodynamics* (John Wiley & Sons, Hoboken, 2005). https://doi.org/10.1002/0471727903

22. A. Monin, A. Yaglom, *Statistical Fluid Mechanics, Volume II: Mechanics of Turbulence*. Dover Books on Physics, vol. 2 (Dover Publications, Mineola, 2013). https://store.doverpublications. com/products/9780486458915

23. H. Tennekes, J.L. Lumley, *A First Course in Turbulence* (MIT Press, Cambridge, 1972). https:// doi.org/10.7551/mitpress/3014.001.0001

24. J.P. Minier, Phys. Rep. **665**, 1 (2016). https://doi.org/10.1016/j.physrep.2016.10.007

# Reduced Statistical Descriptions and the Probabilistic Framework

**4**

**Abstract**

There are basically two steps in dispersed two-phase flow modeling. The first one deals with the selection of the variables entering the particle state vector and with the construction of the stochastic processes used to model particle dynamics. This step encompasses most of the issues related to physics and provides answers to the questions: How do we describe a mechanical system? How do we represent its dynamical evolution? The second step concerns the probabilistic framework that is needed to guide us from particle stochastic models to the statistics of interest in practical situations. This step is more mathematical in nature and provides answers to the questions: What are the main stochastic processes? How do we handle them to stay clear of mathematical pitfalls? To concentrate on the physical issues in the following chapters while relying on a safe probabilistic framework, we give here an outline of the mathematical background.

**Chapter Content** We start with a reminder of the PDF methodology which leads from Lagrangian stochastic models to mean fields in Sect. 4.1. The definition and essential characteristics of stochastic processes are addressed in Sect. 4.2 before focusing on jump-diffusion processes and the related Fokker-Planck equations in Sect. 4.3. The dynamical Monte Carlo method and its applications are then discussed in Sect. 4.4. This provides the background to clarify the issues related to the choice of a particle state vector which are brought out in Sect. 4.5.

© The Author(s) 2025
J.-P. Minier et al., *Understanding Turbulent Systems*,
Lecture Notes in Physics 1039, https://doi.org/10.1007/978-3-031-84466-9_4

## 4.1   From Microscopic to Macroscopic Descriptions

### 4.1.1   Mass Density Functions and Mean Fields

The PDF approach to single-phase and dispersed two-phase turbulent flows has reached a mature level and detailed presentations can be found in [1–6]. The formalism bears similarities with the formulation in statistical physics based on distribution functions but there are also differences. First, the normalization is implemented in terms of the mass of the fluid or particle phases instead of the number of particles. Second, this PDF framework grew out of the probabilistic description of turbulent single-phase flows where the 'microscopic level of description' (i.e., the Navier-Stokes equations) is made up by fields, and of dispersed two-phase flows where it is a combination of particles and fields. As noted in Sect. 3.4.1, this is in contrast with classical statistical physics where the microscopic level involves discrete particles (typically atoms/molecules) whereas the macroscopic one corresponds to fields (the hydrodynamical level). For this reason, a specific distinction is made between Lagrangian and Eulerian PDFs to avoid confusions between descriptions where the position is a variable (Lagrangian) or a parameter (Eulerian).

The methodology can be developed for general $N$-particle PDF descriptions and state vectors (whose selections are addressed later on) but, to keep simple notations, we consider essentially the one-particle PDF approach. This allows to handle a large number of particles as samples instead of a large number of pairs of particles for two-particle PDF models, etc. Adopting therefore a Lagrangian standpoint and focusing on the particle phase, we introduce the decomposition of the particle state vector $\mathbf{Z}_p = (\mathbf{X}_p, \mathbf{Z}_c)$ where the particle location $\mathbf{X}_p$ is always present and where $\mathbf{Z}_c$ is the complementary part of the state vector. In sample space, the corresponding variables are noted $\mathbf{z}_p = (\mathbf{y}_p, \mathbf{z}_c)$. Starting from the unconditional Lagrangian PDF $p(t; \mathbf{y}_p, \mathbf{z}_c)$ and using $M_p$ for the total mass of the discrete particles in the domain, the Lagrangian and Eulerian mass density functions (MDFs) for the dispersed phase are defined by the relations

$$\mathcal{F}_p^L(t; \mathbf{y}_p, \mathbf{z}_c) = M_p \, p(t; \mathbf{y}_p, \mathbf{z}_c), \tag{4.1a}$$

$$\mathcal{F}_p^E(t, \mathbf{x}; \mathbf{z}_c) = \mathcal{F}_p^L(t; \mathbf{y}_p = \mathbf{x}, \mathbf{z}_c) = \int \mathcal{F}_p^L(t; \mathbf{y}_p, \mathbf{z}_c) \delta(\mathbf{y}_p - \mathbf{x}) \, d\mathbf{y}_p. \tag{4.1b}$$

It follows from Eq. (4.1b) that the Eulerian MDF is derived from the Lagrangian one. The formulation in Eq. (4.1b), where values at the same location are taken as equal, should not hide that this position is a parameter for the Eulerian MDF (the position is before the double comma in the list of arguments for $\mathcal{F}_p^E$)) while it is a variable in the Lagrangian MDF (the position is after the double comma in the list of arguments for $\mathcal{F}_p^L$). The superscripts E and L are used to emphasize this distinction.

The key relation between the Eulerian and Lagrangian descriptions is [3, 7]

$$\mathcal{F}_{\mathrm{p}}^{E}(t, \mathbf{x}; \mathbf{z}_{\mathrm{c}}) = \int p(t; \mathbf{x}, \mathbf{z}_{\mathrm{c}} \mid t_0; \mathbf{x}_0, \mathbf{z}_{\mathrm{c},0})\, \mathcal{F}_{\mathrm{p}}^{E}(t_0, \mathbf{x}_0; \mathbf{z}_{\mathrm{c},0})\, \mathrm{d}\mathbf{x}_0\, \mathrm{d}\mathbf{z}_{\mathrm{c},0}\,, \qquad (4.2)$$

showing that the Lagrangian transition PDF is the propagator of the information, here the Eulerian MDF from an initial state $(t_0; \mathbf{x}_0, \mathbf{z}_{\mathrm{c},0})$ to the present one $(t; \mathbf{x}, \mathbf{z}_{\mathrm{c}})$. This central property indicates that there is a one-way street from Lagrangian models to Eulerian ones and explains why most of the physics is contained in the Lagrangian transition PDF $p(t; \mathbf{y}_{\mathrm{p}}, \mathbf{z}_{\mathrm{c}} \mid t_0; \mathbf{y}_{\mathrm{p},0}, \mathbf{z}_{\mathrm{c},0})$ and the unconditional one, on which we focus from now on (without having to use a superscript L). This leads directly to Eulerian PDFs being obtained as conditional Lagrangian ones [2, 3] but the important connection is between the Lagrangian PDF and the Eulerian MDF.

Once the Eulerian MDF is obtained, particle mean field properties are extracted as follows: for a particle variable noted $H_{\mathrm{p}}(t; \mathbf{Z}_{\mathrm{c}})$, its average $\langle H_{\mathrm{p}} \rangle$ is defined as

$$\alpha_{\mathrm{p}}(t, \mathbf{x})\, \rho_{\mathrm{p}} \langle H_{\mathrm{p}} \rangle(t, \mathbf{x}) = \int H_{\mathrm{p}}(t; \mathbf{z}_{\mathrm{c}}) \mathcal{F}_{\mathrm{p}}^{E}(t, \mathbf{x}; \mathbf{z}_{\mathrm{c}})\, \mathrm{d}\mathbf{z}_{\mathrm{c}}\,, \qquad (4.3)$$

where $\alpha_{\mathrm{p}}(t, \mathbf{x})$ is the mean particle volumetric fraction. Note that the mass fractions, e.g. $\alpha_{\mathrm{p}}\rho_{\mathrm{p}}$, can vary in space making fluid and particle phases akin to a compressible fluid. In a discrete sense, when we introduce $N_{\mathrm{p}}$ stochastic particles or samples, the definitions of Lagrangian and Eulerian MDFs given in Eq. (4.1) become

$$\mathcal{F}_{\mathrm{p},N_{\mathrm{p}}}^{L}(t; \mathbf{y}_{\mathrm{p}}, \mathbf{z}_{\mathrm{c}}) = \sum_{i=1}^{N_{\mathrm{p}}} m_{\mathrm{p}}^{[i]} \delta(\mathbf{y}_{\mathrm{p}} - \mathbf{x}_{\mathrm{p}}^{[i]})\, \delta(\mathbf{z}_{\mathrm{c}} - \mathbf{Z}_{\mathrm{c}}^{[i]}) \qquad (4.4\mathrm{a})$$

$$\mathcal{F}_{\mathrm{p},N_{\mathrm{p}}}^{E}(t, \mathbf{x}; \mathbf{z}_{\mathrm{c}}) = \mathcal{F}_{\mathrm{p},N_{\mathrm{p}}}^{L}(t; \mathbf{y}_{\mathrm{p}} = \mathbf{x}, \mathbf{z}_{\mathrm{c}}) \qquad (4.4\mathrm{b})$$

where $m_{\mathrm{p}}^{[i]}$ is the mass of the particle labeled $[i]$. This shows that in a small volume $\mathcal{V}^{(\mathbf{x})}$ around a point $\mathbf{x}$ where averages are obtained from Monte Carlo estimations, that is as the ensemble averages over the $N_{\mathrm{p}}^{(\mathbf{x})}$ particles present in that volume, we get the equivalent of Favre, or mass-weighted, averages

$$\langle H_{\mathrm{p}} \rangle(t, \mathbf{x}) \simeq \langle H_{\mathrm{p}} \rangle_{N_{\mathrm{p}}} = \frac{\sum_{i=1}^{N_{\mathrm{p}}^{(\mathbf{x})}} m_{\mathrm{p}}^{[i]} H_{\mathrm{p}}(t; \mathbf{z}_{\mathrm{c}}^{[i]}(t))}{\sum_{i=1}^{N_{\mathrm{p}}^{(\mathbf{x})}} m_{\mathrm{p}}^{[i]}}\,. \qquad (4.5)$$

## 4.1.2  A Unified Approach to Statistical Operators

In Eq. (4.5), we retrieved the traditional ensemble averaging applied in the Reynolds-averaged Navier-Stokes (RANS) equations in turbulence modeling, using the relation $\mathcal{F}_p^L = \langle \mathcal{F}_{p,N_p}^L \rangle \simeq \mathcal{F}_{p,N_p}^L$ for large $N_p$. This is an echo of the all-important role of the fine-grained PDF defined as $\delta(\mathbf{Z}(t) - \mathbf{z})$, since we have $p(t; \mathbf{z}) = \langle \delta(\mathbf{Z}(t) - \mathbf{z}) \rangle$, where we use a general notation for the stochastic process $\mathbf{Z}$ to encompass the fluid and particle phases in the description. In recent decades, another commonly encountered approach is the LES method in which a spatial filtering operation is applied instead of an averaging one. For single-phase turbulent flows where we handle fields, e.g. $\mathcal{Q}(t, \mathbf{x})$, this means that we consider the statistical operator noted $\langle \cdot \rangle_{ls}$

$$\langle \mathcal{Q} \rangle_{ls}(t, \mathbf{x}) = \int \mathcal{Q}(t, \mathbf{x}') \, G(\mathbf{x}' - \mathbf{x}) \, d\mathbf{x}' \,, \tag{4.6}$$

where the spatial filter $G$ is taken as a spatially and temporally invariant positive function with a compact support and such that $\int G(\mathbf{x}) \, d\mathbf{x} = 1$. In a series of papers, Pope and co-workers [8–10] demonstrated that the PDF road can still be followed provided that we consider the filtered density function (FDF) defined as $p_{ls}(t; \mathbf{z}) = \langle \delta(\mathbf{Z}(t) - \mathbf{z}) \rangle_{ls}$ (note that this makes sense since the position is always present in the state vector, whether we consider fluid particles or discrete ones). This is referred to as the FDF approach. Its original formulation consisted in handling field quantities only but, given the upstream role of the Lagrangian MDF, this suggests to extend the FDF approach to a system made up by either fields or particles by manipulating the discrete Lagrangian MDF. In [5, section 7], this idea to derive FDF models for LES from a Lagrangian description in terms of particles by starting from the discrete Lagrangian Mass Density Function (LMDF) was developed.

We can combine the RANS and LES descriptions in a single formulation. Indeed, if we consider the discrete LMDF in Eq. (4.4a)

$$\widetilde{\mathcal{F}}_p^L(t; \mathbf{y}_p, \mathbf{z}_c) = \sum_{i=1}^{N_p} m_p^{[i]} \, \delta(\mathbf{y}_p - \mathbf{x}_p^{[i]}(t)) \, \delta(\mathbf{z}_c - \mathbf{Z}_c^{[i]}(t)) \,, \tag{4.7}$$

we remark that the LMDF needed for RANS formulations and the Lagrangian Filtered Density Function (LFDF) needed for LES are retrieved by applying the different statistical operators, $\langle \cdot \rangle$ and $\langle \cdot \rangle_{ls}$ respectively. This general approach based on the discrete LMDF as the 'parent function' is represented in Fig. 4.1. For more detailed accounts, readers are referred to [5, 6].

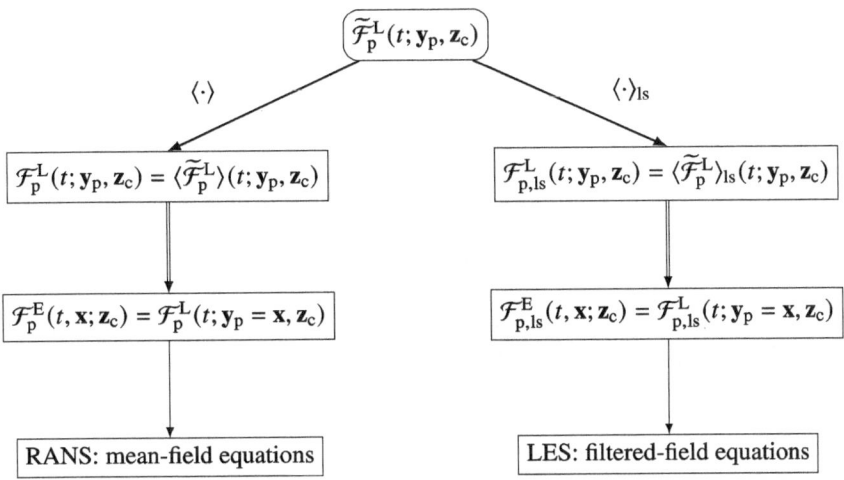

**Fig. 4.1** Sketch of the central role played by the discrete LMDF $\widetilde{\mathcal{F}}_p^L$ from which the application of different statistical operators yields the Lagrangian and Eulerian MDFs and the resulting macroscopic descriptions

The great benefit from such a unified formulation is that we do not have to bother anymore about which framework we are evolving in (RANS/LES) or which statistical averages to apply (ensemble or spatial filter). We can concentrate on how the particle state vector $\mathbf{Z}_p$ is represented by a stochastic processes.

## 4.2    Stochastic Processes

In the introduction to this chapter, we indicated that we limited ourselves to providing only an overview of the mathematical background. Actually, this is not suggested to newcomers in this subject who, on the contrary, are encouraged to study in-depth the mathematical aspects of stochastic processes. There is a vast literature on the subject with several books dedicated to the mathematical construction of stochastic processes and the analysis of their characteristics [11–14]. For physically-oriented readers with an interest for rigorous presentations, a highly-recommended textbook is [15] which does a very good job at introducing stochastic processes and whose tracks we follow in the account given below. As a side remark, it is telling that half of this book on polymer models is dedicated to the mathematical properties of stochastic processes and, in particular, of stochastic diffusion processes. This can be nicely complemented by the slightly less rigorous account in [16] but which is more widespread in its interests and covers interestingly jump processes.

## 4.2.1 Definition and Key Characteristics

### 4.2.1.1 Definition and Key Characteristics

A stochastic process, noted as $\mathbf{Z}$ or $(\mathbf{Z}(t), t \geq 0)$, is a family of random variables indexed by a parameter which is usually time (with $t \in [0\,; T_{\max}]$, where $T_{\max}$ is an end time with the possible value $T_{\max} = +\infty$). In the following, we consider vector-valued stochastic processes in $\mathfrak{R}^d$ (where $d$ is the dimension) and denote the values in the corresponding sample space by $\mathbf{z}$. To capture the key characteristics of a stochastic process, it is necessary to resort to the mathematical definition according to which $\mathbf{Z}$ is a family of measurable functions on an underlying probability space $\Omega$ equipped with a $\sigma$-algebra $\Sigma$ and a probability measure $\mathbb{P}$. This gives that $\mathbf{Z}$ is a measurable function of two variables:

$$[0\,; T_{\max}] \times (\Omega, \Sigma, \mathbb{P}) \longrightarrow (\mathfrak{R}^d, \mathbf{B}^d)$$
$$(t, \omega) \longmapsto \mathbf{Z}(t, \omega)\,, \tag{4.8}$$

where $\mathbf{B}^d$ is the Borel $\sigma$-algebra for $\mathfrak{R}^d$ obtained as the $d$-tensor-product of the Borel $\sigma$-algebra for $\mathfrak{R}$.

In some presentations, this rigorous definition is skipped and stochastic processes are introduced directly in the image space $(\mathfrak{R}^d, \mathbf{B}^d)$. This turns out to be unfortunate. A first reason is that the introduction of $\sigma$-algebras allows to quantify the intuitive notions of the information content and of fine- or coarse-grained descriptions which correspond to different embedded $\sigma$-algebras, or sub-$\sigma$-algebras, embodying the information resolved by one description. A second reason is that the correspondence between two ways to characterize a stochastic process may be overlooked. On the one hand, we can consider the family of vector-valued random variables $\mathbf{Z}(t)$ at fixed times $t$. This corresponds to the 'PDF point of view' where the aim is to derive the PDF equation satisfied by $p(t\,; \mathbf{z})$ in sample space. On the other hand, we can consider a family of 'elementary events', represented by different values of $\omega$ and handle a (large) number of time functions $t \longmapsto \mathbf{Z}_\omega(t)$. This corresponds to the 'trajectory point of view' where the objective is to write the time-evolution equations of these 'particles' (samples of the PDF to be used in Monte Carlo estimations). For a stochastic process, there is more information contained in its trajectories than in its PDF. However, since we are interested in approximating statistics from a stochastic process, we can refer to an equivalence between the PDF and trajectory points of view in a weak sense. Thus, considering for instance Langevin type of equations in physical space for a number of trajectories has the same status as the PDF equation in sample space, as revisited below in Sect. 4.4.

### 4.2.1.2 The Law of a Stochastic Process

Knowing a stochastic process is not equivalent to knowing the family of its one-time PDFs $(p(t\,; \mathbf{z}), t \geq 0)$ since it implies knowing also the joint laws of joint random variables $\mathbf{Z}(t)$ and $\mathbf{Z}(t')$ at any times $t$ and $t'$, etc. Using a discrete time setting, this means that the knowledge of the law of a stochastic process is equivalent to the knowledge of the joint PDF $p((t_1; \mathbf{z}_1); (t_2, \mathbf{z}_2); \ldots; (t_N; \mathbf{z}_N))$ for any set of $N$ times and for any values of the chosen times $(t_1, t_2, \ldots, t_N)$. It is thus clear that the amount of information required is huge and, in particular, much larger than the sole access to the one-time PDF functions $p(t; \mathbf{z})$. In practice, we need to restrict ourselves to more tractable sub-classes of stochastic processes, such as Markov processes.

### 4.2.2 Markovian Processes

In the world of ODEs, knowledge of an initial condition (at a time $t_0$) and of the rate of change of a deterministic dynamical system is enough to predict its future. A classical example is Lagrangian/Hamiltonian analytical mechanics. The counterpart of this notion for probabilistic approaches leads to define Markov processes for which knowledge of the present (in a probabilistic sense) and of the law of conditional increments is enough to predict the future (still in a probabilistic sense). To capture this idea, it is sufficient to translate the Markov property in a discrete time setting

$$p(t_{n+1}; \mathbf{z}_{n+1} \mid ((t_n; \mathbf{z}_n); (t_{n-1}; \mathbf{z}_{n-1}); \ldots; (t_1; \mathbf{z}_1))) = p(t_{n+1}; \mathbf{z}_{n+1} \mid t_n; \mathbf{z}_n) ,$$
(4.9)

where $t_n$ represents the present time, $t_{n+1}$ the future and $(t_{n-1}, \ldots, t_1)$ the past, while $\mathbf{z}_i$ is the value of the process at $t = t_i$ (i.e. $\mathbf{Z}(t_i) = \mathbf{z}_i$). An important element is that the condition entering the conditional PDF, written as $(t_n; \mathbf{z}_n)$ in Eq. (4.9), represents the information known at the present time $t = t_n$.

The fundamental property of Markov processes is that the law of the stochastic process is determined from the knowledge of only two functions: the initial PDF $p(t_0; \mathbf{z}_0)$ and the transition PDF $p(t; \mathbf{z} \mid t_0; \mathbf{z}_0)$ from $\mathbf{z}_0$ at $t = t_0$ to $\mathbf{z}$ at a later time $t$. Indeed, all the $N$-time PDFs are reconstructed from the chain-rule [15–17]

$$p((t_n; \mathbf{z}_n); (t_{n-1}; \mathbf{z}_{n-1}); \cdots ; (t_1; \mathbf{z}_1); (t_0; \mathbf{z}_0)) =$$
$$p(t_n; \mathbf{z}_n \mid t_{n-1}; \mathbf{z}_{n-1}) \cdots p(t_1; \mathbf{z}_1 \mid t_0; \mathbf{z}_0) p(t_0; \mathbf{z}_0) \qquad (4.10)$$

demonstrating that information on the complete law of the process is derived. In particular, we obtain the Chapman-Kolmogorov equation

$$p(t; \mathbf{z} \mid t_0; \mathbf{z}_0) = \int p(t; \mathbf{z} \mid t_1; \mathbf{z}_1) \, p(t_1; \mathbf{z}_1 \mid t_0; \mathbf{z}_0) \, \mathrm{d}\mathbf{z}_1 , \qquad (4.11)$$

which forms the basis of path-integral formulations [18].

In a weak sense, Markov processes are characterized by the infinitesimal operator

$$\mathcal{L}_t g(\mathbf{z}) = \lim_{\mathrm{d}t \to 0} \frac{\langle (g(\mathbf{Z}_{t+\mathrm{d}t}) | \mathbf{Z}_t = \mathbf{z}) \rangle - g(\mathbf{z})}{\mathrm{d}t} , \qquad (4.12)$$

which measures the effect of a conditional increment of the Markov process over a test function $g$. It can then be shown [11, 15, 16] that the transitional PDF $p(t; \mathbf{z} \mid t_0; \mathbf{z}_0)$ is the solution of two equations depending on whether we fix the end condition $(t; \mathbf{z})$ or the initial one $(t_0; \mathbf{z}_0)$. In the first case, this corresponds to the Kolmogorov backward equation written as

$$
\begin{cases}
\dfrac{\partial\, p(t; \mathbf{z} \mid t_0; \mathbf{z}_0)}{\partial t_0} + \mathcal{L}_{t_0} \left[\, p(t; \mathbf{z} \mid t_0; \mathbf{z}_0)\, \right] = 0 \,, \\[2ex]
p(t; \mathbf{z} \mid t_0; \mathbf{z}_0) = \delta(\mathbf{z} - \mathbf{z}_0) \quad t_0 \to t \,.
\end{cases}
\tag{4.13}
$$

In the second case, the evolution equation is the Kolmogorov forward equation

$$
\begin{cases}
\dfrac{\partial\, p(t; \mathbf{z} \mid t_0; \mathbf{z}_0)}{\partial t} = \mathcal{L}_t^{\perp} \left[\, p(t; \mathbf{z} \mid t_0; \mathbf{z}_0)\, \right] \,, \\[2ex]
p(t; \mathbf{z} \mid t_0; \mathbf{z}_0) = \delta(\mathbf{z} - \mathbf{z}_0) \quad t \to t_0 \,,
\end{cases}
\tag{4.14}
$$

where $\mathcal{L}_t^{\perp}$ is the adjoint operator of $\mathcal{L}_t$. In physics, we often consider the time evolution of a dynamical system from an initial state and we are mostly concerned with the Kolmogorov forward equation but the backward one is also of great interest, as recalled in Sect. 4.4.3. When dealing with Markov processes, these equations are central since all the needed information is generated by the transition PDF. Note that the same Kolmogorov forward equation is obtained for the one-time PDF $p(t; \mathbf{z})$ by integrating the transition PDF over all initial conditions.

The subclass of Markov stochastic processes is the only one for which we can develop a complete description. For non-Markov processes, it is possible to write PDF equations for the one-time PDF but we cannot reconstruct the law of the process. As shown later in Chap. 5, more serious troubles lie ahead for non-Markov processes.

### 4.2.3    The Two Building Blocks: The Wiener and Poisson Processes

#### 4.2.3.1 The Poisson Process

For random discrete events, the reference model is the Poisson process where there is a very small probability to have one event but where each event implies a discontinuous evolution represented by a jump (multiple jumps at one time are not possible). The trajectories of the Poisson process are therefore piecewise constant with jumps (having a step of one unity in the standard Poisson process) occurring at random times [16, 19] as illustrated in Fig. 4.2.

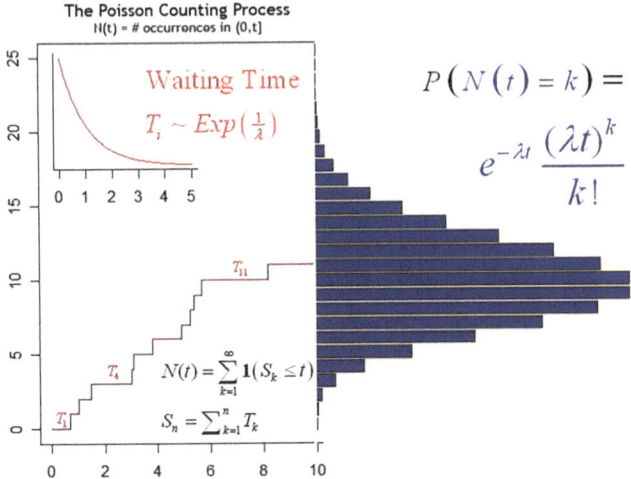

**Fig. 4.2** Properties of the Poisson process $N(t)$: one trajectory of the process jumping at random times $T_i$ which follows an exponential distribution and the PDF of these waiting times between jumps as well as the PDF of the Poisson process. Reprinted with permission from [20]. © 2014 CISM, Udine. All rights reserved

A Poisson process is characterized by its intensity $\lambda$, which is the mean value of the number of events per unit time. In a time interval $[t\,,\,t + \Delta t]$, the number of events $\Delta N(t) = N(t + \Delta t) - N(t)$, is a Poisson random variable

$$\mathbb{P}[\,\Delta N(t) = k\,] = \frac{(\lambda \Delta t)^k}{k!} e^{-\lambda \Delta t} \tag{4.15}$$

from which it follows that the mean and variance of $\Delta N(t)$ are linear in $\Delta t$

$$\langle \Delta N(t) \rangle = \lambda \Delta t \qquad \langle (\Delta N(t) - \langle \Delta N(t) \rangle)^2 \rangle = \lambda \Delta t. \tag{4.16}$$

In any finite interval, the times at which the Poisson process jumps are uniformly distributed while the increments $\Delta N(t)$ are independent. In practice, this property explains that deviations of measured particle distributions from the Poisson law are regarded as manifesting an underlying order. There are two ways to simulate a Poisson process. The first method is based on waiting times (the time intervals between successive random jumps) and, since values are constant between jumps, the idea is to go directly from one event to the next one. These waiting times are random variables following an exponential distribution with the same intensity $\lambda$:

$$\mathbb{P}[\,T_i = t\,] = \lambda e^{-\lambda t} \qquad \Rightarrow \langle T_i \rangle = \frac{1}{\lambda}\,, \tag{4.17}$$

and are generated before applying the unit-one jump or more general events when considering generalized Poisson processes (see Sect. 4.3). This method is often used to simulate collision and agglomeration events in molecular dynamics (cf. Sect. 12.2) but imposes variable time steps. The second way is to generate the number of events occurring in fixed time intervals (cf. Sect. 12.5). When the time interval $\Delta t$ is much smaller than the timescale $1/\lambda$ (i.e. $\lambda \Delta t \ll 1$), the statistics of the increments, which follow the Poisson distribution given in Eq. (4.15), become:

$$\mathbb{P}[\,\Delta N(t) = 0\,] \simeq 1 - \lambda \Delta t \,, \quad \mathbb{P}[\,\Delta N(t) = 1\,] \simeq \lambda \Delta t \,, \tag{4.18}$$

while higher-order values are negligible, i.e. $\mathbb{P}[\,\Delta N(t) = k\,] \simeq 0\ (k \geq 2)$.

### 4.2.3.2 The Wiener Process

The Wiener process is the canonical model of Brownian motion [15, 16] and can be defined by the following three properties [13]:

(i) The process has independent increments, i.e. $(W(t_3) - W(t_2))$ and $(W(t_1) - W(t_0))$ are independent when $t_0 < t_1 < t_2 < t_3$;
(ii) The trajectories of the process are continuous functions (almost everywhere);
(iii) The increments of the Wiener process $(W(t_2) - W(t_1))$ are centered Gaussian random variables and with a variance equal to $(t_2 - t_1)$.

From these characteristics, it results that the Wiener process $W$ is a Gaussian, Markov process, with zero mean and a covariance equal to $\langle W(t)W(t')\rangle = \min(t, t')$. Its trajectories are continuous and represent continuous evolutions where there is a near-one chance that modifications happen but with small changes. The most important properties are that the trajectories of the Wiener process are non-differentiable at any point and have even unbounded total variations in any interval. Furthermore, the increments $dW = W(t+dt) - W(t)$ are stationary and independent with moments

$$\langle dW \rangle = 0, \quad \langle (dW)^2 \rangle = dt, \quad \langle (dW)^n \rangle = o(dt) \quad \forall n > 2 \,. \tag{4.19}$$

The linear variation of $\langle (dW)^2 \rangle$ with respect to $dt$ is of paramount importance in the development of stochastic calculus [11, 14, 19] and explains the special care with which stochastic integrals must be defined. One trajectory of a Wiener process is displayed in Fig. 4.3 illustrating the wild variations at any scale.

## 4.3    Jump-diffusion Processes and Fokker-Planck Equations

To explain the shift from ODEs to stochastic differential equations (SDEs), we consider a one-dimensional dynamical system $Z(t)$ under the influence of Gaussian white-noise $\zeta(t)$ (with two functions $A$ and $B$ whose physical meaning will be

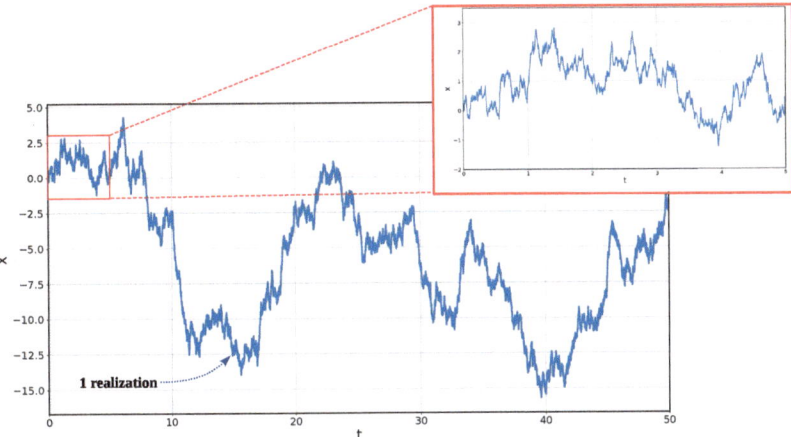

**Fig. 4.3** One trajectory of the Wiener process showing a continuous but ragged profile. The window, with a zoom of the trajectory, reveals the self-similar nature and the fast fluctuations of the trajectories of the Wiener process at all scales

introduced later in Sect. 4.3.2)

$$\frac{dZ(t)}{dt} = A(t, Z(t)) + B(t, Z(t))\,\zeta(t)\,. \tag{4.20}$$

To arrive at a well-posed definition, we can benefit from the smoothing properties of the integration using the identification of $\int_0^t \zeta(s)\,ds$ with a white-noise process, i.e. $\zeta_t\,dt = dW(t)$. Instead of Eq. (4.20), we therefore try to give meaning to

$$Z(t) = Z(t_0) + \underbrace{\int_{t_0}^t A(s, Z(s))\,ds}_{\text{regular integral}} + \underbrace{\int_{t_0}^t B(s, Z(s))\,dW(s)}_{\text{stochastic integral}}. \tag{4.21}$$

Due to the infinite total variation in any finite interval of the trajectories of the Wiener process $W$, we cannot obtain the stochastic integral as the limit of classical Riemann-Stieltjes sums. For a non-anticipating process $Z$, the stochastic integral is then defined in the Ito sense as

$$\int_{t_0}^t B(s, Z(s))\,dW(s) = \text{ms-}\lim_{N\to\infty}\sum_{i=1}^N B(t_i, Z(t_i))\,(W(t_{i+1}) - W(t_i)) \tag{4.22}$$

where $(t_i)_{i=1,N}$ represents a partition of the interval $[t_0 ; t]$ (with $t_1 = t_0$ and $t_{N+1} = t$) and where the limit is to be understood in the mean-square (ms) sense [11, 14, 15] and not as a convergence trajectory by trajectory. In the physics literature, these equations are written in incremental form as a short-hand notation and are referred to as 'Langevin equations'

$$dZ(t) = A(t, Z(t)) \, dt + B(t, Z(t)) \, dW(t) \,. \tag{4.23}$$

The Ito definition of the stochastic integral given in Eq. (4.22) is an essential point in stochastic calculus. On the one hand, it implies that two fundamental properties are always satisfied. The first one, called the zero-mean property, states that for any non-anticipating process $Z$ we have

$$\left\langle \int_{t_0}^{t} Z(s) \, dW(s) \right\rangle = 0 \,, \tag{4.24}$$

which indicates that the stochastic term can be truly regarded as noise (in the mathematical literature, this is an echo of the martingale property). The second fundamental property, called the isometry property, ensures that for any two non-anticipating processes $Z_1$ and $Z_2$ we can write that

$$\left\langle \left( \int_{t_0}^{t_2} Z_1(s) \, dW(s) \right) \left( \int_{t_1}^{t_3} Z_2(s) \, dW(s) \right) \right\rangle = \int_{t_1}^{t_2} \langle Z_1(s) \, Z_2(s) \rangle \, ds \,, \tag{4.25}$$

when $t_0 \le t_1 \le t_2 \le t_3$. On the other hand, the classical rules of ordinary differential calculus no longer apply and must be replaced by Ito stochastic calculus. This is expressed by the Ito formula: if $Z(t)$ is a stochastic process solution of the SDE given in Eq. (4.22), then the SDE satisfied by $g(t, Z(t))$ for any smooth function $g$ (at least twice continuously differentiable) is

$$dg(t, Z(t)) = \frac{\partial g}{\partial t}(t, Z(t)) \, dt + \frac{\partial g}{\partial z}(t, Z(t)) \, dZ(t) + \frac{1}{2} \left[ B^2 \frac{\partial^2 g}{\partial z^2} \right] (t, Z(t)) \, dt \,. \tag{4.26}$$

The extra term, which is the last one on the right-hand side of Eq. (4.26), is due to the linear-in-$dt$ variation of the variance of its increments, i.e. $\langle (dW)^2 \rangle = dt$, as shown by performing a Taylor expansion.

Using Ito calculus for a set of such test functions $g$, it is easy to show that the corresponding PDF equation in sample space is the Fokker-Planck (FP) equation

$$\frac{\partial \, p(t; z \mid t_0; z_0)}{\partial t} =$$

$$- \frac{\partial [\, A(t, z) \, p(t; z \mid t_0; z_0) \,]}{\partial z} + \frac{1}{2} \frac{\partial^2 [\, B^2(t, z) \, p(t; z \mid t_0; z_0) \,]}{\partial z^2} \,. \tag{4.27}$$

The FP equation is a forward Kolmogorov equation, cf. Eq. (4.14), corresponding to the following infinitesimal operator $\mathcal{L}_t$

$$\mathcal{L}_t[\cdot] = A(t, z)\frac{\partial [\cdot]}{\partial z} + \frac{1}{2}B^2(t, z)\frac{\partial^2 [\cdot]}{\partial z^2} . \qquad (4.28)$$

Since the FP equation has the form of a convection-diffusion equation in sample space, these processes are called stochastic diffusion processes and the functions $A$ and $B$ are referred to as the drift and diffusion coefficients, respectively. Note that the diffusion coefficient in the PDF equation is $B^2$ and is always positive or null.

The correspondence between Langevin and Fokker-Planck equations is easily extended to the multi-dimensional case. For instance, the SDEs written for a $d$-dimensional process $\mathbf{Z} = (Z_1, \ldots, Z_d)$ are

$$dZ_i(t) = A_i(t, \mathbf{Z}(t))\, dt + B_{ij}(t, \mathbf{Z}(t))\, dW_j(t) , \qquad (4.29)$$

with $(W_j)_{j=1,d}$ a set of $d$ independent Wiener processes. In Eq. (4.29), the drift $\mathbf{A} = (A_i)$ is a vector while the diffusion $\mathbf{B} = (B_{ij})$ is a matrix. The Ito formula is

$$dg(t, \mathbf{Z}(t)) = \frac{\partial g}{\partial t}(t, \mathbf{Z}(t))\, dt + \frac{\partial g}{\partial z_i}(t, \mathbf{Z}(t))\, dZ_i(t)$$

$$+ \left[\frac{1}{2} D_{ij}\frac{\partial^2 g}{\partial z_i \partial z_j}\right](t, \mathbf{Z}(t))\, dt , \qquad (4.30)$$

and the corresponding Fokker-Planck equation is

$$\frac{\partial\, p(t; \mathbf{z}\,|\, t_0; \mathbf{z}_0)}{\partial t} =$$

$$- \frac{\partial\, [\, A_i(t, \mathbf{z})\, p(t; \mathbf{z}\,|\, t_0; \mathbf{z}_0)\,]}{\partial z_i} + \frac{1}{2}\frac{\partial^2\, [\, D_{ij}(t, \mathbf{z})\, p(t; \mathbf{z}\,|\, t_0; \mathbf{z}_0)\,]}{\partial z_i \partial z_j} , \qquad (4.31)$$

where $\mathbf{D} = \mathbf{BB}^\perp$ (or $D_{ij} = B_{ik}B_{jk}$) is a symmetric definite-positive matrix. As in the one-dimensional case, the positivity of $\mathbf{D}$ justifies the reference to a diffusive nature induced by the rapidly-varying terms, $B_{ij}(t, \mathbf{Z}(t))\, dW_j(t)$, in Eq. (4.29). The fact that several matrices $\mathbf{B}$ can correspond to the same diffusion matrix $\mathbf{D}$ translates the equivalence between the trajectory and PDF points of view in a weak sense.

### 4.3.1   A Note on Stratonovich Stochastic Calculus

The Ito choice made in Eq. (4.22) is not the only way to define a stochastic integral and another possibility is the Stratonovich approach according to which the stochastic integral in Eq. (4.21) is defined by

$$
\int_{t_0}^{t} B_{ij}(s, \mathbf{Z}(s)) \circ dW_j(s) =
$$

$$
\text{st-} \lim_{N \to \infty} \sum_{i=1}^{N} \frac{1}{2} \left[ B(t_i, Z(t_i)) + B(t_{i+1}, Z(t_{i+1})) \right] (W(t_{i+1}) - W(t_i)) . \tag{4.32}
$$

The symbol $\circ$ indicates that the stochastic integral is defined in the Stratonovich sense and the limit is here to be understood in the stochastic (st) sense [11, 14, 15]. The properties of stochastic integrals in the Stratonovich definition are opposite to those defined in the Ito sense: the two fundamental properties (the zero-mean and the isometry ones) are no longer true but the rules of classical differential calculus remain valid. The fact that the zero-mean property no longer applies (unless the diffusion coefficient is constant) is not a moot point and suggests that a stochastic integral in the Stratonovich sense does not necessarily represent true noise. Actually, the Ito and Stratonovich definitions of a SDE with the same form correspond to two different stochastic processes or, conversely, a given diffusion stochastic process can be expressed in the Ito or in the Stratonovich contexts but with different SDEs. This is revealed by working out the correspondence between these two interpretations. Indeed, it can be shown [11, 14] that a SDE written in the Stratonovich sense as

$$
dZ_i(t) = A_i(t, \mathbf{Z}(t)) \, dt + B_{ij}(t, \mathbf{Z}(t)) \circ dW_j(t), \tag{4.33}
$$

is equivalent to the following SDE written in the Ito sense,

$$
dZ_i(t) = A_i(t, \mathbf{Z}(t)) \, dt + \left[ \frac{1}{2} B_{kj} \frac{\partial B_{ij}}{\partial z_k} \right] (t, \mathbf{Z}(t)) \, dt + B_{ij}(t, \mathbf{Z}(t)) \, dW_j(t).
$$

$$
\tag{4.34}
$$

An interesting argument giving credit to the Stratonovich integral comes from a theorem by Wong and Zakai (see a discussion for physics-oriented readers in [15, section 3.3.6]). In a layman formulation, this theorem states that if we consider the white-noise term expressed by the Wiener process as the limit of a sequence of processes, say $\widetilde{W}^{(n)}$ when $n \to +\infty$, where each $\widetilde{W}^{(n)}$ has continuous trajectories with bounded variations, then the series made up by the differential equations when the increments of $\widetilde{W}^{(n)}$ are used in the stochastic part of Eq. (4.21) (note that, in that case, Eq. (4.21) is an ODE) converges towards the SDE in the Stratonovich sense, that is Eq. (4.33). This theorem is thus in line with the fact that the rules of ordinary differential calculus apply in the Stratonovich world. In a modeling perspective, the central point is, however, that the Ito and Stratonovich definitions are similar to two different languages with which one message (one modeled stochastic process) can be expressed. We can switch from one expression to another one, using the correspondence between Eqs. (4.33) and (4.34), provided that we do not forget that we are handling the same physical object modeled as a stochastic process. The choice of which language to resort to is a matter of practicability. Given the major interest of being able to use the two fundamental properties given in Eqs. (4.24)– (4.25), we choose to follow the Ito definition and calculus from this point onwards. On a few occasions, we refer to Stratonovich calculus, typically when handling terms such as $u \circ d\mathcal{M}$. This is done for the sake of convenience to avoid having to distinguish between the cases when $\mathcal{M}$ is a differentiable process or when it stands for a Wiener process. This short-cut notation appears, for instance, in Sects. 7.3 and 10.4.3.

## 4.3.2   A Physical Interpretation

We can now shed light on the physical characteristics of the terms defining a stochastic diffusion process. Indeed, the physical meaning of $A$ and $B$ in the Langevin and Fokker-Planck equations are revealed by considering the statistics of the conditional increments (using a one-dimensional version)

$$\langle \Delta Z \mid Z(t) = z \rangle = A(t, z)\,\Delta t \,, \tag{4.35a}$$

$$\langle (\Delta Z)^2 \mid Z(t) = z \rangle = B^2(t, z)\,\Delta t \,. \tag{4.35b}$$

As shown in Fig. 4.4, the drift term, $A(t, z)$, governs the mean evolution of the conditional increments while the diffusion coefficient, $B(t, z)$, governs the spread of the conditional increments around its mean value.

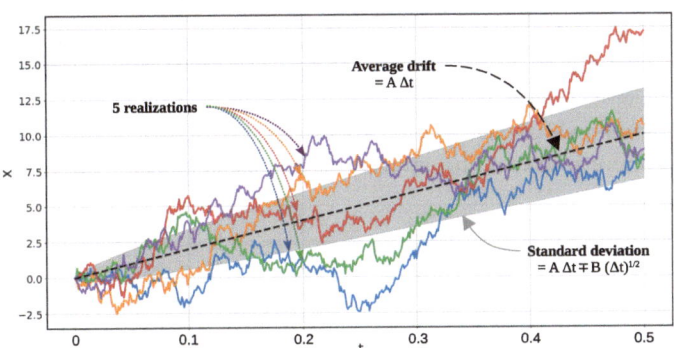

**Fig. 4.4** Statistics over small time increments of the conditional trajectories of a diffusion process illustrating the meaning of the drift and diffusion coefficients

There is often some confusion about the nature of the Gaussian hypothesis in SDEs. As expressed by Eq. (4.35), the Gaussian hypothesis applies only to the conditional increments and, in the general situation when $B(t, z)$ is not constant, the resulting PDF of the stochastic diffusion process $Z$ can deviate from Gaussianity. Though this point has been repeatedly clarified [3, 5], it is worth emphasizing that the observation of non-Gaussian distributions does not invalid models based on stochastic diffusion processes.

Apart from the case of Brownian motion itself (where $A = 0$ and $B = 1$), the simplest example of a stochastic diffusion process is the Ornstein-Uhlenbeck (OU) process whose trajectories are

$$dZ(t) = -\frac{Z(t)}{\tau}\, dt + K\, dW(t)\,, \tag{4.36}$$

where the timescale $\tau$ and the diffusion coefficient $K$ are constant. From Ito calculus [16], it is trivial to show that $K$, $\tau$ and $\langle Z^2 \rangle$ are related through $K^2 = 2\langle Z^2 \rangle / \tau$ which corresponds to a fluctuation-dissipation theorem [21]. Another immediate property is that the auto-correlation function of the stationary process $\mathcal{R}(t' - t) = \langle Z(t)Z(t') \rangle / \langle Z^2 \rangle$ is $\mathcal{R}(t' - t) = \exp(-(t' - t)/\tau)$, revealing that the relaxation timescale $\tau$ is also the integral timescale since $\tau = \int_0^{+\infty} \mathcal{R}(s)\, ds$. This is the first hint that timescales play a central role when formulating stochastic models, as exemplified on several occasions in later chapters.

### 4.3.3  Another Note on McKean Diffusion Processes

The preceding presentation of stochastic diffusion processes was made by considering that the drift and diffusion coefficients are functions of the value of the process, $A(t, Z)$ and $B(t, Z)$, so that the Fokker-Planck equation is a linear equation with respect to $p(t, z)$. In many physical applications and, in particular, in the ones considered in this book, we need to extend this well-established framework to cases where the drift and diffusion coefficients become functions of averages of the process, which we write as $A(t, Z, \langle f(Z(t)) \rangle)$ and $B(t, Z, \langle g(Z(t)) \rangle)$, where $f$ and $g$ represent general functions.

As an example, we extend an OU process, cf. Eq. (4.36), to situations where the drift term is a return-to-equilibrium towards an unsteady mean value,

$$dZ = \frac{d\langle Z \rangle(t)}{dt}\, dt - \frac{Z - \langle Z \rangle(t)}{\tau}\, dt + K\, dW\,, \tag{4.37}$$

so that the drift term depends on the average $\langle Z \rangle(t) = \int z\, p(t, z)\, dz$ and, therefore, on the PDF of the process.

The theory developed for the definition of the SDEs as well as the Fokker-Planck equation remain valid but this equation, which can be written as

$$
\begin{cases}
\dfrac{\partial p}{\partial t} = -\dfrac{\partial[\, A(t, z, f(p))\, p\,]}{\partial z} + \dfrac{1}{2}\dfrac{\partial^2[\, B^2(t, z, g(p))\, p\,]}{\partial z^2} \\[2mm]
p(t; z \mid t_0; z_0) = \delta(z - z_0) \ \text{ when } \ t \to t_0.
\end{cases}
\tag{4.38}
$$

to manifest the dependence of the drift and diffusion coefficients with $p$, is now a non-linear equation with respect to the transitional PDF of the process. Correspondingly, the evolution equations for the trajectories of the process take a more general form

$$
dZ(t) = A(t, Z(t), \langle f(X(t)) \rangle)\, dt + B(t, Z(t), \langle g(Z(t)) \rangle)\, dW.
\tag{4.39}
$$

In the mathematical literature, these processes are referred to as McKean diffusion processes or as processes with mean-field interactions [15]. When we use Monte Carlo methods to obtain estimates of these averaged quantities, we arrive at a formulation of the SDEs where the different stochastic particles are (weakly) coupled instead of being strictly independent as in the well-defined SDEs given above and as in the classical Monte Carlo method in which samples are independent. This is seen by writing the SDE for one stochastic particle, labeled as $[i]$, which becomes

$$
dZ^{[i]} = A\!\left(t, Z^{[i]}, \frac{1}{N}\sum_{j=1}^{N} f(Z^{[j]})\right) dt + B\!\left(t, Z^{[i]}, \frac{1}{N}\sum_{j=1}^{N} g(Z^{[j]})\right) dW
\tag{4.40}
$$

with $\mathrm{ms} - \lim_{n\to\infty} \frac{1}{N}\sum_{j=1}^{N} f(X_t^{(j)}) = \langle f(X_t) \rangle$. Interestingly, this last property about the correct limit of the ensemble averages in the context of weakly interacting stochastic particles is called the propagation of chaos (to be understood here as meaning that the weakly interacting samples behave nearly as independent ones in that limit, see an interesting discussion in physical terms in [15, section 3.3.4]). In the rest of this book, we treat extended SDEs of the McKean type, cf. Eq. (4.39), as well-defined and handle them without further precautions compared to the original version, cf. Eq. (4.23). It is nevertheless better to be aware of the difference and to remember that things rest perhaps on safer grounds when the mean-field interactions are represented by smooth functions rather than noisy estimates.

### 4.3.4  Jump-diffusion Processes

So far, we have built SDEs based on the Wiener process for diffusion processes with continuous trajectories. Similar steps can be made with the Poisson process to introduce sudden jumps. Still using the trajectory point of view, the SDEs for a

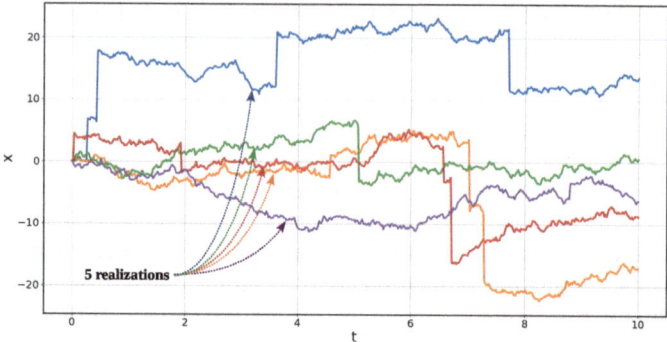

**Fig. 4.5** Five trajectories of a jump-diffusion process with their diffusive behavior in between sudden jumps having variable magnitudes

jump-diffusion process are written for a one-dimensional process

$$dZ(t) = A(t, Z(t)) \, dt + B(t, Z(t)) \, dW(t) + C(t, Z(t)) \, dN(t) \,, \qquad (4.41)$$

where $N$ is a Poisson process with intensity $\lambda$ and $C(t, Z(t))$ represents the amplitude of the jumps. The statistics of the conditional increments over a time $\Delta t$ become

$$\langle \Delta Z \mid Z(t) = z \rangle = A(t, z) \, \Delta t + C(t, z) \lambda \, \Delta t \,, \qquad (4.42a)$$

$$\langle (\Delta Z)^2 \mid Z(t) = z \rangle = B^2(t, z) \, \Delta t + C^2(t, z) \lambda \, \Delta t \,. \qquad (4.42b)$$

The introduction of jumps is illustrated in Fig. 4.5, revealing the discontinuous nature of the trajectories and their diffusive behavior between successive jumps.

From the properties of the Poisson process, we know that, over an infinitesimal time increment $dt$ and retaining only the first order contributions, the increments of $N(t)$ can take only two possible values

$$\mathbb{P}[\, dN(t) = k \,] = (1 - \lambda dt) \, \delta_{k,0} + \lambda \, dt \, \delta_{k,1} \Rightarrow \langle (dN(t))^m \rangle = \lambda \, dt \quad \forall m. \qquad (4.43)$$

The Poisson jumps contribute to any order in $dt$ (whereas the increments of the Wiener process contributes only to the second order in $dt$). The SDEs of a jump-diffusion process can be generalized by considering random amplitudes for the jumps (this is referred to as a compound Poisson process), which writes

$$dZ(t) = A(t, Z(t)) \, dt + B(t, Z(t)) \, dW(t) + C(t, Z(t), Q) \, dN(t) \,, \qquad (4.44)$$

where $Q$ is an independent random variable. The law of the random jumps becomes

$$W(y \mid t, z) = W(y \mid Z(t) = z) = \mathbb{P}[\, C = y \mid Z(t) = z \,] \lambda \,, \qquad (4.45)$$

where the probability $\mathbb{P}$ refers to the law of the independent random variable $Q$. In sample space, the equation for the transitional PDF $p(t, z \mid t_0, z_0)$ is

$$\frac{\partial p(t; z \mid t_0; z_0)}{\partial t} = -\frac{\partial [A(t, z) \, p(t; z \mid t_0; z_0)]}{\partial z} + \frac{1}{2}\frac{\partial^2 [B^2(t, z) \, p(t; z \mid t_0; z_0)]}{\partial z^2}$$

$$+ \int [W(z \mid t, y) \, p(t; y \mid t_0; z_0) - W(y \mid t, z) \, p(t; z \mid t_0; z_0)] \, dy. \quad (4.46)$$

This form is referred to as the Chapman-Kolmogorov PDF equation in [16]. While jump processes are not needed when modeling discrete particle transport in turbulent flows in the next chapters (from Chaps. 5 to 11) since the random motions of turbulent flows induce continuous changes in particle dynamics, they resurface in Sect. 12.5 to account for sudden velocity jumps due to particle collisions.

## 4.4 Dynamical Monte Carlo Methods and Solutions of Partial Differential Equations

We have already mentioned in Sect. 4.2.1 the correspondence between the PDF and the trajectory points of view. Clarifying the nature of this correspondence requires however to define the measure according to which a stochastic process, or a series of them, is said to converge towards another one. In many physical applications and for the developments presented in this book, we are concerned with the approximation in the weak sense, which means that we are essentially trying to capture statistics derived from a stochastic process. Among the four typical models of convergence (see definitions and analyses in [14–16, 20]), this corresponds to a convergence in law where the PDF is characterized in the weak sense (as for distributions). In that case, we can then speak of the equivalence between the PDF and trajectory points of view. Since we are essentially adopting the trajectory point of the view in the following, it is important to be aware of how things translate for the PDF or, more precisely, for the PDF evolution equation, as well as of the formulation of many partial differential equations (PDE) in terms of the law of a stochastic process. These aspects have been established for some time and are well covered in the above-cited literature on the subject, both in mathematically- and physically-oriented textbooks. Nevertheless, for the sake of a self-contained presentation and for readers who may not be familiar with this mathematical background, we recall in this section the main points of the dynamical Monte Carlo methods for stochastic processes, starting with the basic Monte Carlo method for random variables, and how the solutions of typical PDEs can be expressed directly as functions of the law of a stochastic process.

### 4.4.1   Monte Carlo Method for Random Variables

In the Monte Carlo method, we consider a random variable $Z$ with a PDF $p(z)$, where $z$ are the values in sample space. The statistics of interest can be expressed as

$$I = \langle g(Z) \rangle = \int_{\mathcal{D}} g(z)\, p(z)\, \mathrm{d}z \,, \tag{4.47}$$

where $g$ is a regular and smooth-enough function. The Monte Carlo method consists in generating $N$ identical and independent random variables, $(Z_i)_{1 \le i \le N}$, with the same law $p(z)$ and to approximate the statistics by the ensemble average

$$I \approx I_N = \frac{1}{N} \sum_{i=1}^{N} g(Z_i). \tag{4.48}$$

It follows that, while $\langle g(Z) \rangle$ is a number, the discrete sums $I_N$ are random variables.

We are yet to be ensured that $I_N$ converges to $\langle g(Z) \rangle$, as $N \to +\infty$ and, if so, to determine in what sense. To keep simple notations, we consider a random variable $Z$ instead of $g(Z)$ without any loss of generality (think about using $Y = g(Z)$ which is just a deterministic function of $Z$). The answer to the first question is given by the Law of the large numbers which states that: if $(Z_i)_{1 \le i \le N}$, $N$ are independent random variables with the same law $p(z)$ (i.e. copies of the same random variable $Z$), and if $\langle Z^3 \rangle < \infty$, then, for almost every $\omega$, we have that

$$\langle Z \rangle = \lim_{N \to \infty} \frac{1}{N} \sum_{i=1}^{N} Z_i(\omega). \tag{4.49}$$

It is thus seen that convergence is guaranteed and is even to be understood in the strongest possible mode of convergence, namely the almost-sure mode (or for 'every $\omega$'). To characterize in what sense $I_N$ converges to $\langle g(Z) \rangle$, it is convenient to introduce the approximation error of the Monte Carlo method $\epsilon_N$, defined as

$$\epsilon_N = \langle Z \rangle - \frac{1}{N} \sum_{i=1}^{N} Z_i(\omega) \,. \tag{4.50}$$

The way with which $\epsilon_N$ tends towards 0 is precised by the central limit theorem (CLT) which states that: let $(Z_i)_{1 \le i \le N}$, $N$ independent random variables with the same law $p(z)$ as a random variable $Z$, and such that $\sigma^2 = \langle Z^2 \rangle - \langle Z \rangle^2 < \infty$, then

$$\frac{\sqrt{N}}{\sigma} \epsilon_N \to \mathcal{N}(0, 1) \,, \tag{4.51}$$

where $\mathcal{N}(0, 1)$ denotes the Gaussian distribution of zero mean and unit standard deviation. From the CLT, it is clear that the mode of convergence of the error $\epsilon_N$ is a convergence in law, that is in a weak sense. The CLT provides what is considered one of the hallmarks of the basic Monte Carlo method, namely its rate of convergence in the weak sense as $\epsilon_N \sim N^{-1/2}$, as well as direct estimates of confidence intervals. If the rate of convergence of Monte Carlo estimations is slow, it does not depends on the regularity of the functions $g$ and, more importantly, on the space dimension. Monte Carlo methods are therefore general and, though not necessarily the best choices in some cases in low-dimensions, they are always applicable and remain at the moment the only real possibility for problems in high-dimensional spaces.

It is easily shown that the Monte Carlo method corresponds to an estimation, in the weak sense, of the PDF of the random variable. Indeed, for any test function $\varphi$

$$\forall \varphi, \quad \langle \varphi(Z) \rangle \approx \frac{1}{N} \sum_{i=1}^{N} \varphi(Z_i). \tag{4.52}$$

By introducing the discrete PDF $p_N(z)$ as a sum of Dirac masses for the $N$ samples

$$p_N(z) = \frac{1}{N} \sum_{i=1}^{N} \delta(z - Z_i), \tag{4.53}$$

we have therefore that

$$\forall \varphi, \quad \int \varphi(z) \, p(z) \, dz \approx \int \varphi(z) \, p_N(z) \, dz, \tag{4.54}$$

from which it derives that $p(z) \simeq p_N(z)$ in the weak sense (or as a distribution).

### 4.4.2  Dynamical Monte Carlo Methods for Stochastic Processes

The dynamical Monte Carlo method follows from the weak approximation of the PDF by a sum of Dirac functions by introducing time as we go from a random variable to a stochastic process. It can be applied for any Markov stochastic process and is illustrated by the representation below.

$$
\underbrace{
\left\{
\begin{array}{l}
\mathbf{Z}(0) \text{ random variable} \\[2em]
\text{with a density } p(0; \mathbf{z})
\end{array}
\right.
}_{\text{sample space}}
\quad \Longleftrightarrow \quad
\underbrace{
\left\{
\begin{array}{l}
N \text{ samples} : \mathbf{Z}^{[k]}(0), \ k = 1, \ldots N \\[1em]
p(0; \mathbf{z}) \approx \dfrac{1}{N} \sum_{k=1}^{N} \delta(\mathbf{z} - \mathbf{Z}^{[k]}(0))
\end{array}
\right.
}_{\text{physical space}}
$$

$$\Downarrow (t > 0)$$

| solution of the PDE | advance of the SDEs |
|---|---|
| (1 equation in $(t, \mathbf{z})$ in a $N_d$-space) | ($N$ equations in $t$, in 3D-space) |

$$\overbrace{\qquad\qquad\text{sample space}\qquad\qquad} \qquad\qquad \overbrace{\qquad\qquad\qquad\text{physical space}\qquad\qquad\qquad}$$

$$\begin{cases} \mathbf{Z}(t) \text{ random variable} \\[2em] \text{with a density } p(t; \mathbf{z}) \end{cases} \quad\Longleftrightarrow\quad \begin{cases} N \text{ samples}: \mathbf{Z}^{[k]}(t), \; k = 1, \ldots N \\[1em] p(t; \mathbf{z}) \approx \dfrac{1}{N} \sum_{k=1}^{N} \delta(\mathbf{z} - \mathbf{Z}^{[k]}(t)) \end{cases}$$

At the initial time ($t = 0$), $Z(0)$ is a random variable and we can apply the basic Monte Carlo method and introduce a number of samples, from which the initial PDF is represented by a sum of Dirac masses. As time passes by ($t > 0$), the dynamical aspects can be addressed from two points of view. As shown on the left part of the sketch, we can solve the evolution equation in sample space (this would be the Fokker-Planck equation for a stochastic diffusion process), which means solving one equation but in a space that has potentially a very high dimension. We can also address the same dynamical evolution by following the second point of view which is indicated on the right part of the sketch. This amounts to tracking in time the evolution of the $N$ initial samples or, in other terms, to simulating the trajectories of these $N$ samples or 'stochastic particles' (this would be solving the SDEs for these particles). At a later time $t$, we obtain therefore $N$ updated samples, written as $\mathbf{Z}^{[k]}(t)$ (with $k = 1, \ldots, N$), and we can apply the basic Monte Carlo method to obtain a weak approximation of the PDF $p(t, \mathbf{z})$.

In this representation, we have used the one-time PDF, $p(t; \mathbf{z})$ to simplify notations but what is really handled is the key distribution for Markov processes, namely the transitional PDF $p(t; \mathbf{z} \,|\, (t_0; \mathbf{z}_0))$ from which the one-time PDF is immediately derived by integration over all initial conditions. We have also assumed that each sample has the same statistical weight, here $1/N$, while a general formulation of the above approximation is to attach a statistical weight $w_{\text{stat}}^{[k]}$ to each particle labeled ($k$) so that the dynamical Monte Carlo estimation of the PDF writes

$$p(t; \mathbf{z}) \approx \left( \frac{1}{\sum_{k=1}^{N} w_{\text{stat}}^{[k]}} \right) \sum_{k=1}^{N} w_{\text{stat}}^{[k]} \delta(\mathbf{z} - \mathbf{Z}^{[k]}(t)) \tag{4.55}$$

Going back to the expressions given in Sect. 4.1.1, it is seen from the definitions of the Lagrangian and Eulerian MDFs in Eq. (4.1) that we retrieve the discrete expressions in Eq. (4.4) with the natural choice in polydispersed two-phase flows to retain the particle mass as the statistical weight attached to that particle. In that sense, the stochastic particles used in the statistical descriptions mimic the actual discrete particles but remain nevertheless 'stochastic particles' to be regarded as samples of the corresponding PDF. In practical applications, a local homogeneous assumption is often made by which the $N_p^{(\mathbf{x})}$ particles located at a time $t$ in a small volume $\mathcal{V}^{(\mathbf{x})}$ around a point $\mathbf{x}$ are considered as equivalent samples of the PDF at that point. This leads to the estimation of mean quantities as mass-weighted averages as in Eq. (4.5).

Thus, applying the dynamical Monte Carlo method is equivalent to making an approximation for the solution of a PDF equation, which is a partial differential equation in sample-space, without apparently 'solving' this PDF equation. This is a first indication that we can solve PDEs with the law of a stochastic process.

## 4.4.3 Probabilistic Solutions of Partial Differential Equations

### 4.4.3.1 The Forward Point of View

The equivalence between the Langevin equations and Fokker-Planck equation is key to express the solutions of PDEs as some statistics extracted from the law of a stochastic diffusion process. For example, the solution of the evolution problem

$$
\begin{cases}
\dfrac{\partial u}{\partial t} = -\dfrac{\partial[\, A(t, x)\, u\,]}{\partial x} + \dfrac{1}{2}\dfrac{\partial^2[\, B^2(t, x)\, u\,]}{\partial x^2} \\
u(0, x) = h(x) \quad \text{when} \ \ t = 0,
\end{cases}
\tag{4.56}
$$

is built from the transitional PDF of the diffusion process $X$ with $A$ and $B$ as the drift and diffusion coefficients (in this section, we use $X$ to denote the stochastic process we are handling since we are referring to a space dimension and also to follow classical notations used in mathematical textbooks). This solution is given by

$$
u(t, x) = \int p(t; x \mid t_0; x_0)\, h(x_0)\, \mathrm{d}x_0 ,
\tag{4.57}
$$

since $p(t; x \mid t_0; x_0)$ is the solution of the corresponding forward Kolmogorov equation (with a Dirac function as an initial condition), and can also be written as

$$
u(t, x) = \langle\, h(X(0)) \mid X_t = x \,\rangle
\tag{4.58}
$$

where the expectation is taken with respect to the trajectories solutions of the SDE

$$
\mathrm{d}X = A(t, X)\, \mathrm{d}t + B(t, X)\, \mathrm{d}W_t
\tag{4.59}
$$

arriving at location $x$ at time $t$.

To translate how things appear in the weak sense, it is illustrative to re-express this result using a particle formulation by considering that we assign a given value, say a 'scalar' variable $\Phi$, to each stochastic particle following the diffusive evolution. The corresponding state-vector is thus extended to include both location and scalar $(X, \Phi)$. The scalar is given an initial value equal to the local value of the field $h$ at the initial location $X(0)$ and remains constant along each particle trajectory. The particle SDEs are therefore

$$
\begin{cases}
X(0) = X_0 \text{ and } \Phi(0) = h(X_0) \text{ at } t = 0 , \\
dX(t) = A(t, X(t)) \, dt + B(t, X(t)) \, dW , \\
d\Phi(t) = 0 .
\end{cases}
\tag{4.60}
$$

From the dynamical Monte Carlo method, the discrete joint PDF is

$$
p_N(t; y, \Psi) = \frac{1}{N} \sum_{k=1}^{N} \delta(y - X^{[k]}(t)) \, \delta(\Psi - \Phi^{[k]}(t)) ,
\tag{4.61}
$$

with $y$ and $\Psi$ the sample space variables for $X$ and $\Phi$, respectively. We define the field $u(t, x) = \langle \Phi \mid X_t = x \rangle$, which represents the averaged value of the scalar for particles arriving at $x$ at time $t$. In the Monte Carlo method, this field is estimated by

$$
u(t, x) = \int \Psi \, \frac{p(t; x, \Psi)}{p(t; x)} \, d\Psi \simeq \frac{1}{N \, p_N(t; x)} \sum_{k=1}^{N} \delta(x - X^{[k]}(t)) \, \Phi^{[k]}(t)
\tag{4.62}
$$

The Dirac functions are then 'smoothed' by considering a small volume $\delta \mathcal{V}^{(x)}$ around $x$ which contains $N^{(x)}$ particles and by using the approximation

$$
\delta(y - x) \simeq \frac{1}{\delta \mathcal{V}^{(x)}} \mathbb{1}_{(y \in \delta \mathcal{V}^{(x)})}, \quad p_N(t, x) \simeq \frac{N^{(x)}}{N \, \delta \mathcal{V}^{(x)}} ,
\tag{4.63}
$$

which gives for the local value of the field $u(t, x)$

$$
u(t, x) \simeq \frac{1}{p_N(t; x)} \frac{1}{N \, \delta \mathcal{V}^{(x)}} \sum_{k=1}^{N^{(x)}} \Phi^{[k]}(t) = \frac{1}{N^{(x)}} \sum_{k=1}^{N^{(x)}} \Phi^{[k]}(t) .
\tag{4.64}
$$

Thus, we retrieve the Monte Carlo estimation of the solution of the PDE since

$$
u(t, x) = \frac{1}{N^{(x)}} \sum_{k=1}^{N^{(x)}} h(X(0)) \simeq \langle h(X(0)) \mid X_t = x \rangle.
\tag{4.65}
$$

### 4.4.3.2 The Backward Point of View

In classical mechanics, the dynamics of physical systems is typically studied with time-marching evolution equations and, consequently, we may regard the forward

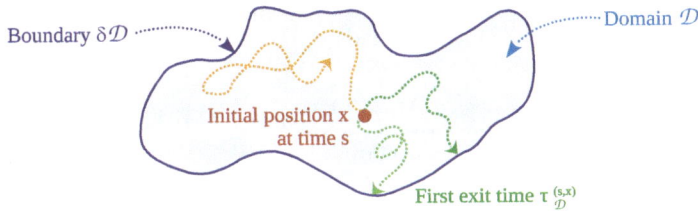

**Fig. 4.6** Illustration of the first exit time $\tau_{\mathcal{D}}^{(s,x)}$ at which a particle reaches the boundary $\delta\mathcal{D}$ of a physical domain $\mathcal{D}$

Kolmogorov point of view, cf. Eq. (4.14), as the most relevant one. However, there is also a direct connection between PDEs and the backward Kolmogorov formulation, cf. Eq. (4.13) which can be worked out with the Dynkin's formula. Indeed, by using Ito calculus to express the increments of $u(t, X)$ where $u$ is a regular function and $X$ a stochastic process whose infinitesimal operator is $\mathcal{L}_t$, we obtain for the evolution of the average value from time $s$ to time $t$ (with $0 \leq s \leq t \leq T$)

$$\langle u(t, X(t)) \rangle = \langle u(s, X(s)) \rangle + \left\langle \int_s^t \left( \frac{\partial u}{\partial s'} + \mathcal{L}_{s'} \right) (s', X(s')) \, ds' \right\rangle . \tag{4.66}$$

In this equation, note that time $t$ is a parameter. In bounded domains $\mathcal{D}$, we need to consider stopping times, $\tau_{\mathcal{D}}^{(s,x)}$, and which are defined as follows: for the trajectories of a stochastic process starting at $x$ at time $s$, $\tau_{\mathcal{D}}^{(s,x)}$ is the first time at which a trajectory hits the boundary $\delta\mathcal{D}$ of the physical domain $\mathcal{D}$, as sketched in Fig. 4.6.

Various boundary conditions (rebound, killing the trajectory, etc.) can be then applied but it is important to realize that $\tau_{\mathcal{D}}^{(s,x)}$ is now a random variable for any given initial conditions $(s, x)$. Nevertheless, it can be demonstrated [14] that the Dynkin's formula remains valid and retains the same expression, so that we can write for a stochastic process evolving in a bounded domain that

$$\langle u(\tau_{\mathcal{D}}^{(s,x)}, X(\tau_{\mathcal{D}}^{(s,x)})) \rangle_{(s,x)} = u(s, X(s))$$

$$+ \left\langle \int_s^{\tau_{\mathcal{D}}^{(s,x)}} \left( \frac{\partial u}{\partial s'} + \mathcal{L}_{s'} \right) (s', X^{(s,x)}(s')) \, ds' \right\rangle_{(s,x)} , \tag{4.67}$$

where $\langle . \rangle_{(s,x)} = \langle . \mid (X(s) = x) \rangle$ denotes the conditional expectation on $X(s) = x$ at time $s$ and $X^{(s,x)}$ the corresponding conditional process.

The backward Kolmogorov point of view is useful to express the solution of PDEs with an end (or terminal) condition. For example, the solution the PDE

$$\begin{cases} \dfrac{\partial u}{\partial s} + A(s, x) \dfrac{\partial u}{\partial x} + \dfrac{1}{2} B^2(s, x) \dfrac{\partial^2 u}{\partial x^2} = r(s, x) \, u(s, x) , \\[2mm] u(T, x) = g(x) \quad \text{when} \quad s = T , \end{cases} \tag{4.68}$$

is obtained from the transitional PDF of the diffusion process $X$ whose infinitesimal operator is given in Eq. (4.28), which reads

$$u(s, x) = \int g(y) \underbrace{e^{-\int_s^T r(X(t'),X(t'))dt'}}_{w(X(T))} p(T; y \mid s; x) \, dy , \qquad (4.69)$$

which can be written with a more compact form as

$$u(s, x) = \langle \, w(X(T)) \, g(X(T)) \mid X_s = x \, \rangle , \qquad (4.70)$$

where the expectation is taken with respect to trajectories leaving the location $x$ at time $s$. In this formula, it is seen that $w(X(T))$ plays the role of a statistical weight attached to each trajectory, similar to a sort of death-birth process along each trajectory. In the case when $r = 0$, the statistical weights remain equal to 1 for all particles and we retrieve a formula which appears as the 'adjoint' of the one developed above for the initial value problem. The expression in Eq. (4.69) played an historical role in the development of stochastic models and of path-integral formulations: this is the celebrated Feynman-Kac formula!

Another classical example for boundary-value problems is the solution of an elliptic PDE with boundary conditions, for instance, with Dirichlet boundary conditions. We are now considering that the drift and diffusion coefficients do not depend on time anymore. Using the above results, it follows that the solution of the PDE

$$\begin{cases} \mathcal{L}u(x) + f(x) = 0 & x \in \mathcal{D} , \\ u(x) = g(x) & x \in \delta\mathcal{D} , \end{cases} \qquad (4.71)$$

is given by

$$u(x) = \left\langle g(X(\tau_{\mathcal{D}}^{(s,x)})) \mathbb{1}_{\tau_{\mathcal{D}}^{(s,x)} < +\infty} \right\rangle_{(0,x)} + \left\langle \int_0^{\tau_{\mathcal{D}}^{(s,x)}} f(X(s))ds \right\rangle_{(0,x)} , \qquad (4.72)$$

with $X$ the stochastic diffusion having $\mathcal{L}$ as its infinitesimal operator (the initial condition is chosen at a time noted $t = 0$ by convention since $\tau_{\mathcal{D}}^{(s,x)}$ is actually the elapsed time or the time interval between the moment when the trajectory is initiated and when it reaches the boundary $\delta\mathcal{D}$ of the domain $\mathcal{D}$). Such expressions are interesting when we want to know the solutions at a limited number of points without having to solve the PDE in the entire domain. These results are well-established and a vast literature on the subject has been developed over past decades [14, 19].

## 4.5    A Double Hierarchy: One- or N-particle PDF and Particle-attached Variables

Having set up the probabilistic framework, we return to the first step mentioned at the beginning of this chapter. Yet, before considering their modeled evolution, we need to define the variables entering the state vector. This includes: (a) choosing a one- or a $N$-particle PDF formulation; (b) determining the relevant variables to describe each particle. Point (a) is a very classical issue in statistical physics in relation with the well-known BBGKY hierarchy for any system with (at least) two-particle interactions. At the moment, two-particle PDF models (not to mention $N$-particle ones, with $N \geq 3$) are not developed enough for general inhomogeneous turbulent flows, especially wall-bounded ones, and are therefore far less tractable [2, 22]. For these reasons and though references are made to such approaches (cf. Chaps. 6 and 12), we retain the one-particle PDF level of description and, correspondingly, a one-point Eulerian PDF description of mean fields. It is then essential to be aware of the inherent loss of information since, for instance, two-point correlations cannot be extracted from one-particle PDF models and particle-particle interactions, such as actions at distance or even collisions, have to be modeled through mean-field formulations. Point (b) turns out to be a challenge to traditional statistical views and requires specific analysis. This is addressed in Chap. 5.

## References

1. S.B. Pope, Progress Energy Combustion Sci. **11**(2), 119 (1985). https://doi.org/10.1016/0360-1285(85)90002-4
2. S. Pope, *Turbulent Flows* (Cambridge University Press, Cambridge, 2000). https://doi.org/10.1017/CBO9780511840531
3. J.P. Minier, E. Peirano, Phys. Rep. **352**(1–3), 1 (2001). https://doi.org/10.1016/S0370-1573(01)00011-4
4. D.C. Haworth, Progress Energy Combustion Sci. **36**(2), 168 (2010). https://doi.org/10.1016/j.pecs.2009.09.003
5. J.P. Minier, Progress Energy Combustion Sci. **50**, 1 (2015). https://doi.org/10.1016/j.pecs.2015.02.003
6. J.P. Minier, Phys. Rep. **665**, 1 (2016). https://doi.org/10.1016/j.physrep.2016.10.007
7. R. Balescu, *Statistical Dynamics: Matter Out of Equilibrium* (Published by Imperial College Press and Distributed by World Scientific Publishing Co., London, 1997). https://doi.org/10.1142/p036
8. L.Y.M. Gicquel, P. Givi, F.A. Jaberi, S.B. Pope, Phys. Fluids **14**(3), 1196 (2002). https://doi.org/10.1063/1.1436496
9. M.R.H. Sheikhi, T.G. Drozda, P. Givi, S.B. Pope, Phys. Fluids **15**(8), 2321 (2003). https://doi.org/10.1063/1.1584678
10. M.R.H. Sheikhi, P. Givi, S.B. Pope, Phys. Fluids **19**(9) (2007). https://doi.org/10.1063/1.2768953
11. L. Arnold, *Stochastic Differential Equations: Theory and Applications (Book)* (Wiley-Interscience, New York, 1974). https://doi.org/10.1142/6453
12. A.N. Borodin, P. Salminen, *Handbook of Brownian Motion-Facts and Formulae* (Springer Science & Business Media, Berlin, 2015). https://doi.org/10.1007/978-3-0348-8163-0

13. I. Karatzas, S. Shreve, *Brownian Motion and Stochastic Calculus*, vol. 113 (Springer Science & Business Media, Berlin, 1991). https://doi.org/10.1007/978-1-4612-0949-2
14. B. Øksendal, *Stochastic Differential Equations* (Springer, Berlin, 2003). https://doi.org/10.1007/978-3-642-14394-6
15. H.C. Öttinger, *Stochastic Processes in Polymeric Fluids: Tools and Examples for Developing Simulation Algorithms* (Springer Science & Business Media, Berlin, 2012). https://doi.org/10.1007/978-3-642-58290-5
16. C. Gardiner, *Stochastic Methods*, 4th edn. (Springer, Berlin, 2009). https://link.springer.com/book/9783540707127
17. H. Risken, *Fokker-Planck Equation* (Springer, Berlin, 1996). https://doi.org/10.1007/978-3-642-61544-3
18. F.W. Wiegel, *Introduction to Path-integral Methods in Physics and Polymer Science* (World Scientific, Singapore, 1986). https://doi.org/10.1142/0178
19. F.C. Klebaner, *Introduction to Stochastic Calculus with Applications* (World Scientific Publishing Company, Singapore, 2012). https://doi.org/10.1142/p821
20. S. Chibbaro, J.P. Minier, *Stochastic Methods in Fluid Mechanics*. CISM International Centre for Mechanical Sciences (Springer, Vienna, 2014). https://doi.org/10.1007/978-3-7091-1622-7
21. H.C. Öttinger, *Beyond Equilibrium Thermodynamics* (John Wiley & Sons, Hoboken, 2005). https://doi.org/10.1002/0471727903
22. S.B. Pope, Ann. Rev. Fluid Mech. **26**(1), 23 (1994). https://doi.org/10.1146/annurev.fl.26.010194.000323

# Statistical Modeling of Particle Transport in Turbulent Flows

**5**

**Abstract**

The Boltzmann picture is dominating the description of transport phenomena to such an extent that the kinetic framework is rarely questioned. Yet, whether kinetic variables are necessarily the ones to retain in the particle state vector is not a point over which to pass too quickly. For discrete particles in turbulent flows, there are typically three situations. The first one corresponds to fully-resolved turbulent flows, where the fluid velocity field is known at every point and every time, which means that the velocity of the fluid seen $U_s$ is similar to an external deterministic force field. It is then accounted for without approximation in Boltzmann-like PDF models based on the kinetic state vector $Z_p^r = (X_p, U_p)$ where we handle the PDF $p^r(t; z_p^r)$. The second situation corresponds to high-inertia particles for which the underlying fluid can be regarded as white-noise, leading to a Langevin equation for the discrete particle velocity and a resulting Fokker-Planck equation for $p^r$. The third situation corresponds to the case of non-fully-resolved turbulent flows, where $U_s$ is neither deterministic nor white-noise but exhibits memory effects. Is the choice of kinetic variables still relevant in that case? And, if not, what guidelines can we follow? These are the questions addressed in this chapter.

**Chapter Content** In Sect. 5.1, it is shown that present kinetic PDF closure proposals result in ill-based equations for non-fully resolved turbulent flows. This leads to reconsider the choice of the variables in the particle state vector and to adopt a formulation based on the Markovian approach, as explained in Sect. 5.2. For turbulent dispersed two-phase flows, this Markovian approach relies on the Kolmogorov theory discussed in Sect. 5.3 which is used to select a suitable state vector.

© The Author(s) 2025
J.-P. Minier et al., *Understanding Turbulent Systems*,
Lecture Notes in Physics 1039, https://doi.org/10.1007/978-3-031-84466-9_5

## 5.1    Shortcomings of Kinetic-like Descriptions

In kinetic-based approaches, the objective is to obtain the equation satisfied in sample space by the PDF $p^r(t; \mathbf{z}_p^r) = p^r(t; \mathbf{y}_p, \mathbf{V}_p)$. In these notations, the superscript r indicates a reduced description compared to $\mathbf{Z}_p = (\mathbf{X}_p, \mathbf{U}_p, \mathbf{U}_s)$ and the PDF $p^r$ is the marginal of $p(t; \mathbf{z}_p) = p(t; \mathbf{y}_p, \mathbf{V}_p, \mathbf{V}_s)$. By manipulating the fine-grained PDF $\mathcal{P}(t; \mathbf{y}_p, \mathbf{V}_p) = \delta(\mathbf{X}_p(t) - \mathbf{y}_p)\delta(\mathbf{U}_p(t) - \mathbf{V}_p)$ and using standard techniques [1] with $p^r(t; \mathbf{y}_p, \mathbf{V}_p) = \langle \mathcal{P}(t; \mathbf{y}_p, \mathbf{V}_p) \rangle$, the kinetic PDF equation is derived from the particle equations keeping only the drag force which is the central point. The unclosed PDF equation is

$$\frac{\partial p^r}{\partial t} + \frac{\partial \left[ V_{p,i}\, p^r \right]}{\partial y_{p,i}} = \left[ \left\langle \frac{V_{p,i}}{\tau_p} \mid (\mathbf{y}_p, \mathbf{V}_p) \right\rangle p^r \right] - \frac{\partial}{\partial V_{p,i}} \left[ \left\langle \frac{U_{s,i}}{\tau_p} \mid (\mathbf{y}_p, \mathbf{V}_p) \right\rangle p^r \right].$$

$$(5.1)$$

It first appears that the complete expression of the particle relaxation time $\tau_p$ cannot be accounted for at this level of description, since it is usually a function of $(\mathbf{U}_p, \mathbf{U}_s)$. In the kinetic approach, it is assumed that we are only dealing with particles in the Stokes regime so that $\tau_p \simeq \tau_p^{St}$ becomes independent of $(\mathbf{U}_p, \mathbf{U}_s)$, which is already a limitation. With this approximation, the unclosed kinetic PDF equation is written as

$$\frac{\partial p^r}{\partial t} + \frac{\partial \left[ V_{p,i}\, p^r \right]}{\partial y_{p,i}} = \frac{\partial}{\partial V_{p,i}} \left[ \frac{V_{p,i}}{\tau_p^{St}}\, p^r \right] - \frac{\partial}{\partial V_{p,i}} \left[ \frac{1}{\tau_p^{St}} \langle U_{f,i} \rangle\, p^r \right]$$

$$- \frac{\partial}{\partial V_{p,i}} \left[ \frac{1}{\tau_p^{St}} \langle u_{s,i}' \mid (\mathbf{y}_p, \mathbf{V}_p) \rangle\, p^r \right] \qquad (5.2)$$

where the velocity of the fluid seen has been decomposed as $\mathbf{U}_s = \langle \mathbf{U}_f \rangle (t, \mathbf{X}_p(t)) + \mathbf{u}_s'$. This does not imply, however, that $\langle \mathbf{u}_s' \rangle = 0$ since the set of velocities of the fluid seen represents only a subset of fluid particle velocities at a given location (unless the so-called well-mixed condition is satisfied), and its mean value is called the drift velocity $\mathbf{U}_d$, which is thus equal to $\mathbf{U}_d = \langle \mathbf{U}_s \rangle - \langle \mathbf{U}_f \rangle (t, \mathbf{X}_p(t))$.

Closure of the open flux in sample space is derived from Furutsu-Novikov-Donsker (FND) relation [2–4], which expresses the correlation between a Gaussian field with zero mean and an arbitrary functional of that field. It is applied by assuming that the fluctuating velocity field $\mathbf{u}_f(t, \mathbf{x})$ is a random Gaussian field and writes for a functional $Q[t; \mathbf{u}_f]$ of $\mathbf{u}_f(t, \mathbf{x})$

$$\langle u_{f,i}(t, \mathbf{x})\, Q[t; \mathbf{u}_f] \rangle = \int_0^t \int_{\mathbf{x}'} R_{ik}(t, \mathbf{x}; t', \mathbf{x}') \left\langle \frac{\delta Q[t; \mathbf{u}_f]}{\delta u_{f,k}(t', \mathbf{x}')} \right\rangle d\mathbf{x}'\, dt' \qquad (5.3)$$

where $R_{ik}(t, \mathbf{x}; t', \mathbf{x}') = \langle u_{f,i}(t, \mathbf{x})\, u_{f,k}(t', \mathbf{x}') \rangle$ is the fluid two-point two-time correlation. Then, applying the FND relation to the fine-grained PDF $\mathcal{P}$ yields the expression of the flux closure as [5]

$$\frac{1}{\tau_p^{St}} \langle U_{s,i} \,|\, \mathbf{z}_p^r \rangle \, p^r = \frac{1}{\tau_p^{St}} \langle U_{f,i} \rangle (t, \mathbf{y}_p) \, p^r + \frac{1}{\tau_p^{St}} U_{d,i}(t, \mathbf{y}_p) \, p^r$$
$$- \lambda_{ij} \frac{\partial [\, p^r \,]}{\partial y_{p,j}} - \mu_{ij} \frac{\partial [\, p^r \,]}{\partial V_{p,j}} \qquad (5.4)$$

with $\lambda_{ij}$ and $\mu_{ij}$ given by

$$\lambda_{ij} = \frac{1}{\tau_p^{St}} \int_0^t \langle \Gamma_{jk}(t, t') \, R_{ik}(t, \mathbf{y}_p; t', \mathbf{X}_p(t')) \rangle_{(\mathbf{y}_p, \mathbf{V}_p)} \, dt' \qquad (5.5a)$$

$$\mu_{ij} = \frac{1}{\tau_p^{St}} \int_0^t \langle \dot{\Gamma}_{jk}(t, t') \, R_{ik}(t, \mathbf{y}_p; t', \mathbf{X}_p(t')) \rangle_{(\mathbf{y}_p, \mathbf{V}_p)} \, dt' \qquad (5.5b)$$

where the notation $\langle . \rangle_{(\mathbf{y}_p, \mathbf{V}_p)}$ indicates the averaged value conditioned on a particle trajectory that 'arrives' at $(\mathbf{y}_p, \mathbf{V}_p)$ at time $t$, which explains that the dispersion tensors are functions of $(\mathbf{y}_p, \mathbf{V}_p)$ even if the eventual dependence on $\mathbf{V}_p$ is often neglected. In these equations, $\Gamma_{jk}(t, t')$ stands for the response function

$$\Gamma_{jk}(t, t') = \frac{\delta X_{p,j}(t)}{\delta u_{f,k}(t', \mathbf{X}_p(t'))} \qquad (5.6)$$

that measures the effect of a perturbation of the fluctuating fluid velocity seen at an earlier time $t'$ on the particle position $\mathbf{X}_p(t)$ at time $t$, and $\dot{\Gamma} = \partial \Gamma / \partial t$. With these expressions, the flux can also be written as

$$\frac{1}{\tau_p^{St}} \langle U_{s,i} \,|\, \mathbf{z}_p \rangle \, p^r = \frac{1}{\tau_p^{St}} \langle U_{f,i} \rangle (t, \mathbf{y}_p) \, p^r + \kappa_i \, p^r - \frac{\partial [\, \lambda_{ij} \, p^r \,]}{\partial y_{p,j}} - \frac{\partial [\, \mu_{ij} \, p^r \,]}{\partial V_{p,j}} \qquad (5.7)$$

with

$$\kappa_i(t; \mathbf{z}_p) = \frac{1}{\tau_p^{St}} U_{d,i}(t, \mathbf{y}_p) + \frac{\partial \lambda_{ij}}{\partial y_{p,j}} + \frac{\partial \mu_{ij}}{\partial V_{p,j}} . \qquad (5.8)$$

Inserting Eq. (5.7) into Eq. (5.2), we obtain the kinetic PDF equation which, in a compact form, is

$$\frac{\partial p^r}{\partial t} = -\frac{\partial}{\partial z_l^r} \left[ A_l^{KE} \, p^r \right] + \frac{1}{2} \frac{\partial^2}{\partial z_l^r \partial z_m^r} \left[ B_{lm}^{KE} \, p^r \right], \qquad (5.9)$$

where the components of the drift vector $(A_l^{KE})_{l=1,6}$ are

$$
A_l^{KE} = \begin{cases} V_{p,l} & l = 1, 3, \\ \dfrac{\langle U_{f,l-3} \rangle - V_{p,l-3}}{\tau_p^{St}} + \kappa_{l-3} & l = 4, 6, \end{cases} \tag{5.10}
$$

and where the symmetrical matrix $B_{lm}^{KE}$ is (using a bloc notation with $i, j = 1, 3$)

$$
\mathbf{B}^{KE} = \begin{pmatrix} 0 & \vline & (\lambda_{ij}) \\ -- & \vline & --- \\ (\lambda_{ji}) & \vline & (\mu_{ij}) + (\mu_{ji}) \end{pmatrix} . \tag{5.11}
$$

As such, Eq. (5.9) looks like a classical convection-diffusion equation with $B_{lm}^{KE}$ in the second-order derivative appearing as a diffusion matrix. However, it is immediate to show that this symmetrical matrix $\mathbf{B}^{KE}$ is negative definite. Indeed, its determinant is $\det(\mathbf{B}^{KE}) = -(\det(\lambda))^2 < 0$ and, consequently, the matrix $\mathbf{B}^{KE}$ has always at least one negative eigenvalue. This point was first put forward in [6] where it was associated with an 'anti-diffusive' behavior and the non-Markovian nature of the reduced state vector $\mathbf{Z}_p^r$. It was later discussed in a comprehensive analysis of kinetic- and dynamic-PDF models in [7] to which readers are referred for a more detailed analysis. The unavoidable consequence is that this kinetic-PDF equation cannot be solved, unless for very special initial conditions, and, therefore, does not qualify as an acceptable PDF model.

Is the failure of kinetic descriptions due to specific closures, such as the FND relation, or to the consideration of discrete particles in non-fully-resolved turbulent flows? Or is it actually related to a poor choice of the variables entering the particle state vector and to the loss of Markovianity? In relation to these questions is the remark that, when considering homogeneous situations (for the sake of avoiding complex notations in general flows, see comprehensive discussions of these relations in [7]), $\lambda_{ij}$ and $\mu_{ij}$ are simply expressed as the correlations between particle positions and velocities and the variable that is eliminated from the reduced PDF description, that is the velocity of the fluid seen:

$$
\lambda_{ij} = \frac{1}{\tau_p^{St}} \langle u_{s,i}(t) \, X_{p,j}(t) \rangle , \tag{5.12a}
$$

$$
\mu_{ij} = \frac{1}{\tau_p^{St}} \langle u_{s,i}(t) \, U_{p,j}(t) \rangle . \tag{5.12b}
$$

We can then wonder if leaving out the velocity of the fluid seen was appropriate. All these issues are now investigated.

## 5.2 Slow and Fast Variables and the Markovian Approach

### 5.2.1 Colored or White Noises and Well-posed PDF Equations

The failure of the kinetic description for discrete particles in non-fully-resolved turbulent flows can be cast in the more general framework of dynamical systems under the influence of 'external noises' [8–11]. To shed light on this issue, we consider a dynamical system characterized by a state vector $\mathbf{Z}(t) = (Z_1(t), Z_2(t), \ldots, Z_n(t))$ with evolution equations

$$\frac{\mathrm{d}Z_i}{\mathrm{d}t} = A_i(t, \mathbf{Z}) + B_{ij}(t, \mathbf{Z})\,\xi_j \tag{5.13}$$

where $\boldsymbol{\xi}(t) = (\xi_j(t))_{j=1,n}$ represents an 'external noise'. For the sake of simplicity since it does not affect results to come, the dependence of $\mathbf{A}$ and $\mathbf{B}$ is simply written as $\mathbf{A}(t, \mathbf{Z})$ and $\mathbf{B}(t, \mathbf{Z})$ instead of a more general dependence including statistics of the stochastic process (such as its mean, variance, etc.).

When the external noise $\boldsymbol{\xi}$ is a Gaussian process with independent values, we are evolving within the well-established framework of Ito SDEs and Fokker-Planck equations, cf. Sect. 4.3. The situation is different for 'colored noises', that is when $\boldsymbol{\xi}$ has a non-zero correlation timescale. This is found, for example, for stationary Gaussian processes simulated as a set of independent OU processes (cf. Sect. 4.3)

$$\mathrm{d}\xi_k = -\frac{\xi_k}{\tau}\,\mathrm{d}t + \sigma\,\mathrm{d}W_k\,, \tag{5.14}$$

where $\tau$ is the timescale of $\xi_k$ and $\sigma$ a constant equal to $\sigma^2 = 2\langle\xi_k^2\rangle/\tau$ from the classical fluctuation-dissipation theorem [12]. For the sake of simplicity, $\tau$ and $\sigma$ are retained for each component $\xi_k$ (differences can be accounted for through the choice of $\mathbf{B}$). The auto-correlation of this process is an exponential function $\langle\xi_k(t)\xi_k(s)\rangle = \langle\xi^2\rangle\exp(-|t-s|/\tau)$ and is therefore not delta-correlated when $\tau \neq 0$. An important element is that for such external noises, the process $\mathbf{Z}$ in Eq. (5.13) is no longer Markovian [10, 11, 13] although a classical remark is to note that Markovianity is retrieved by considering the extended process $(\mathbf{Z}, \boldsymbol{\xi})$ [10, 11, 13]. Even in a non-Markovian case, we can still consider the equation satisfied by the one-time PDF of the process $p(t; \mathbf{z})$, from which a set of PDEs are derived for some relevant statistical moments. It is therefore essential that this equation be mathematically well-posed for the resulting macroscopic descriptions to be regarded as acceptable, regardless of the choices of $\mathbf{A}$ and $\mathbf{B}$ in Eq. (5.13).

#### 5.2.1.1 PDF Equation and the Well-posed Criterion

The PDF equation for Gaussian colored noise is derived by standard techniques [1] from the fine-grained PDF $\mathcal{P}(t; \mathbf{z}) = \delta(\mathbf{Z}(t) - \mathbf{z})$ with $p(t; \mathbf{z}) = \langle\mathcal{P}(t; \mathbf{z})\rangle$. The exact but unclosed PDF equation for $p(t; \mathbf{z})$ is

$$\frac{\partial p}{\partial t} = -\frac{\partial}{\partial z_i} [A_i(t, \mathbf{z}) \, p] - \frac{\partial}{\partial z_i} [B_{ik}(t, \mathbf{z})\langle \xi_k \,|\, (t, \mathbf{z})\rangle \, p] \tag{5.15}$$

which, using $\langle X\mathcal{P}\rangle = \langle X|(t, \mathbf{z})\rangle p(t, \mathbf{z})$, can be written as

$$\frac{\partial p}{\partial t} = -\frac{\partial}{\partial z_i} [A_i(t, \mathbf{z}) \, p] - \frac{\partial}{\partial z_i} [B_{ik}(t, \mathbf{z})\langle \xi_k \mathcal{P}\rangle] . \tag{5.16}$$

In Eq. (5.16), the open flux $\langle \xi_k \mathcal{P}\rangle$ can be written as $\langle \xi_k Q[t; \boldsymbol{\xi}]\rangle$, where $Q[\cdot]$ stands for a functional dependence since $\mathcal{P}$ can be seen as a functional of the Gaussian centered process $\boldsymbol{\xi}$. As for derivation of the kinetic PDF, we can apply the Furutsu-Novikov-Donsker (FND) relation [2–4]

$$\langle \xi_k(t)Q[t; \boldsymbol{\xi}]\rangle = \int_0^t \langle \xi_k(t)\xi_l(t')\rangle \left\langle \frac{\delta Q[t; \boldsymbol{\xi}]}{\delta \xi_l(t')} \right\rangle dt' . \tag{5.17}$$

Then, using

$$\frac{\delta \mathcal{P}(t; \mathbf{z})}{\delta \xi_l(t')} = -\frac{\delta Z_j(t)}{\delta \xi_l(t')} \frac{\partial \mathcal{P}(t; \mathbf{z})}{\partial z_j} \tag{5.18a}$$

$$= -\frac{\partial}{\partial z_j} \left[ \frac{\delta Z_j(t)}{\delta \xi_l(t')} \mathcal{P} \right] + \left( \frac{\partial}{\partial z_j} \left[ \frac{\delta Z_j(t)}{\delta \xi_l(t')} \right] \right) \mathcal{P} \tag{5.18b}$$

and applying the averaging operator, we obtain

$$\langle \xi_k \mathcal{P}\rangle = \alpha_k(t, \mathbf{z}) \, p(t; \mathbf{z}) - \frac{\partial \left[ \lambda_{kj}(t, \mathbf{z}) \, p(t; \mathbf{z}) \right]}{\partial z_j} \tag{5.19}$$

with

$$\alpha_k(t, \mathbf{z}) = \int_0^t \langle \xi_k(t)\xi_l(t')\rangle \left\langle \frac{\partial}{\partial z_j} \left[ \frac{\delta Z_j(t)}{\delta \xi_l(t')} \right] |(t, \mathbf{z}) \right\rangle dt' \tag{5.20}$$

and where $\lambda_{kj}$ is given by

$$\lambda_{kj}(t, \mathbf{z}) = \int_0^t \langle \xi_k(t)\xi_l(t')\rangle \left\langle \frac{\delta Z_j(t)}{\delta \xi_l(t')} |(t, \mathbf{z}) \right\rangle dt' . \tag{5.21}$$

On the other hand, applying directly the FND relation to $Z_j(t)$ in Eq. (5.17) leads to

$$\langle \xi_k Z_j(t)\rangle = \int \lambda_{kj}(t, \mathbf{z}) p(t; \mathbf{z}) d\mathbf{z} = \int_0^t \langle \xi_k(t)\xi_l(t')\rangle \left\langle \frac{\delta Z_j(t)}{\delta \xi_l(t')} \right\rangle dt' . \tag{5.22}$$

Combining Eqs. (5.19) and (5.16) gives the closed form for the one-time PDF equation

$$\frac{\partial p}{\partial t} = -\frac{\partial}{\partial z_i} \left[ (A_i + B_{ik}\alpha_k) \, p \right] + \frac{\partial}{\partial z_i} \left[ B_{ik} \frac{\partial \left( \lambda_{kj} \, p \right)}{\partial z_j} \right] , \tag{5.23}$$

which can be re-arranged as

$$\frac{\partial p}{\partial t} = -\frac{\partial}{\partial z_i} \left[ \widetilde{A}_i \, p \right] + \frac{1}{2} \frac{\partial^2}{\partial z_i \partial z_j} \left[ \widetilde{D}_{ij} \, p \right] , \tag{5.24}$$

with $\widetilde{A}_i = A_i + B_{ik}\alpha_k + \lambda_{kj} \partial B_{ik}/\partial z_j$ and $\widetilde{D}_{ij}$ the symmetrical matrix

$$\widetilde{D}_{ij}(t, \mathbf{z}) = B_{ik}(t, \mathbf{z})\lambda_{kj}(t, \mathbf{z}) + B_{jk}(t, \mathbf{z})\lambda_{ki}(t, \mathbf{z}) . \tag{5.25}$$

The well-posed nature of the PDF equation, Eq. (5.24), is determined by the matrix $\widetilde{D}_{ij}$ in Eq. (5.25). Indeed, if $\widetilde{D}_{ij}$ is not positive definite (there is at least one negative eigenvalue), this implies that a marginal of the one-time PDF appears as the 'solution' of an 'anti-diffusion' PDE which are ill-posed in that they can only be solved for very special initial conditions. Hence, the well-posed criterion is to require that $\widetilde{D}_{ij}(t, \mathbf{z})$ have only positive, or null, eigenvalues, $\forall \mathbf{z}$.

*Analysis of the Linear Case*  In [14, section 9.2], the well-posed property of such PDF equations is investigated, especially for a dynamical system with a general structure containing kinetic descriptions, and it is shown that the positive nature of $D_{ij}$ is only obtained when taking the white-noise limit. In the present context, it is sufficient to consider a simpler situation where the drift vector is linear in $\mathbf{Z}$ (or linearized around a given point in sample space). Then, Eq. (5.13) becomes

$$\frac{dZ_i}{dt} = -G_{ik}Z_k + B_{ik}\xi_k \tag{5.26}$$

where $\mathbf{G}$ is a constant matrix representing return-to-equilibrium effects and where the colored noise $\boldsymbol{\xi}$ is a set of independent stationary OU processes as in Eq. (5.14). In the linear case, the 'response functions' $\delta Z_i(t)/\delta \xi_l(t')$ in Eq. (5.21) are independent of the sample space value $\mathbf{z}$, showing that $\lambda_{ji} = \langle Z_i \xi_j \rangle$. In the stationary state where $\lambda_{ji}$ reach constant values, the correlations are easily derived through

$$\frac{d\lambda_{ji}}{dt} = 0 \implies (\delta_{ik} + \tau G_{ik}) \langle Z_k \xi_j \rangle = B_{ij}\tau \langle \xi^2 \rangle , \tag{5.27}$$

which can be inverted to give $\lambda_{ji} = \tau \langle \xi^2 \rangle \widetilde{G}_{ik}^{-1} B_{kj}$ with $\widetilde{\mathbf{G}} = \mathbb{1} + \tau \mathbf{G}$.

To show that the positive-definite property of $\widetilde{D}_{ij}$ is not automatically satisfied, it is sufficient to consider a specific counter-example. Taking for instance a simple two-dimensional situation where $\mathbf{Z} = (Z_1, Z_2)$ with an isotropic noise term, i.e. $\mathbf{B} = B\,\mathbb{1}$, and with a return-to-equilibrium matrix of the form

$$\mathbf{G} = \begin{pmatrix} 1 & -\kappa \\ -\kappa & 1 \end{pmatrix} . \tag{5.28}$$

The evolution equations for this system are therefore

$$\frac{dZ_1}{dt} = -Z_1 + \kappa Z_2 + B\,\xi_1 , \tag{5.29a}$$

$$\frac{dZ_2}{dt} = -Z_2 + \kappa Z_1 + B\,\xi_2 , \tag{5.29b}$$

and from Eq. (5.25) we obtain that the determinant of the $(2 \times 2)$ matrix $\widetilde{D}_{ij}$ is

$$\det(\widetilde{\mathbf{D}}) = \frac{4B^2 \left(\tau\langle\xi^2\rangle\right)^2}{(1+\tau)^2 - \tau^2\kappa^2} . \tag{5.30}$$

As soon as $\tau \neq 0$, this determinant is negative for large values of $\kappa$ and, hence, such a system is ill-posed. In other words, as soon as the two components $Z_1$ and $Z_2$ are strongly coupled, the resulting probabilistic description becomes ill-based even for such a trivial system. Note for small values of $\tau$, the inverse matrix can be approximated by $\widetilde{\mathbf{G}}^{-1} = \mathbb{1} - \tau\mathbf{G} + O(\tau^2)$, which shows that, for the general formulation in Eq. (5.26), the correlations $\lambda_{ji}$ can be written as

$$\lambda_{ji} = \tau\langle\xi^2\rangle B_{ij} - \tau^2\langle\xi^2\rangle G_{ik}B_{kj} + O(\tau^2) . \tag{5.31}$$

Using this approximation in Eq. (5.25), the symmetrical matrix $\widetilde{D}_{ij}$ is obtained as

$$\widetilde{D}_{ij} = 2\tau\langle\xi^2\rangle(BB^\perp)_{ij} - \tau^2\langle\xi^2\rangle C_{ij} + O(\tau^2) , \tag{5.32}$$

where $\mathbf{C} = (BB^\perp)G^\perp + G(BB^\perp)$ is a symmetrical matrix. The first term on the right-hand side of Eq. (5.32) constitutes a positive-definite matrix but the same conclusion does not necessarily hold for the second term. The important point is that this second term is explicitly dependent upon $\mathbf{G}$. The only possibility to ensure that the resulting matrix $\widetilde{D}_{ij}$ remains definite positive whatever the choice of $\mathbf{G}$ is to take the limit $\tau \to 0$. Yet, in order to retain a non-zero $\widetilde{D}_{ij}$ matrix, this limit must be taken as: $\tau \to 0$ with $\langle\xi^2\rangle \to \infty$, such that $\lim_{\tau \to 0} \left(\tau\langle\xi^2\rangle\right) = K$ where $K$ is a positive constant. This corresponds to the Markovian approximation which is now introduced.

## 5.2.2   The Markovian Approach

It follows from the preceding discussions that there are two main modeling philosophies. The first one consists in keeping the same set of variables, for instance the kinetic variables $\mathbf{Z}_p^r = (\mathbf{X}_p, \mathbf{U}_p)$, whatever the context. We have then to handle colored noises when 'external forcing' involves non-zero time or space correlations with the risk of ending up with ill-posed PDF formulations, as in the case for discrete particles in non-fully-resolved turbulent flows. The second modeling philosophy consists in adjusting the particle state vector by including additional variables until the eliminated degrees of freedom and/or the 'external forcing' can be treated as white-noise terms on the now-extended mechanical system under consideration. The technical derivations of the eigenvalues of the diffusion matrix should not hide the physical issues at stake. In a thermodynamic formulation, the existence of a negative eigenvalue indicates that one is trying to describe a system whose contact with the 'external world' cannot be treated as a contact with a heat bath since it contains a (negative) correlation and, thus, an underlying order that needs to be taken into account. On the other hand, positive eigenvalues of the second-order matrix $\widetilde{D}_{ij}$ means that the corresponding effects can be regarded as the sum of uncorrelated 'pure noise' perturbations, leading to real diffusive actions on the system. In this second situation, we can regard the extended system as being in contact with the equivalent of 'heat bathes' and we are now dealing with Markovian systems. To underline the importance of this notion, we quote [15, page 213] who gave an excellent presentation of this principle: "One needs to keep sufficiently many variables and the appropriate non-linearities for achieving a realistic description of a system by *Markovian time-evolution equations*. We here insist on avoiding explicit memory effects and take the standpoint that memory effects always indicate the existence of unrecognized variables relevant to the definition of a proper system for understanding certain phenomena of interest."

The application of this principle consists in classifying the degrees of freedom of a system as slow and fast variables with respect to an observation time $\Delta t$ which needs to be introduced (in a discrete time version, this observation time interval corresponds to the time step). Variables whose auto-correlation timescale are larger than $\Delta t$ are defined as slow variables while variables with an auto-correlation timescale smaller than $\Delta t$ are defined as fast variables. As such, this is not sufficient since the eliminated fast variables would appear as colored noises in the evolution equations of the retained slow variables. The search is therefore for a scale separation which allows to treat fast variables as white-noise effects while slow variables have not changed appreciably (this is the essence of the 'slaving principle').

### 5.2.2.1  An Enlightening Toy Model

To exemplify the slaving principle, it is instructive to consider a toy model involving a variable $X_{slow}$ whose time-rate-of-change is $X_{fast}$, i.e. $dX_{slow}/dt = X_{fast}$. We regard then $X_{fast}$ as a centered process having reached its equilibrium distribution, which can be conditioned (or not) on a given value of the slow variable $X_{\text{slow}} = x$. Therefore, $X_{\text{fast}}$ is a stationary process with variance $\langle X_{\text{fast}}^2 \rangle$ and an auto-correlation function $\mathcal{R}_{X_{\text{fast}}}$, defined such that $\langle X_{\text{fast}}^2 \rangle \mathcal{R}_{X_{\text{fast}}}(t' - t) = \langle X_{\text{fast}}(t) X_{\text{fast}}(t') \rangle$, which depends only on the time difference $t' - t$ and whose integral timescale is $T_{X_{\text{fast}}}$. Straightforward calculations show that for $X_{\text{slow}}(0) = 0$

$$\frac{\mathrm{d}\langle X_{\text{slow}}^2 \rangle}{\mathrm{d}t} = 2\langle X_{\text{fast}}^2 \rangle \int_0^t \mathcal{R}_{X_{\text{fast}}}(s)\,\mathrm{d}s \; . \tag{5.33}$$

Since by definition $T_{X_{\text{fast}}} = \int_0^{+\infty} \mathcal{R}_{X_{\text{fast}}}(s)\,\mathrm{d}s$, we get that for 'long-enough time lapses'

$$t \gg T_{X_{\text{fast}}} \implies \langle X_{\text{slow}}^2 \rangle(t) \simeq \left( 2\langle X_{\text{fast}}^2 \rangle T_{X_{\text{fast}}} \right) t \; , \tag{5.34}$$

which is the linear behavior for the second-order moment of $X_{\text{slow}}$ characterizing the diffusive regime. If we wish to obtain the same result at any time $t$ (of the order of the observation time), we need to take the limit of vanishing timescale $T_{X_{\text{fast}}}$ and, to retain a non-zero diffusive coefficient for the evolution of $\langle X_{\text{slow}}^2 \rangle$ in Eq. (5.34), we are led to assume that we have

$$T_{X_{\text{fast}}} \to 0 \text{ and } \langle X_{\text{fast}}^2 \rangle \to +\infty \quad \text{such that} \quad \langle X_{\text{fast}}^2 \rangle T_{X_{\text{fast}}} \to \mathcal{D} \; , \tag{5.35}$$

where $\mathcal{D}$ is the diffusion coefficient for $X_{\text{slow}}$. This corresponds to the white-noise limit for $X_{\text{fast}}$ and, in terms of the equation for the trajectories of $X_{\text{slow}}$, we have replaced an ODE by a SDE

$$\mathrm{d}X_{\text{slow}} = X_{\text{fast}}\,\mathrm{d}t \implies \mathrm{d}X_{\text{slow}} = \sqrt{2\mathcal{D}}\,\mathrm{d}W \; , \tag{5.36}$$

where $W$ is a Wiener process. Note that the resulting diffusion coefficient $\mathcal{D}$ can be written as

$$\mathcal{D} = \int_0^\tau \langle X_{\text{fast}}(t) X_{\text{fast}}(0) \rangle \mathrm{d}t \; , \tag{5.37}$$

where $\tau$ is an intermediate timescale separating $X_{\text{slow}}$ from the rapidly-varying part of its time derivative, which is here $X_{\text{fast}}$ (as long as $\tau \gg T_{X_{\text{fast}}}$, the upper limit of the integral does not modify significantly the integral compared to $\tau = +\infty$). This is actually a Green-Kubo expression [15]. Following the slaving principle, an heuristic formulation consists in keeping the same results conditioned on a given value $x$ of $X_{\text{slow}}$, so that we obtain $\mathcal{D}(x)$ as in Eq. (5.37) with the averaging operator being $\langle \cdot \mid X_{\text{slow}} = x \rangle$. This turns out to be valid for the diffusion coefficient resulting from the elimination of fast variables but the shift from an ODE to a SDE, as in Eq. (5.36), may involve additional drift terms when $T_{X_{\text{fast}}}(x)$ is an explicit function of the slow variable. In the course of the following sections, these questions resurface regularly and they are addressed more rigorously in Sect. 10.2. Going back to the well-posed criterion for PDF equations involving colored noises, it is seen that the criterion needed to guarantee the positive nature of the diffusion matrix $D_{ij}$ in Eq. (5.32) is the same one with $\tau$, $\langle \xi^2 \rangle$ and $K$ being replaced here by $T_{X_{\text{fast}}}$, $\langle X_{\text{fast}}^2 \rangle$ and $\mathcal{D}$, respectively. The conclusion is that well-posed formulations are obtained when taking the white-noise limit based on a scale separation to distinguish between slow and (very) fast variables.

### 5.2.2.2 Particle State Vector and Markovian Modeling Steps

With the benefit of hindsight, we can revisit the shortcomings of the kinetic approach demonstrated in Sect. 5.1 to emphasize that the way out of such difficulties consists in selecting adequate variables in the particle state vector so that the Markovian approach can be applied, similarly to what is done in the toy model above. It is important to be aware that including additional variables is not enough as such and, furthermore, that there is nothing wrong with the FND relation either. What matters is to tailor the choice of the particle state vector so as to end up with a system influenced by fast variables that can then be safely replaced by white-noises in order to obtain well-based PDF formulations.

To bring out the significance of combining these modeling principles, we take up an example introduced in [7]. In this example, the stochastic model is reduced to its bare essentials by retaining a 1D formulation without mean terms and with constant coefficients so that it is straightforward to derive physically-relevant expressions. For this reason, we consider the following system in which the velocity of the fluid seen $U_s$ is included in the particle state vector but is modeled as being influenced by a colored noise, noted $\xi_s$, instead of a white-noise

$$\frac{dX_p}{dt} = U_p \, , \tag{5.38a}$$

$$\frac{dU_p}{dt} = \frac{U_s - U_p}{\tau_p} \, , \tag{5.38b}$$

$$\frac{dU_s}{dt} = -\frac{U_s}{T_L} + \xi_s \, . \tag{5.38c}$$

As mentioned, the timescales $\tau_p$ and $T_L$ are constant and, since $\xi_s$ is a colored noise, $U_s$ is differentiable which explains the derivative form in Eq. (5.38c). For $\xi_s$, we consider a simple stationary Gaussian process with a non-zero correlation timescale, that is an OU process

$$d\xi_s = -\frac{\xi_s}{\tau} \, dt + \sigma \, dW \, , \tag{5.39}$$

where $\tau$ and $\sigma$ are constant (with $\sigma^2 = 2\langle \xi_s^2 \rangle / \tau$ from the classical fluctuation-dissipation theorem). As already indicated in Sect. 5.2.1, the auto-correlation of $\xi_s$ is an exponential function

$$\langle \xi_s(t) \xi_s(t') \rangle = \langle \xi_s^2 \rangle \exp(-|t - t'|/\tau) \qquad (5.40)$$

and thus not delta-correlated when $\tau \neq 0$. From the form in Eq. (5.38), it appears that, though the fluid velocity seen $U_s$ is included, the particle state vector $\mathbf{Z}_p = (X_p, U_p, U_s)$ does not constitute a Markov process (a Markov process would be retrieved by considering the extended state vector $(X_p, U_p, U_s, \xi_s)$). Thus, the PDF equation for $p(t; y_p, V_p, V_s)$ is open and is

$$\frac{\partial p}{\partial t} + \frac{\partial [V_p \, p]}{\partial y_p} = -\frac{\partial}{\partial V_p} \left[ \left( \frac{V_s - V_p}{\tau_p} \right) p \right] + \frac{\partial}{\partial V_s} \left[ \frac{V_s}{T_L} p \right]$$

$$- \frac{\partial}{\partial V_s} \left[ \langle \xi_s \mid (y_p, V_p, V_s) \rangle \, p \right] . \qquad (5.41)$$

Since the external noise $\xi_s$ is Gaussian, we can work out an exact closed PDF equation by applying the same FND relation used in Sect. 5.1, cf. Eq. (5.3). For the jointly-Gaussian process $(X_p, U_p, U_s, \xi_s)$, this yields that

$$\langle \xi_s \mid (y_p, V_p, V_s) \rangle \, p = -\Lambda_{X_p} \frac{\partial p}{\partial y_p} - \Lambda_{U_p} \frac{\partial p}{\partial V_p} - \Lambda_{U_s} \frac{\partial p}{\partial V_s} \qquad (5.42)$$

where the coefficients $\Lambda_{X_p}$, $\Lambda_{U_p}$ and $\Lambda_{U_s}$ represent the correlations between the variables kept in the state vector and the noise that is eliminated, as in Eq. (5.12). In our example, this gives the following expressions

$$\Lambda_{U_s} = \langle U_s \, \xi_s \rangle, \quad \Lambda_{U_p} = \langle U_p \, \xi_s \rangle, \quad \Lambda_{X_p} = \langle X_p \, \xi_s \rangle . \qquad (5.43)$$

With the help of stochastic calculus, we obtain from the system in Eqs. (5.38) and (5.39) that these dispersion coefficients (in the PDF equation) scale as the successive powers of the timescale $\tau$ multiplied by the 'noise energy' $\langle \xi_s^2 \rangle$

$$\Lambda_{U_s} = \frac{\tau \, T_L}{\tau + T_L} \langle \xi_s^2 \rangle , \qquad (5.44a)$$

$$\Lambda_{U_p} = \frac{\tau}{\tau + \tau_p} \langle U_s \, \xi_s \rangle = \frac{\tau^2 \, T_L}{(\tau + T_L)(\tau + \tau_p)} \langle \xi_s^2 \rangle , \qquad (5.44b)$$

$$\Lambda_{X_p} = \tau \langle U_p \, \xi_s \rangle = \frac{\tau^3 \, T_L}{(\tau + T_L)(\tau + \tau_p)} \langle \xi_s^2 \rangle . \qquad (5.44c)$$

For the moment, we are concerned with the structure of the resulting PDF equation obtained by inserting Eq. (5.42) in Eq. (5.41), which gives

$$\frac{\partial p}{\partial t} = -\frac{\partial}{\partial z_{p,l}} \left[ \widetilde{A}_l \, p \right] + \frac{1}{2} \frac{\partial^2}{\partial z_{p,l} \partial z_{p,m}} \left[ \widetilde{B}_{lm} \, p \right] \qquad (5.45)$$

where the drift vector $(\widetilde{A}_l)_{l=1,3}$ follows directly from Eq. (5.41) and $z_p = (y_p, V_p, V_s)$. From the analysis developed in Sects. 5.1 and 5.2.1, we know that the well-posed property of this PDF equation is determined by the second-order tensor

$$\widetilde{\mathbf{B}} = \begin{pmatrix} 0 & 0 & \Lambda_{X_p} \\ 0 & 0 & \Lambda_{U_p} \\ \Lambda_{X_p} & \Lambda_{U_p} & 2\Lambda_{U_s} \end{pmatrix} . \qquad (5.46)$$

The characteristic polynomial of this tensor, $\mathbb{P}_{\widetilde{\mathbf{B}}}(x) = \det(x\mathbb{1} - \widetilde{\mathbf{B}})$, is

$$\mathbb{P}_{\widetilde{\mathbf{B}}}(x) = x \left( x^2 - 2x\Lambda_{U_s} - \Lambda_{U_p}^2 - \Lambda_{X_p}^2 \right) , \qquad (5.47)$$

from which it follows that its eigenvalues are

$$e_{v1} = 0 , \qquad (5.48a)$$

$$e_{v2} = \Lambda_{U_s} - \sqrt{\Lambda_{U_s}^2 + \Lambda_{U_p}^2 + \Lambda_{X_p}^2} , \qquad (5.48b)$$

$$e_{v3} = \Lambda_{U_s} + \sqrt{\Lambda_{U_s}^2 + \Lambda_{U_p}^2 + \Lambda_{X_p}^2} . \qquad (5.48c)$$

The inescapable outcome is that $e_{v2} < 0$ and, thus, that there is always a negative eigenvalue involved, which entails that Eq. (5.45) is an ill-posed equation.

The conclusion from this analysis is that the sole inclusion of the velocity of the fluid seen in the particle state vector does not change the situation for an external colored Gaussian noise and we end up again with an incomplete and ill-based PDF formulation. It can easily be seen that continuing to include further variables would only shift the problem at a higher, or deeper, level but would not change the structure of the second-order tensor in the resulting PDF equation. The dimension of the corresponding sample space would only increase but the PDF equation would still be ill-posed. The only way out of this predicament is to apply the methodology used with the toy model described above and to isolate a sufficiently-fast variable that can be replaced by a white-noise. For the system in Eq. (5.38), this is illustrated if we consider that $\tau$ becomes small (with respect to $T_L$) but with the total energy, $\tau \langle \xi_s^2 \rangle$, remaining constant, as in Eq. (5.35), so that we obtain a proper SDE involving white-noise increments, as in Eq. (5.36). Indeed, when we consider the two limits $\tau \to 0$ with $\langle \xi_s^2 \rangle \to \infty$, such that $\lim_{\tau \to 0} \left( \tau \langle \xi_s^2 \rangle \right) = K$ where $K$ is a positive constant, the dispersion coefficients tend to

$$\Lambda_{U_s} \xrightarrow[\tau \to 0]{} K, \ \Lambda_{U_p} \xrightarrow[\tau \to 0]{} 0, \ \Lambda_{X_p} \xrightarrow[\tau \to 0]{} 0 . \tag{5.49}$$

In that limit, the second-order tensor $\widetilde{\mathbf{B}}$ becomes

$$\lim_{\tau \to 0} \widetilde{\mathbf{B}} = \begin{pmatrix} 0 & 0 & 0 \\ 0 & 0 & 0 \\ 0 & 0 & 2K \end{pmatrix} \tag{5.50}$$

with eigenvalues equal to $e_{v1} = e_{v2} = 0$ while $e_{v3} > 0$. Thus, $\widetilde{\mathbf{B}}$ is now degenerate but with a strictly positive sub-matrix, here the positive coefficient $K$, and we retrieve a well-posed FP formulation for the PDF equation in Eq. (5.45). The SDEs are

$$dX_p = U_p \, dt \tag{5.51a}$$

$$dU_p = \frac{U_s - U_p}{\tau_p} \, dt \tag{5.51b}$$

$$dU_s = -\frac{U_s}{T_L} \, dt + \sqrt{2K} \, dW \tag{5.51c}$$

which can also be obtained, using a less rigorous but short-cut manner, by writing directly that $\lim_{\tau \to 0} \xi_s \, dt \simeq \sqrt{2K} \, dW$ in the original trajectory equations, Eq. (5.38).

As a side remark, we can observe that, in this small-$\tau$ limit, the dispersion coefficients in Eq. (5.44) appear as successive powers of $\tau$ since

$$\langle U_s \, \xi_s \rangle \simeq K \sim O(1), \quad \langle U_p \, \xi_s \rangle \simeq \tau K , \quad \langle X_p \, \xi_s \rangle \simeq \tau^2 K . \tag{5.52}$$

It turns out that similar scaling laws are at the core of the developments based on fast-variable elimination techniques (for small $T_L$) in Chap. 10 and for the formulation of the source terms representing particle back effects on the fluid in the tracer-particle limit (for small $\tau_p$) in Chap. 11.

### 5.2.2.3 Summary of the Markovian Approach

For discrete particles in non-fully-resolved turbulent flows, the Markovian approach leads to include the velocity of the fluid seen in the particle state vector (having a similar timescale as $\mathbf{U}_p$ for relatively low-inertia particles, $\mathbf{U}_s$ is clearly a slow variable). However, do we need additional variables with, for instance, the time derivative of $\mathbf{U}_s$? Answers to this question are provided by the Kolmogorov theory.

## 5.3 The Kolmogorov Theory and Lagrangian Models

In spite of some limitations, the Kolmogorov description of turbulent flows remains the reference theory in turbulence modeling. The first theory was presented in 1941 (the K41 theory) and later refined to account for intermittency in 1962 (the refined or K62 theory) [16–18]. Interestingly, A. N. Kolmogorov, who was one of the most brilliant mathematicians of the twentieth century, chose a rather qualitative approach based on the image of an energy cascade from which statistical predictions are derived. As in Richardson's first pictorial description in 1922 ("Big whorls have little whorls, which feed on their velocity, and little whorls have lesser whorls, and so on until viscosity"), the energy cascade corresponds to a description in terms of what is loosely defined as 'an eddy' (or a velocity fluctuating 'component'), characterized by its 'size' $l$, velocity scale $\delta u_f(l)$ and timescale $\tau_f(l)$: energy is produced at the large scales imposed by the geometry of the flow domain or the boundary conditions and energy is transferred through the inertial range (eddies for which energy is neither created nor dissipated) until it is dissipated at the smallest scales by viscous motions, see Fig. 5.1. Given its central role, the Kolmogorov theory has made its way in classical textbooks on turbulence and extensive accounts are available [1, 16, 17]. Consequently, we only give a brief outline of its main characteristics with a view towards its application for Lagrangian stochastic models.

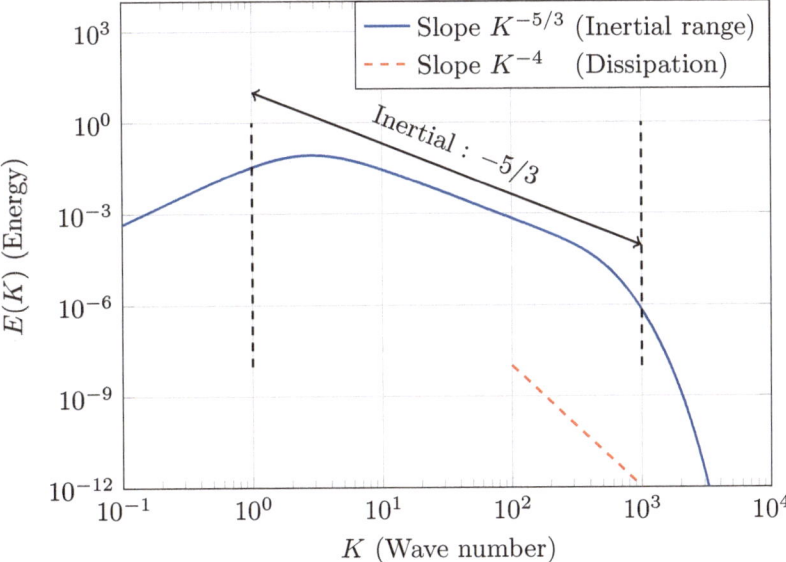

**Fig. 5.1** The Kolmogorov picture of energy cascade in turbulence: energy is produced at the large scales and is transferred through the inertial range before being dissipated by viscous motions at the same rate $\langle \epsilon_f \rangle$

Since traditional presentations tend to concentrate on spatial correlations, it is worth recalling that the fundamental K41 theory is more general and proceeds from a Lagrangian vision. In that sense, the most comprehensive description remains the one given in [16, chapter 8]. It defines the notion of locally isotropic turbulence by considering the fields relative to a chosen fluid particle in a small space-time region around that moving particle. For velocity statistics, this means that we consider the relative field, $\delta \mathbf{v}_f(\tau, \mathbf{r}) = \mathbf{U}_f(t_0 + \tau, \mathbf{x}) - \mathbf{U}_f(t_0, \mathbf{x}_0)$, in the reference frame moving with the velocity of a chosen fluid particle, $\mathbf{U}_f(t_0, \mathbf{x}_0)$, and where the space separation $\mathbf{r}$ is $\mathbf{r} = \mathbf{x} - \mathbf{x}_0 - \mathbf{U}_f(t_0, \mathbf{x}_0) \tau$. Based on this description, the K41 theory states that, for high-Reynolds-number turbulent flows and for small enough $r = |\mathbf{r}|$ and $\tau$, turbulence is locally isotropic (in the small space-time region defined by $\mathbf{r}$ and $\tau$ around the moving fluid particle). Then, the K41 first similarity hypothesis is that statistics within that small space-time region are uniquely determined by the mean rate of dissipation of the fluid kinetic energy $\langle \epsilon_f \rangle$ (which represents also the mean rate of energy transfer in the inertial range), the fluid viscosity $v_f$ and the space or time coordinates $\mathbf{r}$ and $\tau$, but not by the specific velocity of the 'observation fluid particle' $\mathbf{U}_f(t_0, \mathbf{x}_0)$.

The first outcome of the Kolmogorov theory is the expression of the length, velocity and time scales of the smallest scales of turbulence

$$\eta_K = \left( \frac{v_f^3}{\langle \epsilon_f \rangle} \right)^{3/4}, \quad u_K = (v_f \langle \epsilon_f \rangle)^{1/4}, \quad \tau_K = \left( \frac{v_f}{\langle \epsilon_f \rangle} \right)^{1/2}. \tag{5.53}$$

If we introduce $L_f$ and $u_f$ the length and velocity of the large-scale motions and use the estimation $\langle \epsilon_f \rangle \sim u_f^3 / L_f$, we obtain the ratios between the length and time scales of the largest to the smallest scales

$$\frac{L_f}{\eta_K} \sim \text{Re}^{3/4}, \quad \frac{T_f}{\tau_K} \sim \text{Re}^{1/2}, \tag{5.54}$$

with $T_f = L_f / u_f$ and $\text{Re} = u_f L_f / v_f$ the Reynolds number based on the large scales. This provides a way to assess the high level of complexity involved in turbulence. Indeed, a complete spatial resolution of the velocity field (i.e., capturing all the turbulent eddies) in a three-dimensional flow implies that we must handle a number of degrees of freedom that scales as $\text{Re}^{9/4}$ which need, moreover, to be tracked in time (to simulate one large-eddy turnover time $T_f$ we must adopt a time resolution of the order of $\tau_K$ implying another $\text{Re}^{1/2}$ factor in the effort required). Given that many flows have Reynolds numbers of the order of $10^6 - 10^8$ (for example, in the atmospheric boundary layer), this estimation demonstrates that we are dealing with huge numbers of degrees of freedom so that statistical reduced descriptions are unavoidable even with today's computational resources.

The Kolmogorov scales allow to properly define the inertial range as the space and time domain where $\eta_K \ll r \ll L_f$ and $\tau_K \ll \tau \ll T_f$. Then, the second Kolmogorov similarity hypothesis assumes that, in the inertial range, statistics do not depend anymore on the fluid viscosity $\nu_f$. This yields scaling relations for the eddy velocity and time scales in the inertial range

$$\delta u_f(l) = (\langle \epsilon_f \rangle \, l)^{1/3} \sim u_f \, (l/L_f)^{1/3} \;, \; \tau_f(l) = \left( l^2/\langle \epsilon_f \rangle \right)^{1/3} \sim \frac{L_f}{u_f} \, (l/L_f)^{2/3} \;. \tag{5.55}$$

### 5.3.1 Eulerian Statistics of Velocity Differences

The most usual application is for the Eulerian fluid velocity correlations which are characterized by the tensor

$$D_{f,ij}(t, \mathbf{x}, \mathbf{r}) = \langle \, \delta v_{f,i}(0, \mathbf{r}) \, \delta v_{f,j}(0, \mathbf{r}) \, \rangle \;,$$
$$= \langle \, \left[ U_{f,i}(t, \mathbf{x} + \mathbf{r}) - U_{f,i}((t, \mathbf{x}) \right] \left[ U_{f,j}(t, \mathbf{x} + \mathbf{r}) - U_{f,j}((t, \mathbf{x}) \right] \rangle. \tag{5.56}$$

When isotropy prevails, this velocity-structure tensor does not depend on $\mathbf{x}$ anymore and is written as [1, 16]

$$D_{f,ij}(t, \mathbf{r}) = D_{f, \mathrm{NN}}(t, r) \delta_{ij} + \left[ D_{f, \mathrm{LL}}(t, r) - D_{f, \mathrm{NN}}(t, r) \right] \frac{r_i r_j}{r^2} \;, \tag{5.57}$$

where the two scalar functions $D_{f, \mathrm{LL}}(t, r)$ and $D_{f, \mathrm{NN}}(t, r)$ are the longitudinal and transverse structure functions, respectively. Moreover, the continuity constraint implies that $D_{f, \mathrm{NN}}$ is uniquely determined by $D_{f, \mathrm{LL}}$ and $D_{f,ij}(t, \mathbf{r})$ is therefore fully characterized by one scalar function $D_{f, \mathrm{LL}}(t, r)$. In the inertial range, the second Kolmogorov similarity hypothesis yields then that

$$D_{f, \mathrm{LL}}(t, r) = C \, (\langle \epsilon_f \rangle \, r)^{2/3} \;, \quad D_{f, \mathrm{NN}}(t, r) = \frac{4}{3} C \, (\langle \epsilon_f \rangle \, r)^{2/3} \;, \tag{5.58}$$

where $C$ is a constant. This is the same result as the one given in Eq. (5.55) since $D_{f, \mathrm{LL}}$ represents $(\delta u_f(l))^2$ (using $\langle \epsilon_f \rangle \sim u_f^3/L_f$). The Fourier transform of $D_{f, \mathrm{LL}}(t, r)$ gives the (longitudinal) kinetic energy spectrum with the well-known $-5/3$ variation in the inertia range (see Fig. 5.1). This point has been the subject of numerous experimental and numerical studies ever since the 1960s and detailed discussions can be found in the references mentioned before [16–18]. When $\nu_f \to 0$ (with fixed $u_f$ and $L_f$), $\mathrm{Re} \to +\infty$ but the kinetic energy dissipation rate $\langle \epsilon_f \rangle$ tends towards a finite value which is the only remaining trace of the vanishing viscosity. Smaller and smaller scales are generated and the energy spectrum is stretched to larger and larger wave numbers but with the same slope.

### 5.3.2  Statistics of Temporal Velocity Increments

The Eulerian velocity difference at a fixed point $\mathbf{x}_0$ and at two instants $t_0$ and $t_0 + \tau$ can be expressed in terms of the velocity field $\delta \mathbf{v}_f$ defined in the reference system moving with the velocity $\mathbf{U}_f(t_0, \mathbf{x}_0)$ as

$$\delta \mathbf{U}_f^{(t_0,\mathbf{x}_0)}(\tau) \simeq \delta \mathbf{v}_f(\tau, -\mathbf{U}_f(t_0, \mathbf{x}_0)\tau). \tag{5.59}$$

However, the situation is more complicated than for the Eulerian increments since $\delta \mathbf{U}_f^{(t_0,\mathbf{x}_0)}(\tau)$ depends explicitly on the reference velocity $\mathbf{U}_{f,0} = \mathbf{U}_f(t_0, \mathbf{x}_0)$. Therefore, we can only conclude from the Kolmogorov theory that there is a conditional probability distribution for $\delta \mathbf{U}_f^{(t_0,\mathbf{x}_0)}(\tau)$. For a given value of $\mathbf{U}_{f,0}$, the tensor $D_{f,ij}^{(t_0,\mathbf{x}_0)}(\tau) = \langle \delta U_{f,i}^{(t_0,\mathbf{x}_0)}(\tau) \delta U_{f,j}^{(t_0,\mathbf{x}_0)}(\tau) \rangle$ depends on the two functions $D_{f,\parallel}^{(t_0,\mathbf{x}_0)}$ and $D_{f,\perp}^{(t_0,\mathbf{x}_0)}$ which correspond to the longitudinal and transverse directions, as in Eq. (5.57). These two functions depend on $\tau$ as well as on $\mathbf{r} = \mathbf{U}_{f,0}\tau$ and in the inertial range we have

$$D_{f,\parallel}^{(t_0,\mathbf{x}_0)} = \langle \epsilon_f \rangle \, \tau \, \alpha_\parallel \left( \frac{|\mathbf{U}_{f,0}|^2}{\langle \epsilon_f \rangle \tau} \right), \quad D_{f,\perp}^{(t_0,\mathbf{x}_0)} = \langle \epsilon_f \rangle \, \tau \, \alpha_\perp \left( \frac{|\mathbf{U}_{f,0}|^2}{\langle \epsilon_f \rangle \tau} \right), \tag{5.60}$$

where $\alpha_\parallel$ and $\alpha_\perp$ are universal functions which have to be specified. The dependence on $\mathbf{U}_{f,0}$ and the conditional nature of the previous results can be removed by resorting to the Taylor, or frozen turbulence, hypothesis. According to this hypothesis, the turbulent fluctuations are much smaller than the mean velocity, i.e. $\mathbf{U}_{f,0} \simeq \langle \mathbf{U}_{f,0} \rangle$. Over a small time interval $\tau$, the turbulent fluctuations at a fixed point are regarded as being transported without modification at a constant velocity $\langle \mathbf{U}_{f,0} \rangle$. This frozen-turbulence assumption removes the dependence on $\mathbf{U}_{f,0}$ and allows to write the statistics of $\delta \mathbf{U}_f^{(t_0,\mathbf{x}_0)}(\tau)$ in terms of the velocity-structure functions derived for the Eulerian space increments by replacing $\mathbf{r}$ by $\langle \mathbf{U}_{f,0} \rangle \tau$. This leads to

$$D_{f,\parallel}^{(t_0,\mathbf{x}_0)} = D_{f,\text{LL}}(t_0, \langle \mathbf{U}_{f,0} \rangle \tau), \quad D_{f,\perp}^{(t_0,\mathbf{x}_0)} = D_{f,\text{NN}}(t_0, \langle \mathbf{U}_{f,0} \rangle \tau), \tag{5.61}$$

with $D_{f,\text{LL}}$ and $D_{f,\text{NN}}$ as in Eq. (5.58), which gives in the inertial range that

$$D_{f,\parallel}^{(t_0,\mathbf{x}_0)} = C \left( \langle \epsilon_f \rangle \langle \mathbf{U}_{f,0} \rangle \, \tau \right)^{2/3}, \quad D_{f,\perp}^{(t_0,\mathbf{x}_0)} = \frac{4}{3} C \left( \langle \epsilon_f \rangle \langle \mathbf{U}_{f,0} \rangle \, \tau \right)^{2/3}. \tag{5.62}$$

### 5.3.3  Lagrangian Statistics for Particle Velocity Increments

The Kolmogorov theory can be directly applied to the velocity increments of a fluid particle $d\mathbf{U}_f(\tau) = \mathbf{U}_f(t+\tau) - \mathbf{U}_f(t)$ which, by choosing this particle as the reference one in the Kolmogorov approach, are written as $d\mathbf{U}_f(\tau) = \delta \mathbf{v}_f(\tau, 0)$. The locally isotropic nature of turbulent flows implies that, for $\tau \ll T_f$, the Lagrangian second-order velocity structure function $\langle dU_{f,i}(\tau) dU_{f,j}(\tau) \rangle$ has an isotropic form

$$\langle dU_{f,i}(\tau) dU_{f,j}(\tau) \rangle = D_f^L(\tau) \delta_{ij} \ . \tag{5.63}$$

For time differences in the inertial range, i.e. $\tau_K \ll \tau \ll T_f$, the second Kolmogorov hypothesis gives that

$$D_f^L(\tau) = C_0 \langle \epsilon_f \rangle \tau \ , \tag{5.64}$$

with $C_0$ the Kolmogorov constant. This is a significant result since the linear variation in $\tau$ of the second-order moment of velocity increments is a signature of white-noise effects and of a diffusive behavior of fluid particle velocities.

While there is no well-marked separation in terms of length scales, it is important to note that there is one in terms of timescales. More precisely, it is seen that $\delta u_f(l)$ diminishes as $l$ becomes smaller but without any sharp decrease between two comparable scales (the decrease is continuous) whereas there is a clear-cut distinction between the correlation timescales of a fluid particle velocity $U_f$ and its acceleration $A_f$. Fluid particle velocities are governed by the large-scale motions of a turbulent flow and scale as $(U_f)^2 \sim u_f^2$ with a timescale $T_L \sim T_f = u_f^2/\langle \epsilon_f \rangle$, while fluid particle accelerations are governed by the small-scale motions and scale as $(A_f)^2 \sim \langle \epsilon_f \rangle / \tau_K$ with a timescale $\tau_A$ which is of the order of the Kolmogorov timescale $\tau_A \simeq \tau_K$. This situation plays a key role in the analysis of stochastic models. Indeed, we can use the result obtained with the toy model presented above, with $X_{slow}$ being the fluid particle velocity $U_f$ and $X_{fast}$ its acceleration $A_f$ (in a one-dimensional notation). In high-Reynolds-number flows where $\tau_A \ll T_L$ and with $\langle \epsilon_f \rangle$ remaining finite as indicated above, we retrieve the white-noise limit since we have

$$\tau_A \to 0 \text{ and } \langle A_f^2 \rangle \to +\infty \ , \quad \text{such that} \quad \tau_A \langle A_f^2 \rangle \to \langle \epsilon_f \rangle \ . \tag{5.65}$$

This is in line with the prediction already given in Eq. (5.64) and this indicates that, in the inertial range, we expect fluid particle velocity increments to be described by a stochastic model containing a white-noise term such as $\sqrt{C_0 \langle \epsilon_f \rangle} \, d\mathbf{W}$, as will be seen in more details in Chap. 6.

## 5.3.4  Intermittency and Refined Hypothesis

So far, this outline of the Kolmogorov theory has essentially followed the K41 picture in that the rate of kinetic transfer and dissipation $\langle \epsilon_f \rangle$ is taken as a parameter rather than as a random variable. This is related to the intermittency of the flow produced by the inherent transfer processes of turbulence, referred to as 'inner intermittency' (to be distinguished from external intermittency due, for example, to large-scale mixing between laminar and turbulent flows) and we have assumed that $\epsilon_f \simeq \langle \epsilon_f \rangle$. While the above picture can be kept substituting $\langle \epsilon_f \rangle$ with $\epsilon_f$, especially for second-order moments where the impact of intermittency is small, a refined description based on a log-normal distribution for the dissipation of kinetic energy was developed in the K62 theory. For this question, the basic presentation of the refined similarity hypothesis is [16, section 25.2] and this question has been investigated in several studies, in particular in [17]. Without addressing in detail this issue, related comments are proposed in connection with coherent structures later in Sect. 8.4.

### 5.3.5 Particle State Vectors and Lagrangian Models

From these results, it appears that the Kolmogorov theory provides clear indications as to which state vectors should be selected to follow the Markovian approach. For tracer particles, the relevant state vector is $\mathbf{Z}_f = (\mathbf{X}_f, \mathbf{U}_f)$ while for discrete particles the velocity of the fluid seen should be included, leading to $\mathbf{Z}_p = (\mathbf{X}_p, \mathbf{U}_p, \mathbf{U}_s)$, since we expect the 'acceleration' of the fluid seen (or its time-rate-of-change) to be also a fast variable. In terms of the trajectories of the stochastic process $\mathbf{Z}_p$, the general structure of one-particle PDF models for discrete particles in turbulent flows is

$$d\mathbf{X}_p = \mathbf{U}_p \, dt \, , \tag{5.66a}$$

$$d\mathbf{U}_p = \frac{\mathbf{U}_s - \mathbf{U}_p}{\tau_p} \, dt + \mathbf{F}_{f \to p} \, dt \, , \tag{5.66b}$$

$$d\mathbf{U}_s = (\text{stochastic model}) \, , \tag{5.66c}$$

where $\mathbf{F}_{f \to p}$ in Eq. (5.66b) stands for hydrodynamic forces other than drag. The task ahead is to work out appropriate stochastic models for $\mathbf{U}_s$ in Eq. (5.66c).

### References

1. S. Pope, *Turbulent Flows* (Cambridge University Press, Cambridge, 2000). https://doi.org/10.1017/CBO9780511840531
2. K. Furutsu, J. Res. Natl. Bureau Standards – D Radio Propag. **67**, 303 (1963). https://doi.org/10.6028/NBS.MONO.79
3. E.A. Novikov, Soviet J. Exp. Theor. Phys. **20**(5), 1290 (1965). https://ui.adsabs.harvard.edu/abs/1965JETP...20.1290N
4. M.D. Donsker, *Analysis in Function Space*, ed. by W.T. Martin, I. Segal (The MIT Press, Cambridge, 1964), pp. 17–30
5. A. Bragg, D. Swailes, R. Skartlien, Phys. Rev. E **86**(5), 056306 (2012). https://doi.org/10.1103/PhysRevE.86.056306
6. J. Pozorski, J.P. Minier, Phys. Rev. E **59**(1), 855 (1999). https://doi.org/10.1103/PhysRevE.59.855

7. J.P. Minier, C. Profeta, Phys. Rev. E **92**(5), 053020 (2015). https://doi.org/10.1103/PhysRevE.92.053020
8. P. Häunggi, P. Jung, Adv. Chem. Phys. **89**, 239 (1994). https://doi.org/10.1002/9780470141489
9. N.G. Van Kampen, J. Stat. Phys. **54**, 1289 (1989). https://doi.org/10.1007/BF01044716
10. N.G. van Kampen, Braz. J. Phys. **28**, 90 (1998). https://doi.org/10.1590/S0103-97331998000200003
11. N.G. Van Kampen, *Stochastic Proceses in Physics and Chemistry*, 3rd edn. (Elsevier, Amsterdam, 2007). https://doi.org/10.1016/B978-0-444-52965-7.X5000-4
12. C. Gardiner, *Stochastic Methods*, 4th edn. (Springer, Berlin, 2009). https://link.springer.com/book/9783540707127
13. H. Risken, *Fokker-Planck Equation* (Springer, Berlin, 1996). https://doi.org/10.1007/978-3-642-61544-3
14. J.P. Minier, Phys. Rep. **665**, 1 (2016). https://doi.org/10.1016/j.physrep.2016.10.007
15. H.C. Öttinger, *Beyond Equilibrium Thermodynamics* (John Wiley & Sons, Hoboken, 2005). https://doi.org/10.1002/0471727903
16. A. Monin, A. Yaglom, *Statistical Fluid Mechanics, Volume II: Mechanics of Turbulence*. Dover Books on Physics, vol. 2 (Dover Publications, Mineola, 2013). https://store.doverpublications.com/products/9780486458915
17. U. Frisch, *Turbulence: The Legacy of A. N. Kolmogorov* (Cambridge University Press, Cambridge, 1995). https://doi.org/10.1017/CBO9781139170666
18. J.C. Hunt, O.M. Phillips, D. Williams, *Turbulence and Stochastic Processes: Kolmogorov's Ideas 50 Years on* (Royal Society, London, 1991). https://doi.org/10.1017/S0022112094212417

# Modeling the Velocity of the Fluid Seen: Current Formulations

**6**

**Abstract**

Having isolated the velocity of the fluid seen as the key variable to include in the particle state vector, we need to devise a stochastic model to capture its dynamical evolution. The purpose of this chapter is therefore to explain the rationale behind its representation as a stochastic diffusion process and to introduce the state-of-the-art formulation. Following a bottom-up approach, we start with the analysis of the physics of particle dispersion in turbulent flows to work out how relevant timescales can be expressed. A second step consists in revisiting Kolmogorov theory to show that a stochastic diffusion model has physical support, though further approximations have to be made compared to the case of fluid particles. A stochastic model for the velocity of the fluid seen is then built by relying on present Langevin models for fluid particles, which are extended to accommodate the specific timescales of the velocity of the fluid seen. At the moment, such a construction respecting a set of well-established criteria has been achieved only for one model whose characteristics are worth presenting if we are to overcome this limitation with improved formulations in the following chapters.

**Chapter Content** The physics of particle dispersion is first discussed in Sect. 6.1. The state-of-the-art Langevin models for fluid particles are presented in Sect. 6.2 before introducing the reference two-phase Langevin model in Sect. 6.3. Then, applications in the LES context are addressed in Sect. 6.4. Finally, consistency issues and potential limitations are brought forward in Sect. 6.5.

© The Author(s) 2025
J.-P. Minier et al., *Understanding Turbulent Systems*,
Lecture Notes in Physics 1039, https://doi.org/10.1007/978-3-031-84466-9_6

## 6.1    The Physics of Particle Dispersion

### 6.1.1    The Crossing-trajectory Effect

In a nutshell, particle dispersion in turbulent flows is due to the fluctuating fluid velocities encountered by discrete particles and is thus governed by the statistical properties of the velocity of the fluid seen $U_s(t)$. Consequently, our purpose in this section is to bring out the main physical phenomena at play and the relevant variables that need to be accounted for when modeling $U_s(t)$.

The problem can be described as follows (see also Fig. 6.1). At time $t$, we consider a discrete particle (P) located at a position $X_p(t)$, with a velocity $U_p(t)$ and a velocity of the fluid seen $U_s(t)$. Then, after a small time interval $\Delta t$, the discrete particle has a probability to move to a downstream location $X_p(t + \Delta t)$, while a fluid particle (F) starting from $X_p(t)$ with the velocity $U_s(t)$ at time $t$ has a probability to move to another position $X_f(t+\Delta t)$. The modeling issue is to estimate $U_s(t + \Delta t)$, which is the velocity of another fluid particle than (F), say (F'), at time $t + \Delta t$. In most cases, due to particle inertia and/or external forces inducing mean velocity slips $\langle U_r \rangle$ (with $U_r = U_s - U_p$ the relative velocity), the trajectories of the discrete particle (P) and of the fluid particle (F) located at the same point at time $t$ separate during the time interval $\Delta t$. This is referred to as the crossing-trajectory effect (CTE). As such, the CTE is related to particle inertia and the various effects mentioned above but its precise definition in the present modeling context will be narrowed down below. When we re-express the velocity of the fluid seen in terms of the instantaneous fluid velocity field $U_f(t, x)$, it appears that the modeling issue is to simulate $U_s(t + \Delta t) = U_f(t + \Delta t, X_p(t + \Delta t))$ knowing $U_s(t) = U_f(t, X_p(t))$. This clearly involves a two-time two-point conditional fluid velocity correlation, which reveals that it can only be treated without approximation at the level of two-particle PDF models, i.e. when following pairs of fluid particles or when generating spatial information about the fluid velocity field in the vicinity of every discrete particle being tracked. At the level of one-particle PDF models, this constitutes an open issue and additional assumptions have to be made.

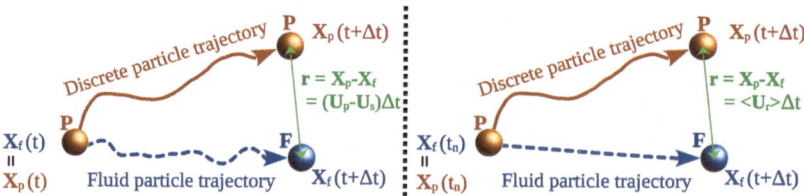

**Fig. 6.1** Illustration of the CTE and its modeled representation. The CTE occurs when there is a non-zero relative velocity between discrete particles and the fluid, due to particle inertia and mean velocity drifts (left side). In present formulations, it is accounted for only due to mean relative velocities (right side)

To pave the way for the more precise developments to come, we first outline the expected statistical characteristics of the velocity of the fluid seen. As pictured in Fig. 6.1, two main physical processes are involved, namely transport and relaxation processes. However, once $\mathbf{U}_s(t)$ is properly defined (as the fluid velocity sampled along discrete particle trajectory), transport processes are implicitly accounted for and we can concentrate on describing how $\mathbf{U}_s(t)$ relaxes towards mean fluid velocities (assuming, therefore, that it does). This means that we essentially consider the relaxation timescale(s) of $\mathbf{U}_s(t)$, which is taken as its integral timescale noted $T_L^*$. In non-homogeneous situations, relaxation is expressed by a matrix in the return-to-equilibrium term but, in the present context, it is sufficient to handle an isotropic matrix involving only one coefficient whose inverse is the correlation timescale $T_L^*$. We need therefore to assess how $T_L^*$ varies as a function of $T_L$ and $T_E$, the fluid Lagrangian and Eulerian timescales, respectively, and particle-related statistics, such as $\langle \mathbf{U}_r \rangle$, $\langle \mathbf{U}_p^2 \rangle$, $\langle \mathbf{U}_p \mathbf{U}_s \rangle$, etc. To that effect, it proves useful to follow the analysis carried out in [1, 2] and to evaluate separately the effects of particle inertia on $T_L^*$: (a) in the absence of a mean velocity slip; and (b) when a mean drift is present.

(a) Even in the absence of mean velocity differences, a CTE is present due to particle inertia, measured by the Stokes number $St = \tau_p / T_L$ which is the ratio of the particle relaxation timescale and the Lagrangian timescale of fluid velocities. In the simple situation of steady homogeneous turbulent flows, the equilibrium particle velocity statistics can be easily derived (see the derivation of the so-called Tchen's formulas in [1, section 7.5.5]) as a function of the Stokes number, leading to $\langle \mathbf{U}_r^2 \rangle \simeq \langle \mathbf{U}_f^2 \rangle St/(1 + St)$ and $\langle \mathbf{U}_p^2 \rangle = \langle \mathbf{U}_p \mathbf{U}_s \rangle \simeq \langle \mathbf{U}_f^2 \rangle/(1 + St)$. Such results are not to be taken as granted in general non-homogeneous turbulent flows but they can be regarded as reference ones and, in that sense, are useful to point out typical particle velocity properties. Based on these guidelines, we can infer that, as depicted in Fig. 6.2, low-inertia discrete particles tend to follow the fluid, which means that we have $T_L^* \simeq T_L$ when $St \ll 1$. Since $\langle \mathbf{U}_r^2 \rangle \neq \langle \mathbf{U}_f^2 \rangle$ when $St > 0$, there is always a separation when particle inertia is not negligible

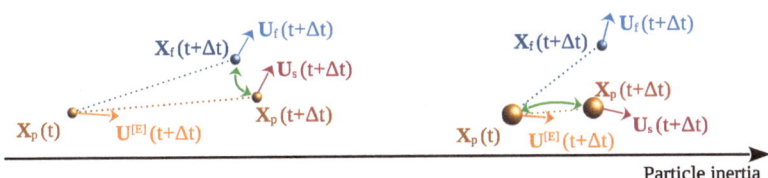

**Fig. 6.2** Estimations of the integral timescale of the velocity of the fluid seen with respect to the Lagrangian and Eulerian timescales in the absence of a mean velocity drift between particles and the fluid. Low-inertia particles tend to follow fluid particle trajectories so that the statistics of $\mathbf{U}_s(t)$ are close to those of a fluid particle velocity $\mathbf{U}_f(t)$, meaning that $T_L^* \simeq T_L$ (left side). High-inertia particles do not move very much over a time interval $\Delta t$ so that the statistics of $\mathbf{U}_s(t)$ are close to those of the fluid velocity $\mathbf{U}^{[E]}(t)$ at a fixed location, meaning that $T_L^* \simeq T_E$ (right side)

and this has led to the introduction of an additional step to correlate the velocities of the two fluid particles located at $\mathbf{X}_f(t + \Delta t)$ and at $\mathbf{X}_p(t + \Delta t)$ at time $t + \Delta t$ (this is referred to as the Eulerian step) once the velocity of the first fluid particle, i.e. $\mathbf{U}_f(t + \Delta t)$, is generated from $\mathbf{U}_s(t)$ (this is referred to as the Lagrangian step). This is, however, risky since we would be building a model based on one sample (we cannot actually say that 'the fluid particle goes there' or 'has this velocity' but simply that it has a probability to do so) and, more importantly, these two steps are not independent. This was nevertheless treated as such in earlier works which led to spurious de-correlation effects in the resulting timescale $T_L^*$. This is revealed by considering the other limit case of high-inertia particles. In that case, we expect discrete particles not to travel a great distance compared to nearby fluid particles since $\langle \mathbf{U}_p^2 \rangle \ll \langle \mathbf{U}_f^2 \rangle$ and, thus, to be essentially correlated with the fluid velocity at time $t + \Delta t$ at the initial discrete particle location $\mathbf{X}_p(t)$ which is noted $\mathbf{U}^{[E]}(t + \Delta t)$ in Fig. 6.2. In short, in that case, the velocity of the fluid seen is similar to Eulerian fluid velocities at a given position and we expect that $T_L^* \simeq T_E$ when $St \gg 1$. In between these two limits, we can derive from this description that $T_L^*$ remains of the order of $T_L$ and $T_E$, and is a function of the Stokes number without being able to work out the exact dependency in the one-particle PDF framework. Note that this does not imply that $T_L^*$ is a monotonous function of $St$ between the two limit values of $T_L$ and $T_E$ (discrete particles can be trapped in specific coherent structures whose timescales are different, cf. Sect. 6.1.3). Although detailed information is rarely available, it is estimated that $T_L$ and $T_E$ are often comparable. As a first approximation and leaving out the possible effects of coherent structures (which, if need be, can be introduced explicitly as discussed in Sect. 8.4), we therefore retain the simplified result that, in the absence of mean velocity slips, we have $T_L^* \simeq T_L$.

(b) When a mean velocity slip $\langle \mathbf{U}_r \rangle$ is present at the discrete particle location at time $t$, the potential trajectories of the discrete particle and the fluid ones starting from $\mathbf{X}_p(t)$ at time $t$ with the velocity $\mathbf{U}_s(t)$ always separate, cf. the right-hand side in Fig. 6.1. Moreover, since these fluid particles are not affected by the mean velocity drifts, a systematic de-correlation is induced between the velocity of the fluid seen at the time $t + \Delta t$, $\mathbf{U}_s(t + \Delta t)$, and the 'equivalent fluid particle' whose velocity timescale is between $T_L$ and $T_E$ (as a result of the first step) and which is represented by (F) at time $t + \Delta t$ on the right-hand side in Fig. 6.1. The essential point is that, since it is due to mean velocity differences, this effect is independent of the first one (the 'inertia effect') and can be introduced successively in a step-by-step model construction.

Based on this approach in terms of, first, particle-inertia effects in the absence of mean velocity drifts and, second, mean-velocity-slip effects inducing a de-correlation between the velocities of discrete particles and nearby fluid ones, the emerging picture is a two-step process where each step can be addressed independently. We deduce from this analysis that the Lagrangian timescale of the velocity of the fluid seen can be written as $T_L^* = T_L^*(\langle \mathbf{U}_r \rangle = 0) \times f(\langle \mathbf{U}_r \rangle; T_L; T_E, k_f; \cdots)$ where $f$ is a correcting factor accounting for the effects of mean drifts and is a function of particle and fluid statistics (thus, with $f(\langle \mathbf{U}_r \rangle = 0; T_L; T_E; k_f, \cdots) = 1$). As indicated above, a further simplification consists in neglecting the influence of particle inertia on $T_L^*(\langle \mathbf{U}_r \rangle = 0)$ and in writing $T_L^*(\langle \mathbf{U}_r \rangle = 0) \simeq T_L$. This means that in the absence of mean velocity slips, the statistics of the velocity of the fluid seen are the same as those of a fluid particle and that the CTE is considered from now on as being due only to the existence of mean velocity drifts, cf. Fig. 6.1. Since $f = T_L^*/T_L$ is a non-dimensional function, it is best expressed as a function of non-dimensional variables, i.e. $f = f(\langle \mathbf{U}_r \rangle / u_f; C_T; \cdots)$, with $u_f = (2/3k_f)^{1/2}$ the characteristic fluid turbulent velocity (with $k_f$ the fluid turbulent kinetic energy to be properly defined in Sect. 6.2.2) and where $C_T$ is the ratio of the Lagrangian to Eulerian timescales, $C_T = T_L/T_E$. Since the difference between $T_L$ and $T_E$ is neglected in $T_L^*(\langle \mathbf{U}_r \rangle = 0)$, it would be consistent to take $C_T = 1$. Yet, to capture the correct behavior in the limit of high mean velocity drifts when the frozen-turbulence hypothesis is applied with (small) differences between $T_L$ and $T_E$, we keep the parameter $C_T$, with $C_T \sim O(1)$.

## 6.1.2 Stochastic Diffusion Models for the Velocity of the Fluid Seen

The description of the crossing-trajectory effect in terms of the mean velocity slip between particles and the fluid allows us to recast the issue of modeling the velocity of the fluid seen into the framework of the Kolmogorov theory. Indeed, the increments of $\mathbf{U}_s$ over a small time interval $dt$ can be written as

$$dU_s = \delta \mathbf{v}_f(dt, \langle \mathbf{U}_r \rangle \, dt), \tag{6.1}$$

where $\delta \mathbf{v}_f(t, \mathbf{x})$ is the fluid velocity field relative to the motion of a fluid particle that was located at the particle location at time $t$ (cf. Sect. 5.3), as sketched on the right-hand side of Fig. 6.1.

According to the Kolmogorov theory, the statistics of $dU_s$ do not depend on the instantaneous velocity of the fluid particle selected to express $dU_s$ through the observation velocity field $\delta \mathbf{v}_f(t, \mathbf{x})$ in Eq. (6.1) but only on some statistics. In high-Reynolds-number flows and for a time increment $dt$ in the inertial range, it follows from the locally isotropic hypothesis that the second-order structure functions of the velocity of the fluid seen, $D_{s,ij}(dt) = \langle dU_{s,i} dU_{s,j} \rangle$, is determined by the two scalars functions $D_{s,\parallel}$ and $D_{s,\perp}$ through

$$D_{s,ij} = D_{s,\perp} \delta_{ij} + \left[ D_{s,\parallel} - D_{s,\perp} \right] \widehat{r}_i \widehat{r}_j , \tag{6.2}$$

with the separation unit vector $\widehat{\mathbf{r}}$ being in the direction of the mean relative velocity, i.e. $\widehat{\mathbf{r}} = \langle \mathbf{U}_r \rangle / |\langle \mathbf{U}_r \rangle|$. The functions $D_{s,\parallel}$ (resp. $D_{s,\perp}$) represents the correlations for the velocity components aligned with the separation vector $\mathbf{r}$ (resp. transverse to $\mathbf{r}$), similarly to the situation with Eulerian velocity differences [3]. For a time increment in the inertial range, the Kolmogorov theory implies that $D_{s,ij}$ depends only on the mean dissipation rate $\langle \epsilon_f \rangle$, $dt$ and $\langle U_r \rangle$. Then, in the same manner as what was done in Sect. 5.3, we obtain

$$D_{s,||}(dt) = \langle \epsilon_f \rangle \, dt \, \alpha_{||} \left( \frac{|\langle U_r \rangle|^2}{\langle \epsilon_f \rangle dt} \right), \qquad D_{s,\perp}(dt) = \langle \epsilon_f \rangle \, dt \, \alpha_{\perp} \left( \frac{|\langle U_r \rangle|^2}{\langle \epsilon_f \rangle dt} \right),$$

$$(6.3)$$

where $\alpha_{||}(x)$ and $\alpha_{\perp}(x)$ are regarded as two universal functions of a single parameter $x = |\langle U_r \rangle|^2 / (\langle \epsilon_f \rangle dt)$, whose form can be obtained in two limit cases. First, when the mean relative velocity is small, the statistics of the velocity of the fluid seen should be close to the fluid ones which means that we have

$$\frac{|\langle U_r \rangle|^2}{\langle \epsilon_f \rangle dt} \ll 1 \quad \Longrightarrow \quad \alpha_{||} \simeq \alpha_{\perp} \simeq C_0. \tag{6.4}$$

Second, when the relative mean velocity is large, we can use the frozen turbulence hypothesis to obtain, similarly to the case of temporal velocity increments (cf. Sect. 5.3.2), that

$$\frac{|\langle U_r \rangle|^2}{\langle \epsilon_f \rangle dt} \gg 1 \quad \Longrightarrow \quad D_{s,||}(dt) \simeq C(\langle \epsilon_f \rangle \langle U_r \rangle \, dt)^{2/3},$$

$$D_{s,\perp}(dt) \simeq \frac{4}{3} C(\langle \epsilon_f \rangle \langle U_r \rangle \, dt)^{2/3}, \tag{6.5}$$

showing that, in this limit, the two functions $\alpha_{||}(x)$ and $\alpha_{\perp}(x)$ vary as $x^{1/3}$ ($C$ is simply a constant). Apart from the limit case of fluid particles, the variation of these functions and their dependencies on the time increment imply that $D_{s,||}(dt)$ and $D_{s,\perp}(dt)$ are not linear in $dt$ and, therefore, do not exhibit the typical signature of stochastic diffusion processes. An approximation can, however, be worked out by freezing the values of the functions $\alpha_{||}$ and $\alpha_{\perp}$ at a certain value of the time interval, say $\Delta t_r$, which gives

$$D_{s,||}(dt) \simeq \langle \epsilon_f \rangle \, dt \, \alpha_{||} \left( \frac{|\langle U_r \rangle|^2}{\langle \epsilon_f \rangle \Delta t_r} \right), \qquad D_{s,\perp}(dt) \simeq \langle \epsilon_f \rangle \, dt \, \alpha_{\perp} \left( \frac{|\langle U_r \rangle|^2}{\langle \epsilon_f \rangle \Delta t_r} \right).$$

$$(6.6)$$

A physically-sound choice for $\Delta t_r$ is the Lagrangian timescale which is the timescale over which fluid velocities remain correlated. Using the simple estimation $\Delta t_r \simeq T_L \simeq k_f / \langle \epsilon_f \rangle$, we get

$$D_{s,||}(dt) \simeq \langle \epsilon_f \rangle \, dt \, \alpha_{||} \left( \frac{|\langle U_r \rangle|^2}{k_f} \right), \qquad D_{s,\perp}(dt) \simeq \langle \epsilon_f \rangle \, dt \, \alpha_{\perp} \left( \frac{|\langle U_r \rangle|^2}{k_f} \right).$$

$$(6.7)$$

As for the case of fluid particles, the linear-in-time variation of the second-order moments of the increments of the velocity of the fluid seen in Eq. (6.7) suggests to model $U_s$ by a stochastic diffusion process. It is also seen that this modeling step has less support than in the fluid case since additional assumptions have to be made, especially concerning how spatial correlations are accounted for. In spite of these limitations, it appears nevertheless that there is sufficient support to model $U_s$ by a stochastic diffusion process. With this choice, the Langevin model for the velocity of the fluid seen can be written as

$$dU_{s,i} = A_{s,i}(t, \mathbf{Z}_p, \langle \mathscr{f}(\mathbf{Z}_p) \rangle, \mathbf{\Phi}_f(\mathbf{X}_p))\, dt + B_{s,ij}(t, \mathbf{Z}_p, \langle \mathscr{g}(\mathbf{Z}_p) \rangle, \mathbf{\Phi}_f(\mathbf{X}_p))\, dW_j, \tag{6.8}$$

where the drift vector $\mathbf{A}_s$ and the diffusion matrix $\mathbf{B}_s$ have to be modeled. As in Sect. 4.3.3, a general notation is used in Eq. (6.8) to indicate that the drift and diffusion coefficients depend not only on the particle state vector $\mathbf{Z}_p$ but also on the value of mean fluid fields at the particle positions noted $\mathbf{\Phi}_f(\mathbf{X}_p)$ as well as on statistics derived from the particle set, written $\langle \mathscr{f}(\mathbf{Z}_p) \rangle$ and $\langle \mathscr{g}(\mathbf{Z}_p) \rangle$ (e.g., the mean particle velocity field). Yet, for the sake of simplicity, these functional dependencies are omitted from now on. The selection of a Langevin model for $U_s$ means that the particle state vector $\mathbf{Z}_p$ is modeled as a stochastic diffusion process with the form

$$dX_{p,i} = U_{p,i}\, dt, \tag{6.9a}$$

$$dU_{p,i} = A_{p,i}\, dt, \tag{6.9b}$$

$$dU_{s,i} = A_{s,i}\, dt + B_{s,ij}\, dW_j, \tag{6.9c}$$

where the particle acceleration is often limited to $\mathbf{A}_p = (\mathbf{U}_s - \mathbf{U}_p)/\tau_p + \mathbf{g}$ (i.e. considering only contributions from hydrodynamic drag and gravity). This formulation is equivalent to a Fokker-Planck equation for the corresponding PDF $p(t; \mathbf{y}_p, \mathbf{V}_p, \mathbf{V}_s)$ in sample space, which is

$$\frac{\partial p}{\partial t} + \frac{\partial}{\partial y_{p,i}} \left[ V_{p,i}\, p \right] =$$

$$- \frac{\partial}{\partial V_{p,i}} \left[ A_{p,i}\, p \right] - \frac{\partial}{\partial V_{s,i}} \left[ A_{s,i}\, p \right] + \frac{1}{2} \frac{\partial^2}{\partial V_{s,i} \partial V_{s,j}} \left[ (B_s B_s^T)_{ij}\, p \right]. \tag{6.10}$$

It can be noted that the approximations leading to Eq. (6.7) are not only useful to point to stochastic diffusion processes in order to model $U_s$ but also to bring out the non-dimensional mean velocity slip, $|\langle \mathbf{U}_r \rangle|^2 / k_f$, which enters typical closures of the Lagrangian timescale of the velocity of the fluid seen, such as the Csanady's formulas to be introduced in Sect. 6.3 and in new proposals presented in Sect. 8.3.

### 6.1.3 Remarks on the Particle Preferential Concentration Effect

Although this point is developed in more details in Sect. 7.3, it is useful to mention already that, in a particle-based approach, the incompressibility of the flow has a double manifestation for fluid particles or fluid-like elements. It requires that the mean or filtered velocity field derived from particle instantaneous velocities be divergence-free but, also, that the distribution of particle positions be uniform. This is not necessarily the case for discrete particles (either solid particles or bubbles) which can concentrate in certain zones of a turbulent flow, often in connection with the existence of specific flow patterns. This phenomenon is referred to as the particle preferential concentration [4, 5].

This effect has been recognized for some time with, for example, the first observations of particle concentration at the outskirts of the large-scale vortices forming downstream of the separating plate in a turbulent mixing layer (see, for instance, an historical account in [6]). It has then received increasing attention and the literature is vast on the subject particularly since the advent of high-resolution DNS which can capture explicitly unsteady fine-scale flow structures even when they are difficult to measure as in the vicinity of a solid wall. In practice, particle preferential concentration is essentially addressed either in wall boundary layers or in homogeneous isotropic situations. The former is in relation with particle transport in the near-wall region and the practical concern of particle deposition while the latter considers flows in which it was believed that preferential concentration was not likely to occur.

Homogeneous isotropic turbulence (HIT) is an interesting situation to consider since all fluid statistics are uniform and it could be expected that an initially uniform distribution of discrete particles would remain so. As revealed by instantaneous snapshots displayed in Fig. 6.3, concentration buildups can nevertheless be observed in some zones while other areas are nearly devoid of particles. This effect depends on particle inertia which is measured here by the Stokes number $St_{\eta_K} = \tau_p/\tau_K$ based on the Kolmogorov timescale $\tau_K$. As mentioned above, particles with negligible inertia (fluid-like elements) remain uniformly distributed and, at the other extreme of the range of Stokes numbers, heavy particles show the same tendency since they are insensitive to fluid structures and, in that sense, filter out any flow patterns that can be formed. Preferential concentration is most marked for particles with Stokes numbers $St_{\eta_K} = O(1)$, which can be attributed to a 'resonance effect'. Indeed, insightful analyses have revealed that homogeneous fluid flows in the statistical sense can nevertheless contain characteristic flow features, exhibiting an organized pattern locally in space and time and which, for that reason, are called 'coherent structures'. Typical examples of such fluid flow topology are vortex or saddle points. Many efforts have been devoted to characterize the geometry and dynamics of these coherent structures. When discrete elements have the 'right inertia', they can correlate with some specific structures or be expelled from others leading to the observed concentration buildup or depletion. For instance, particles heavier than the fluid are preferentially found in low vorticity and high strain-rate regions while

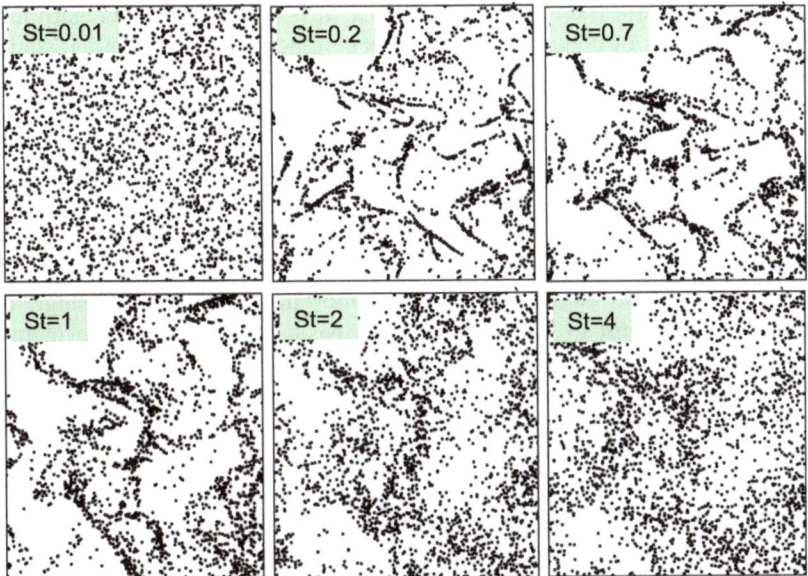

**Fig. 6.3** Snapshots of instantaneous particle positions in HIT obtained from tracking discrete particles in well-resolved DNS for different Stokes numbers based on the Kolmogorov timscale: (a) $St_{\eta_K} = 0.01$; (b) $St_{\eta_K} = 0.2$; (c) $St_{\eta_K} = 0.7$; (d) $St_{\eta_K} = 1$; (e) $St_{\eta_K} = 2$ and (f) $St_{\eta_K} = 4$. Reprinted with permission from [7]. © 2008 Elsevier Ltd. All rights reserved

bubbles are captured in low-pressure zones [4]. A noteworthy consequence is that the statistics of the 'turbulence seen by discrete particles' can be different from the fluid ones (this bias effect is significant for bubbles but is considered less noticeable for small solid particles).

Particle preferential concentration is also interesting in that it challenges traditional statistical views. Once a statistical averaging operator is introduced, it is tempting to regard fluctuations as a manifestation of 'disorder' (noise, heat, etc.) and averages as an expression of 'order' (forces, work, etc.). This turns out to be misleading in some situations such as the one discussed here, which is helpful to remind us that disorder can turn out to be an order whose understanding was eluding us. From a modeling standpoint, the notion of coherent structures is attractive to physicists since they are related to small scales which, according to the Kolmogorov theory, are the only features susceptible to be universal. This is, however, a simplification and these structures cannot be classified as belonging to the small-scale world only. Furthermore, developments are presently hindered by the case-by-case approach mentioned above and a quantitative theory that goes beyond mere observations and that is not limited to specific turbulent flows has yet to emerge. Finally, it is worth recalling that dispersion effects remain governed by the large energy-containing scales which are usually flow dependent and that predicting the mean energy transfer rate or capturing the anisotropy of the Reynolds stresses remain major issues [8]. In that sense, it is perhaps best to regard models developed in terms of coherent structures as complements to traditional statistical formulations. Further comments to that effect are proposed in Sect. 6.5 and, more specifically, in Sect. 8.4, after having presented the state-of-the-art stochastic models based on Langevin equations.

## 6.2    Generalized Langevin Models for Fluid Particles

Since two-phase PDF models are based on the ones developed for single-phase turbulence, it is useful to introduce first the basic features of Langevin models used for single-phase turbulent flows. PDF methods have a rather long history in turbulence (see historical accounts of earlier attempts in [1, 9]) but decisive steps were made by connecting stochastic models with macroscopic formulations for the transport equations for the velocity second-order moments. This was achieved essentially by Pope and co-workers through a series of papers, and detailed presentations can be found in several publications [8–11].

A well-established and, by now, reference stochastic model is the generalized Langevin model (GLM) which represents fluid particle velocities by a stochastic diffusion process with the form [8, 12, 13]

$$dX_{\mathrm{f},i} = U_{\mathrm{f},i}\, dt \,, \tag{6.11a}$$

$$dU_{\mathrm{f},i} = -\frac{1}{\rho_{\mathrm{f}}}\frac{\partial \langle P_{\mathrm{f}}\rangle}{\partial x_i}\, dt + G_{ij}\left(U_{\mathrm{f},j} - \langle U_{\mathrm{f},j}\rangle\right) dt + \sqrt{C_0 \langle \epsilon_{\mathrm{f}}\rangle}\, dW_i \,. \tag{6.11b}$$

where the matrix $G_{ij}$ depends on the particle location and on statistics of the fluid flow but not on the particle velocity, i.e. $G_{ij} = G_{ij}(t, \mathbf{X}_{\mathrm{f}}(t), \mathcal{H}[(\mathbf{X}_{\mathrm{f}}, \mathbf{U}_{\mathrm{f}})])$ where the notation $\mathcal{H}[(\mathbf{X}_{\mathrm{f}}, \mathbf{U}_{\mathrm{f}})]$ refers to fluid mean quantities given or calculated from the set of particles and interpolated at the particle position. Therefore, in homogeneous flows, $G_{ij}$ depends only on time and Eq. (6.11b) is a linear model for fluid particle velocities, while the complete particle system in Eq. (6.11) is non-linear in general non-homogeneous situations [1, 10]. For the sake of simplicity, these dependencies are considered as implicit and are not kept from now onwards. Similarly, the mean terms entering these equations are to be understood as the values at the particle location, e.g. $\langle \mathbf{U}_{\mathrm{f}}\rangle = \langle \mathbf{U}_{\mathrm{f}}\rangle(t, \mathbf{X}_{\mathrm{f}}(t))$. Before detailing up-to-date GLMs, we revisit how we arrive at these equations with a fruitful viewpoint for later developments.

### 6.2.1    The Local Linear Response Theory

In most accounts of the development of PDF methods [8, 12, 13], the above-mentioned connection between Langevin models and the fluid mean-field equations serve not only as a justification but also as a way to close the drift and diffusion coefficients. Given that these velocity-moment equations are the equivalent of the 'macroscopic equations' in the classical terminology of statistical physics, this amounts to following a top-down approach. It does not mean, however, that Langevin models cannot be constructed from a bottom-up perspective. In that sense, it is instructive to consider works where these models are built from first principles or by relying on a step-by-step formulation guided by underlying physical directions [10, 14].

Indeed, if we follow the approach in [10], the elaboration of Langevin models as in Eq. (6.11) starts by applying Kolmogorov theory which suggests to represent the evolution of fluid particle velocities by a stochastic diffusion model of the form

$$\mathrm{d}X_{\mathrm{f},i} = U_{\mathrm{f},i}\,\mathrm{d}t\,,  \tag{6.12a}$$

$$\mathrm{d}U_{\mathrm{f},i} = A_i\,(\mathbf{X}_{\mathrm{f}}, \mathbf{U}_{\mathrm{f}})\,\mathrm{d}t + B_{ij}\,(\mathbf{X}_{\mathrm{f}}, \mathbf{U}_{\mathrm{f}})\,\mathrm{d}W_j\,,  \tag{6.12b}$$

with $\mathbf{A}$ and $\mathbf{B}$ the drift vector and diffusion matrix, respectively. Furthermore, the same Kolmogorov theory tells us that a well-justified closure for the diffusion coefficient is to retain an isotropic matrix with $B_{ij} = \sqrt{C_0 \langle \epsilon_{\mathrm{f}} \rangle}\,\delta_{ij}$. The first modeling step is to consider the situation of homogeneous turbulence where we know that the mean value of the drift term is zero but whose form is still to be worked out. To that end, we use the fact that the resulting equilibrium PDF of fluid particle velocities should be Gaussian. With a SDE as in Eq. (6.12) with a constant diffusion coefficient, it is then straightforward to show that $\mathbf{A}$ must be a linear function of $\mathbf{U}_{\mathrm{f}} - \langle \mathbf{U}_{\mathrm{f}} \rangle$ [15], thus yielding that $\mathbf{A} = \mathbf{G}\,(\mathbf{U}_{\mathrm{f}} - \langle \mathbf{U}_{\mathrm{f}} \rangle)$, where the matrix $\mathbf{G}$ is a constant. When moving to general non-homogeneous flows, the next step consists in adding the mean pressure-gradient, as it should be from the mean Navier-Stokes equation. In the last step of the construction of the stochastic model, the general form of the SDEs is retained, that is we consider that the coefficients $\mathbf{A}$ and $\mathbf{B}$ keep the same expression, but with values of the fluid mean fields which enter these expressions sampled at the particle location. These simple modeling arguments provide therefore the structure of Langevin models, as in Eq. (6.11), with an economy of means that is to be appreciated.

Note that the above formulation amounts to making a locally conditional Gaussian assumption for the increments of fluid particle velocities over a small time increment, conditioned on particles starting from a given position. The resulting ability of Langevin models to capture Gaussian PDFs in homogeneous flows (by construction) as well as deviations from it in general situations has been discussed repeatedly [1, 16] but is still a point not to pass over too quickly.

Far from being just a mere trick that happens to retrieve the fluid mean-field equations, it appears therefore that Langevin models have advanced built-in physical principles. Nevertheless, the above argument used to justify to select a linear drift coefficient may sound like nothing more than a technical requirement from stochastic calculus. To shed new light into its physical content, we propose to recast it in the frame of linear response theory.

For that purpose, we follow the guiding principles of generalized thermodynamics where the formulation of stochastic models is addressed in terms of 'phenomenological forces' $\mathbf{F}(\mathbf{Z})$, defined as the gradient of the potential $\Psi = \log(p_{stat}(\mathbf{z}))$ with $p_{stat}(\mathbf{z})$ the stationary PDF, and corresponding 'fluxes', defined as $\mathrm{d}\langle\mathbf{Z}\rangle/\mathrm{d}t$, resulting from these driving forces. In the spirit of linear irreversible thermodynamics and force-flux relations [15, 17–19], the next step consists in proposing a linear relation between the fluxes and their driving forces, which writes

$$\frac{\mathrm{d}\langle Z_i\rangle}{\mathrm{d}t} = L_{ij}\, F_j(\mathbf{Z})\,, \tag{6.13}$$

where the matrix $\mathbf{L}$ plays a special role, see [15] or other reference textbooks [17–19]). The connection to thermodynamics comes, in particular, from the identification often made in physics of $\Psi/k_B\Theta$ (with $\Theta$ the temperature) with the entropy of the system under consideration. In the present context, $\mathbf{Z} = \mathbf{U}_f$ the velocity of a fluid particle and we can take the equilibrium PDF $p_{stat}$ as Gaussian, which means that the phenomenological force is

$$F_i(\mathbf{U}_f) = -\frac{\partial\Psi(\mathbf{U}_f)}{\partial U_{f,i}} = \left(R_f^{-1}\right)_{ij}\left(U_{f,j} - \langle U_{f,j}\rangle\right)\,, \tag{6.14}$$

with $\mathbf{R}_f$ the tensor of fluid velocity second-order moments (defined as the Reynolds-stress tensor just below). Going back to Eq. (6.12b), we retrieve therefore that the drift vector $\mathbf{A}$ is a linear function of fluid particle velocities, since we have

$$A_i\left(\mathbf{X}_f, \mathbf{U}_f\right) = \left(LR_f^{-1}\right)_{ij}\left(U_{f,j} - \langle U_{f,j}\rangle\right) = G_{ij}\left(U_{f,j} - \langle U_{f,j}\rangle\right)\,, \tag{6.15}$$

where the matrix $\mathbf{G}$ is defined by $\mathbf{G} = \mathbf{LR}_f^{-1}$. This result is for homogeneous situations and, for general non-homogeneous turbulent flows, the next step consists in adding the mean pressure-gradient term as indicated above, leading to the GLM expression of the drift vector as

$$A_i\left(\mathbf{X}_f, \mathbf{U}_f\right) = -\frac{1}{\rho_f}\frac{\partial\langle P_f\rangle}{\partial x_i}\,\mathrm{d}t + G_{ij}\left(U_{f,j} - \langle U_{f,j}\rangle\right)\,. \tag{6.16}$$

Presented as such, we seem to have applied more or less the same arguments as in the first account, albeit in a different order, and it is thus not clear whether we have gained something new or not. A first benefit is nevertheless to bring out that present constructions of Langevin models in turbulence follow in the footsteps of what is currently achieved in non-equilibrium thermodynamics and, consequently, share similar physical validity or limitation. Yet, a real contribution of an original viewpoint is when new information is provided or new questions raised. This is the case here. Indeed, it was mentioned earlier that the matrix $\mathbf{L}$ has special significance in the field of linear irreversible thermodynamics which is manifested by typical characteristics: it is symmetrical and, when diagonalized, contains dynamical material properties such as transport coefficients or relaxation timescales [18]. The symmetry of $\mathbf{L}$ is the translation of the famous Casimir-Onsager relations [15, 17, 18] which express at the level of the macroscopic and usually irreversible equations the underlying reversibility of the behavior of the microscopic degrees of freedom that have been eliminated. This property is thus deeply connected to what is called micro-reversibility or detailed balance in probabilistic descriptions (interested readers are referred to discussions in [18] and to specific presentations and applications for stochastic models in [15, chapter 6.3] involving general symmetric and anti-symmetric variables). When this notion is carried over to the present context of Langevin models for turbulent flows, we are led to wonder whether we should consider that the matrix $\mathbf{G}$ should be such that $\mathbf{GR_f}$ be symmetric. At the moment, this point has received very little attention [14] and such a requirement is not applied in present modeling formulations of $\mathbf{G}$, as seen below, though investigating the possible specific roles played by the symmetrical and anti-symmetrical parts of $\mathbf{L} = \mathbf{GR_f}$ might be an interesting question.

To summarize, we retain that the matrix $\mathbf{G}$ governing the return-to-equilibrium term in Eqs. (6.12b) and (6.11b) is a response function determined locally in space as

$$\frac{1}{\Delta t} \frac{\delta}{\delta \mathbf{U_f}} \langle \Delta \mathbf{U_f}[\mathbf{U_f}] \mid \mathbf{X_f} = \mathbf{x} \rangle \ , \tag{6.17}$$

where we have written $\Delta \mathbf{U_f}[\mathbf{U_f}]$ for the increments of $\mathbf{U_f}$ taken as a functional of the velocity process (indicated by the brackets) over a discrete time increment $\Delta t$, conditioned on a given fluid particle position $\mathbf{x}$. This formulations turns out to be useful for the new formulations of two-phase models for the velocity of the fluid seen which are developed in Sect. 8.1.

## 6.2.2  Current Formulations: SLM and GLM

As indicated by its name, the simplest model is the simplified Langevin model (SLM) where the matrix $\mathbf{G}$ is given by

$$G_{ij} = - \left( \frac{1}{2} + \frac{3}{4} C_0 \right) \frac{\langle \epsilon_f \rangle}{k_f} \delta_{ij} \ , \tag{6.18}$$

with $C_0$ the Kolmogorov constant (cf. Sect. 5.3) as in the diffusion coefficient in Eq. (6.11b). In this expression, $\langle \epsilon_f \rangle$ is the mean rate of dissipation of the turbulent kinetic energy (as already introduced in Sect. 5.3) while $k_f$ is the turbulent kinetic energy defined as half the trace of the Reynolds-stress tensor whose components are $R_{f,ij} = \langle u_{f,i} u_{f,j} \rangle$ with $u_{f,i} = U_{f,i} - \langle U_{f,i} \rangle$ the fluctuating velocity component, which means that $k_f = 1/2 \langle u_{f,i} u_{f,i} \rangle$.

A classical decomposition of the matrix $G_{ij}$ is [10, 13]

$$G_{ij} = -\left(\frac{1}{2} + \frac{3}{4}C_0\right) \frac{\langle\epsilon_f\rangle}{k_f} \delta_{ij} + G_{ij}^a , \qquad (6.19)$$

where the matrix $G_{ij}^a$ represents anisotropic effects and, to be consistent with the kinetic energy budget, is subject to the condition $\mathrm{Tr}(\mathbf{G}^a \cdot \mathbf{R}_f) = 0$ if the diffusion coefficient is written as in Eq. (6.11b). There are, in fact, several ways to express the GLM which are considered as equivalent (they belong to the same class of models) if they yield the same Reynolds-stress equations. For example, relaxing the constraint $\mathrm{Tr}(\mathbf{G}^a \cdot \mathbf{R}_f) = 0$, it is possible to write a stochastic diffusion process for the particle velocity $\mathbf{U}_f$ as

$$dU_{f,i} = -\frac{1}{\rho_f} \frac{\partial \langle P_f\rangle}{\partial x_i} \, dt - \left(\frac{1}{2} + \frac{3}{4}C_0\right) \frac{\langle\epsilon_f\rangle}{k_f} \left(U_{f,i} - \langle U_{f,i}\rangle\right) dt + \sqrt{C_0\langle\epsilon_f\rangle} \, dW_i^{(1)}$$
$$+ G_{ij}^a \left(U_{f,j} - \langle U_{f,j}\rangle\right) dt + \sqrt{-2/3\,\mathrm{Tr}(\mathbf{G}^a\,\mathbf{R}_f)} \, dW_i^{(2)} , \qquad (6.20)$$

where $\mathbf{W}^{(1)}$ and $\mathbf{W}^{(2)}$ are two independent Wiener processes. In a weak formulation, these two increments of Wiener processes can be added to give

$$dU_{f,i} = -\frac{1}{\rho_f} \frac{\partial \langle P_f\rangle}{\partial x_i} \, dt$$
$$+ G_{ij} \left(U_{f,j} - \langle U_{f,j}\rangle\right) dt + \sqrt{\langle\epsilon_f\rangle \left(C_0 - 2/3 \frac{\mathrm{Tr}(\mathbf{G}^a\,\mathbf{R}_f)}{\langle\epsilon_f\rangle}\right)} \, dW_i . \qquad (6.21)$$

For example, if we consider that $G_{ij}^a$ is given by

$$G_{ij}^a = C_2 \frac{\partial \langle U_{f,i}\rangle}{\partial x_j} , \qquad (6.22)$$

where $C_2$ is a constant, we have then

$$dU_{f,i} = -\frac{1}{\rho_f} \frac{\partial \langle P_f\rangle}{\partial x_i} \, dt + G_{ij} \left(U_{f,j} - \langle U_{f,j}\rangle\right) dt + \sqrt{\langle\epsilon_f\rangle \left(C_0 + 2/3 \frac{\mathcal{P}_{k_f}}{\langle\epsilon_f\rangle}\right)} \, dW_i , \qquad (6.23)$$

where $\mathcal{P}_{k_f}$ is the turbulent kinetic energy production term, i.e., $\mathcal{P}_{k_f} = 1/2\,\mathcal{P}_{R_{f,kk}}$ with $\mathcal{P}_{R_{f,ij}}$ the production tensor whose components are the source/sink terms in the transport equation for the Reynolds stress components $R_{f,ij}$ which write

$$\mathcal{P}_{R_{f,ij}} = -\langle u_{f,j}u_{f,k}\rangle \frac{\partial \langle U_{f,i}\rangle}{\partial x_k} - \langle u_{f,i}u_{f,k}\rangle \frac{\partial \langle U_{f,j}\rangle}{\partial x_k} . \qquad (6.24)$$

With the form of $G_{ij}^a$ in Eq. (6.22), the resulting matrix $G_{ij}$ given in Eq. (6.19) retrieves the second-order LRR-IP model [8] (Launder, Reece and Rodi model based on the isotropization of production (IP) formulation). A word of caveat: while the diffusion coefficient in the first formulation, cf. Eq. (6.11b), ensures its realizability (we are taking the square root of a positive quantity), the second formulation in Eq. (6.21) implies that the expression within the square root remains positive (which is generally the case). Therefore, it should be considered that diffusion coefficients written as $\sqrt{D}$ must, in fact, be read as $\sqrt{\min(0, D)}$ for the various functions $D$ that are considered in all the models presented in this section.

Starting from a GLM, written either as in Eqs. (6.11) or (6.20), there is a well-established procedure leading to the transport equations satisfied by the moments of the fluid velocity field. Based on the corresponding FP equation in sample space, the first two velocity moment transport equations are easily derived using the methodology recalled in Sect. 4.1 and detailed in several works [8, 12, 13]. This yields the exact high Reynolds-number form of the mean Navier-Stokes equation

$$\frac{\partial \langle U_{f,i} \rangle}{\partial t} + \langle U_{f,j} \rangle \frac{\partial \langle U_{f,i} \rangle}{\partial x_j} = -\frac{1}{\rho_f} \frac{\partial \langle P_f \rangle}{\partial x_i} - \frac{\partial \langle u_{f,i} u_{f,j} \rangle}{\partial x_j} , \tag{6.25}$$

as well as the second-order transport equations for the components of the Reynolds stress tensor $\langle u_{f,i} u_{f,j} \rangle$ which, using the form of the GLM in Eq. (6.11), write as

$$\frac{\partial \langle u_{f,i} u_{f,j} \rangle}{\partial t} + \langle U_{f,k} \rangle \frac{\partial \langle u_{f,i} u_{f,j} \rangle}{\partial x_k} + \frac{\partial \langle u_{f,i} u_{f,j} u_{f,k} \rangle}{\partial x_k} =$$
$$- \langle u_{f,i} u_{f,k} \rangle \frac{\partial \langle U_{f,j} \rangle}{\partial x_k} - \langle u_{f,j} u_{f,k} \rangle \frac{\partial \langle U_{f,i} \rangle}{\partial x_k}$$
$$+ G_{ik} \langle u_{f,k} u_{f,j} \rangle + G_{jk} \langle u_{f,k} u_{f,i} \rangle + C_0 \langle \epsilon_f \rangle \delta_{ij} . \tag{6.26}$$

Note that Eq. (6.26) is the first level in the hierarchy of transport equations where the modeled part of the GLM appears through the matrix **G** and the diffusion coefficient $\sqrt{C_0 \langle \epsilon_f \rangle}$. For the SLM, the corresponding second-moment closure reads

$$\frac{\partial \langle u_{f,i} u_{f,j} \rangle}{\partial t} + \langle U_{f,k} \rangle \frac{\partial \langle u_{f,i} u_{f,j} \rangle}{\partial x_k} + \frac{\partial \langle u_{f,i} u_{f,j} u_{f,k} \rangle}{\partial x_k} =$$
$$- \langle u_{f,i} u_{f,k} \rangle \frac{\partial \langle U_{f,j} \rangle}{\partial x_k} - \langle u_{f,j} u_{f,k} \rangle \frac{\partial \langle U_{f,i} \rangle}{\partial x_k}$$
$$- \left( 1 + \frac{3}{2} C_0 \right) \frac{\langle \epsilon_f \rangle}{k_f} \left( \langle u_{f,i} u_{f,j} \rangle - \frac{2}{3} k_f \delta_{ij} \right) - \frac{2}{3} \langle \epsilon_f \rangle \delta_{ij} . \tag{6.27}$$

which is the Rotta model involving only a return-to-equilibrium term to represent the redistributive terms between the Reynolds stress components [8, 12, 13].

This relationship between Lagrangian stochastic models and second-moment closures has been used to suggest expressions of the matrix **G** from established formulations of Eq. (6.26). This corresponds to the top-down view mentioned in Sect. 6.2.1. The situation turns out to be different in turbulent dispersed two-phase flows since similar second-moment or macroscopic equations are not known beforehand, which requires that a bottom-up approach be followed instead. Nevertheless, the class of GLMs for single-phase turbulence is used as a building block.

## 6.3    Langevin Models for the Velocity of the Fluid Seen

### 6.3.1    State-of-the-art Two-phase SLM

The current Langevin model for the velocity of the fluid seen is an extension of the SLM for fluid particle velocity and has the form [1,2,20]

$$dU_{s,i} = -\frac{1}{\rho_f} \frac{\partial \langle P_f \rangle}{\partial x_i} \, dt + \left( \langle U_{p,j} \rangle - \langle U_{f,j} \rangle \right) \frac{\partial \langle U_{f,i} \rangle}{\partial x_j} \, dt$$

$$+ G_{ij}^* \left( U_{s,j} - \langle U_{f,j} \rangle \right) dt + B_{s,ij} \, dW_j \, . \qquad (6.28)$$

The matrix $G_{ij}^*$ is built from $G_{ij}$ used in the fluid-SLM and is given by

$$G_{ij}^* = -\left( \frac{1}{2} + \frac{3}{4} C_0 \right) \frac{\langle \epsilon_f \rangle}{k_f} H_{ij}, \qquad (6.29)$$

where the matrix $H_{ij}$ accounts for the crossing-trajectory effect and is expressed by

$$H_{ij} = b_\perp \delta_{ij} + \left[ b_\| - b_\perp \right] \widehat{r}_i \widehat{r}_j, \qquad (6.30)$$

with $(\widehat{r}_i)_{i=1,3}$ the components of the unit vector $\widehat{\mathbf{r}}$ aligned with the mean relative velocity, $\widehat{\mathbf{r}} = \langle \mathbf{U}_r \rangle / |\langle \mathbf{U}_r \rangle|$. The coefficients $b_\|$ and $b_\perp$ represent the Csanady factors which stand for the ratio between the timescale of fluid particle velocities $T_L$ and the timescale of the fluid velocities seen by discrete particles $T_{L,\|}^*$ and $T_{L,\perp}^*$, in the direction parallel to the mean relative velocity or transverse to it, respectively. Using the Csanady formulas for these timescales (see [1])

$$T_{L,\|}^* = \frac{T_L}{\sqrt{1 + C_T^2 \dfrac{|\langle \mathbf{U}_r \rangle|^2}{2k_f/3}}}, \qquad T_{L,\perp}^* = \frac{T_L}{\sqrt{1 + 4C_T^2 \dfrac{|\langle \mathbf{U}_r \rangle|^2}{2k_f/3}}}, \qquad (6.31)$$

(remember that $C_T = T_L/T_E$), the Csanady factors are obtained as $b_\| = T_L/T_{L,\|}^*$ and $b_\perp = T_L/T_{L,\perp}^*$

$$b_\| = \sqrt{1 + C_T^2 \frac{|\langle \mathbf{U}_r \rangle|^2}{2k_f/3}}, \qquad (6.32a)$$

$$b_\perp = \sqrt{1 + 4\,C_T^2 \frac{|\langle U_r \rangle|^2}{2k_f/3}}\,. \tag{6.32b}$$

With the classical expression for $T_L$ from the fluid SLM, whereby

$$T_L = \left( \frac{1}{1/2 + 3/4\,C_0} \right) \frac{k_f}{\langle \epsilon_f \rangle} \tag{6.33}$$

the matrix $G_{ij}^*$ can be re-expressed as

$$G_{ik}^* = -\frac{1}{T_{L,\perp}^*}\,\delta_{ik} - \left[ \frac{1}{T_{L,\|}^*} - \frac{1}{T_{L,\perp}^*} \right] \widehat{r}_i\,\widehat{r}_k\,. \tag{6.34}$$

In Eq. (6.28), the diffusion matrix $B_{s,ij}$ is the square root of the matrix $L_{ij}$ (i.e. $\mathbf{B_s}\,\mathbf{B}_s^T = \mathbf{L}$) given by

$$L_{ij} = L_\perp \delta_{ij} + \left[ L_\| - L_\perp \right] \widehat{r}_i\,\widehat{r}_j, \tag{6.35}$$

where the coefficients $L_\|$ and $L_\perp$ are

$$L_\| = \langle \epsilon_f \rangle \left( C_0 b_\| \widetilde{k}_f / k_f + \frac{2}{3}(b_\| \widetilde{k}_f / k_f - 1) \right), \tag{6.36a}$$

$$L_\perp = \langle \epsilon_f \rangle \left( C_0 b_\perp \widetilde{k}_f / k_f + \frac{2}{3}(b_\perp \widetilde{k}_f / k_f - 1) \right). \tag{6.36b}$$

In these expressions, a new kinetic energy $\widetilde{k}_f$ is introduced and is defined as

$$\widetilde{k}_f = \frac{3}{2} \frac{\mathrm{Tr}(\mathbf{H} \cdot \mathbf{R}_f)}{\mathrm{Tr}(\mathbf{H})}, \tag{6.37}$$

where $\mathrm{Tr}(\mathbf{H}) = H_{ii}$ denotes the trace of the matrix $\mathbf{H}$. As indicated at the end of Sect. 6.2, the eigenvalues of $\mathbf{L}$ must remain positive so that we are actually handling $\min(0; L_\|)$ and $\min(0; L_\perp)$ with $L_\|$ and $L_\perp$ given in Eq. (6.36). Note that the matrix equation $\mathbf{B_s} = \mathbf{L}^{1/2}$ does not yield a unique solution but, since we are only interested in weak solutions, different solutions $\mathbf{B_s}$ give the same statistics and, in a weak sense, are equivalent. It is also worth noting that the decomposition of $\mathbf{L}$ in Eq. (6.35) follows the same one as $\mathbf{G}^*$ in Eqs. (6.29) and (6.30) since we can rewrite Eqs. (6.35) and (6.36) as

$$L_{ij} = \left( 1 + \frac{3}{2}C_0 \right) \frac{\langle \epsilon_f \rangle}{k_f} \frac{\mathrm{Tr}(\mathbf{H} \cdot \mathbf{R}_f)}{\mathrm{Tr}(\mathbf{H})} H_{ij} - \frac{2}{3}\langle \epsilon_f \rangle\,\delta_{ij}\,. \tag{6.38}$$

Therefore, $\mathbf{G}^*$ and $\mathbf{L}$ are directly formulated in terms of the matrix $\mathbf{H}$ which appears as the key operator to go from the fluid-SLM to the two-phase-SLM expressions.

In [20], an analysis was performed to assess various modeling formulations with respect to a set of criteria. It appears that, at present, only the two-phase SLM outlined above meets all the requirements set forth, which is not a satisfactory situation. However, do we need more elaborate models than the SLM? Answers to that question depend on the statistical context in which these models are applied, that is whether we consider LES or RANS approaches.

## 6.4    Applications in LES and Accounting for Viscous Terms

For sufficiently well-resolved LES, not only the large energy-containing scales but also a substantial fraction of the energy spectrum are explicitly resolved, leaving small scales as the unresolved ones. The distinction between the large (resolved) and small (unresolved) scales is not always clear-cut since there is no physically well-justified scale separation (see Sect. 5.3) but we can consider that the unresolved ones are well inside the inertial range where the Kolmogorov theory prevails, at least as the reference framework. This does not mean that we assume these scales to be isotropic but, according to the Kolmogorov theory, we can expect that the driving mechanism towards isotropy is the dominant one. This explains that the SLM is often considered as an adequate model in LES, as reflected by a series of developments for dynamical and reactive flow applications [21–24].

Yet, the same reasoning suggests to consider also viscous terms since the unresolved velocity field can be a low-enough Reynolds-number flow to be impacted by molecular viscosity. The introduction of molecular transport coefficients in PDF/FDF formulations raises specific issues which have, however, not received much attention. To the authors' knowledge, only two proposals have been put forward and it is interesting to discuss the ideas behind these developments and the related challenges. The first proposition [25, 26] consists in introducing a random-walk model added to convective displacements which, with Ito stochastic calculus and further modeling choices, yields the following model for fluid-particle positions and velocities

$$dX_{f,i} = U_{f,i}\, dt + \sqrt{2\, \nu_f}\, dW_i' , \qquad (6.39a)$$

$$dU_{f,i} = -\frac{1}{\rho_f} \frac{\partial \langle P_f \rangle}{\partial x_i}\, dt + \sqrt{2\, \nu_f}\, \frac{\partial \langle U_{f,i} \rangle}{\partial x_k}\, dW_k'$$

$$+ G_{ik} \left( U_{f,k} - \langle U_{f,k} \rangle \right)\, dt + \sqrt{C_0 \langle \epsilon_f \rangle}\, dW_i , \qquad (6.39b)$$

where $\mathbf{W}$ and $\mathbf{W}'$ are two independent vector Wiener processes. An important feature of this model is that the white-noise term, $d\mathbf{W}'$, used in Eq. (6.39a) intervenes also in the particle velocity equation, cf. the second term on the right-hand side of Eq. (6.39b), leading to cross-correlations that need to be properly ascertained by carefully writing the corresponding Fokker-Planck equation [8, 25, 26]. The second proposition [27] retains the random-walk formulation for particle positions but considers that the only remaining viscous effect to account for is the dissipation of the kinetic energy of the mean flow field. This leads to the model

$$dX_{f,i} = U_{f,i}\, dt + \sqrt{2\, \nu_f}\, dW'_i \,, \tag{6.40a}$$

$$dU_{f,i} = -\frac{1}{\rho_f}\frac{\partial \langle P_f \rangle}{\partial x_i}\, dt + \left[A_{\text{vis},ik} + G_{ik}\right] \left(U_{f,k} - \langle U_{f,k} \rangle\right)\, dt + \sqrt{C_0\, \langle \epsilon_f \rangle}\, dW_i \tag{6.40b}$$

where the matrix $\mathbf{A}_{\text{vis}}$ is such that we have

$$A_{\text{vis},ik}\, R_{f,kj} = -\nu_f \left(\frac{\partial \langle U_{f,i} \rangle}{\partial x_k}\frac{\partial \langle U_{f,j} \rangle}{\partial x_k}\right) \,, \tag{6.41}$$

so that this term is written as a classical dissipation term. Both models rely therefore on extending particle convective transport by a viscosity-governed random walk using the exact correspondence between white-noise shifts, $\sqrt{2\, \nu_f}\, d\mathbf{W}$, and the viscous diffusion term, $\nu_f\, \partial^2 p/(\partial x_k \partial x_k)$, in the Fokker-Planck equation. It is worth noting that this in line with similar ideas used in Vortex Methods [28]. From a physical standpoint, it may be considered that we are not dealing anymore with fluid particles but with particles like macro-molecules and that the random-walk model is the remaining non-vanishing trace of molecular collisions.

To retrieve the correct description of velocity first and second-order moments, other formulations are also possible. For example, one can propose the model

$$dX_{f,i} = U_{f,i}\, dt \,, \tag{6.42a}$$

$$dU_{f,i} = -\frac{1}{\rho_f}\frac{\partial \langle P_f \rangle}{\partial x_i}\, dt + F_{\text{vis},i}\, dt + G_{ik} \left(U_{f,k} - \langle U_{f,k} \rangle\right)\, dt + \sqrt{C_0\, \langle \epsilon_f \rangle}\, dW_i \tag{6.42b}$$

where viscous effects are now treated through the new acceleration term $\mathbf{F}_{\text{vis}}$. To provide a closed expression of $\mathbf{F}_{\text{vis}}$, a first possibility is to retain a deterministic expression as a function of $\mathbf{Z}_f = (\mathbf{X}_f, \mathbf{U}_f)$, such as

$$F_{\text{vis},i}(\mathbf{U}_f) = \nu_f \frac{\partial^2 \langle U_{f,i} \rangle}{\partial x_k \partial x_k} + \widetilde{A}_{\text{vis},ik} \left(U_{f,k} - \langle U_{f,k} \rangle\right) \,, \tag{6.43}$$

where the matrix $\widetilde{A}_{\text{vis}}$ satisfies the following equation

$$\widetilde{A}_{\text{vis},ik}\, R_{f,kj} = \frac{\nu_f}{2}\frac{\partial^2 R_{f,ij}}{\partial x_k \partial x_k} \,. \tag{6.44}$$

It is also possible to propose a time-evolution model for $\mathbf{F}_{\text{vis}}$ having the form

$$dF_{\text{vis},i} = -\frac{F_{\text{vis},i} - \langle F_{\text{vis},i} \mid \mathbf{U}_f = \mathbf{V}_f \rangle}{\tau_{\text{vis}}} dt + K \, dW_{\text{vis},i} , \qquad (6.45)$$

with $K$ is a diffusion coefficient and $\tau_{\text{vis}}$ a timescale that we can both tune, $\mathbf{W}_{\text{vis}}$ a vector of independent Wiener processes, and where the conditional mean value noted $\langle F_{\text{vis},i} \mid \mathbf{U}_f = \mathbf{V}_f \rangle$ stands for the right-hand side of Eq. (6.43). When the timescale $\tau_{\text{vis}}$ is very small compared to the Lagrangian one for particle velocity, $\mathbf{F}_{\text{vis}}$ can be regarded as a fast-variable and (at the moment) heuristic manipulations lead to approximating the effect of $\mathbf{F}_{\text{vis}}$ over small time increments as

$$\mathbf{F}_{\text{vis}} \, dt \simeq \langle \mathbf{F}_{\text{vis}} \mid \mathbf{U}_f = \mathbf{V}_f \rangle \, dt + (K \, \tau_{\text{vis}}) \, d\mathbf{W}_{\text{vis}} . \qquad (6.46)$$

This can be seen as an extension of the first expression given in Eq. (6.43) in which $\mathbf{F}_{\text{vis}}$ appears now as a conditional Gaussian random variable.

Although it may be surprising to model viscous transport by a local term, we are here sticking to the notion of macroscopic fluid particles (rather than 'macro-molecules') for which viscous length scales are considered small for moderate- or high-Reynolds-number turbulent flows. It is also important to note that formulations such as the one in Eq. (6.42) are more attractive than the one in Eqs. (6.39) or (6.40) since the unchanged position equation, cf. Eq. (6.42a), makes it easier to carry out developments to the case of discrete particles. In fact, discrete particles are also affected by molecular random motions but through Brownian effects rather than through the same viscosity-governed random walks. Clearly, this is an area where further analysis is needed. In the present context, this formulation is nevertheless already useful in that it introduces discussions in terms of viscous/Brownian effects and local/non-local closures in relation with fast-variable elimination. These points are taken up in Sects. 10.2 and 10.3.

## 6.5    Consistency Issues and Limitations

From the above description, it transpires that present stochastic models for two-phase flows are extensions of the ones for single-phase flows which are retrieved when particle inertia vanishes. In that sense, single-phase PDF models are contained in two-phase PDF ones. This is true for the dynamical variables, such as the velocity of fluid particles and the resulting mean or filtered moments of the velocity field, but not for the kinetic energy dissipation rate which has no equivalent in the set of variables retained in discrete-particle state vectors. It is interesting to note that the difference between the RANS and LES frameworks hinges precisely on how the dissipation rate of the turbulent kinetic energy is modeled.

If the dissipation rate is obtained as the solution of a physically-based transport equation, we are evolving in the RANS framework and the ensemble averages obtained from the particle set correspond to mean field quantities. To obtain a self-contained PDF description (also referred to as stand-alone methods, for obvious reasons), an additional particle-attached variable $\epsilon_f$ is added to the fluid-particle state vector. In practice, models are devised in terms of the turbulent frequency, defined as $\omega_f = \epsilon_f / k_f$, but this does not change the information content ($k_f$ being a mean quantity, $\epsilon_f$ and $\omega_f$ are basically the same stochastic processes, although the physical signification of $\omega_f$ is perhaps clearer). To express the evolution equation of the particle instantaneous frequency, it is natural to rely on the refined K62 theory and model $\omega_f$ so that it follows a log-normal distribution in homogeneous situations while being applicable to general non-stationary and non-homogeneous flows. This leads to the following expression for $\omega_f$ [8, 29]

$$
d\omega_f = -\omega_f \langle \omega_f \rangle \left( S_{\omega_f} + C_\chi \left[ \ln \left( \frac{\omega_f}{\langle \omega_f \rangle} \right) - \left\langle \frac{\omega_f}{\langle \omega_f \rangle} \ln \left( \frac{\omega_f}{\langle \omega_f \rangle} \right) \right\rangle \right] \right) dt
$$
$$
+ \omega_f \sqrt{2 C_\chi \langle \omega_f \rangle \sigma^2} \, dW , \qquad (6.47)
$$

where $\sigma^2$ and $C_\chi$ are constants of the model ($\sigma^2$ is the variance of $\chi = \ln(\omega_f / \langle \omega_f \rangle)$ whose correlation timescale is written $T_\chi^{-1} = C_\chi \langle \omega_f \rangle$). The mean value of the frequency is the solution of the transport equation

$$
\frac{\partial \langle \omega_f \rangle}{\partial t} + \langle U_{f,k} \rangle \frac{\partial \langle \omega_f \rangle}{\partial x_k} = -S_{\omega_f} \langle \omega_f \rangle^2 , \qquad (6.48)
$$

where the driving term $S_{\omega_f}$ in this equation as well as in Eq. (6.47) is specified as

$$
S_{\omega_f} = -C_{\omega_f,1} \frac{\mathcal{P}_{k_f}}{\langle \epsilon_f \rangle} + C_{\omega_f,2} , \qquad (6.49)
$$

with $C_{\omega_f,1}$ and $C_{\omega_f,2}$ two constants of the model. It is worth noting that even if $\epsilon_f$ is included as an instantaneous variable and described with the refined K62 theory, thereby accounting for internal intermittency, the fact that only $\langle \epsilon_f \rangle$ (through $\langle \epsilon_f \rangle = k_f \langle \omega_f \rangle$) enters the GLM equations, as in Eq. (6.11), implies that fluid-particle velocities are modeled with the K41 theory. As such, the role of the log-normal model for the dissipation rate or the turbulent frequency in Eq. (6.47) is merely to provide the mean value $\langle \epsilon_f \rangle$ used in the various Langevin formulations in Sects. 6.2 and 6.3. The evolution toward a velocity-frequency PDF description fully consistent with the refined K62 theory seems straightforward and consists in replacing $\langle \epsilon_f \rangle$ with $\epsilon_f$ in the SDEs, for example by substituting the diffusion coefficient $\sqrt{C_0 \langle \epsilon_f \rangle} \, dW$ with $\sqrt{C_0 \epsilon_f} \, dW$ and, similarly, in the drift term. This leads to the refined Langevin model, as described in [29]. On the one hand, such steps emphasize the great malleability of the PDF modeling framework. On the other hand, to maintain the basic property that the velocity PDF should be Gaussian in homogeneous situations, additional terms must be added to the drift term, increasing the complexity of the model formulation and making it harder to handle [8, 29]. We are thus faced with a first hint that addressing situations of increasing complexity with a single formulation could lead to more intricate model expressions. The philosophy behind the different approaches to complexity is investigated in more details in Sect. 8.4.

If the dissipation rate is obtained from a parameter-based local equation, we are evolving in the LES framework and the ensemble averages obtained from the particle set correspond to filtered field quantities. While the stochastic models for particle velocity have the same expression, the turbulent kinetic energy becomes the residual kinetic energy $k_f^r$ and, more importantly, the dissipation rate noted $\widehat{\epsilon_f}$ is now estimated locally using the cut-off parameter $\Delta$ separating the resolved from the unresolved scales. A model applied in several studies is [21–23]

$$\widehat{\epsilon_f} = C_{\epsilon_f} \left(k_f^r\right)^{3/2} / \Delta , \qquad (6.50)$$

where $C_{\epsilon_f}$ is a constant. In most practical applications, the parameter $\Delta$ is the grid size used in the computations, which means that there is a rather indistinct, if not blurred, separation between the continuous and numerical formulations (a characteristic feature of LES and a subject of discussions on the foundations of the approach [8]). More recent proposals [24, 30] introduce the notion of a cascade time delay between production and dissipation of energy, which leads to (re-)introduce an instantaneous particle-attached value of the dissipation rate, say $\epsilon_f^r$, which is the solution of a time-evolution equation such as

$$d\epsilon_f^r = -\left(\frac{\epsilon_f^r - \widehat{\epsilon_f}}{\tau_{\epsilon_f^r}}\right) dt , \qquad (6.51)$$

where different estimations have been proposed for the timescale $\tau_{\epsilon_f^r}$ governing this delay [24, 30].

These considerations bring us back to the issue mentioned at the beginning, namely the consistency between the description of the fluid phase on the one hand and the description obtained from the dispersed-phase model in the tracer-particle limit on the other hand. When concentrating on PDF models for discrete particles, the mean or filtered fields which characterize the fluid phase are considered as available. Yet, one unfortunate error is to assume that the descriptions retained for each phase are unrelated. Indeed, once a choice is made for the dissipation rate, the corresponding RANS or LES frameworks are selected for both phases. Furthermore, the limit of a two-phase GLM when particle inertia goes to zero is a GLM model for the fluid phase. It follows that, in order to be consistent in the tracer-particle limit, both should be identical. For example, if we select a two-phase SLM coupled to a description of the fluid phase, that description of the fluid-phase must be based on the Rotta second-order model with the same constants [16, 31]. In that sense, it is of importance to note that a stochastic model based on a particle state vector which includes the particle velocity corresponds necessarily to a second-order turbulence model [13] and not to one based on an 'eddy'- or 'turbulent-' viscosity. It is inconsistent to simulate the fluid phase with such a turbulence model and then track discrete particles with a stochastic model for the velocity of the fluid seen. Unfortunately, this is a mistake impacting several simulations of discrete particles in LES since the majority of LES formulations rely on the Smagorinsky model based on the notion of an effective, or sub-grid scale (SGS), viscosity to model the effects of the residual fluid motions. Note that coupling an 'ideal LES' (obtained by filtering a DNS) with a stochastic model designed to mimic SGS fluctuations is still inconsistent since, if the ideal LES is free of modeling error for single-phase predictions, the complete model formulation is not since there is a discrepancy between the 'ideal LES' description and the one corresponding to the tracer-particle limit of the stochastic model used for the velocity of the fluid seen. When using a two-phase GLM for discrete particles in LES, then the components of the SGS tensor (the equivalent of the Reynolds-stress tensor in the context of filtered fields) must be obtained as the solution of their own transport equations based on the corresponding single-phase GLM [21]). On the other hand, when an eddy-viscosity formulation is retained for the fluid phase, it can be wondered which stochastic formulations for discrete particles are consistent. Such questions resurface in Sects. 10.3 and 10.4.

Even when consistency is ensured in the tracer particle limit, it is also useful to be aware of inherent limitations of present Langevin-based stochastic models as revealed, for instance, in studies of particle concentration effects [7, 32]. Actually, Langevin-type of models are devised so as to reproduce essentially one-particle statistics, such as the correct level of subgrid kinetic energy, velocity auto-correlations, dispersion coefficients, etc., but not necessarily instantaneous distributions due to captures by some underlying fluid structures which are not present in the mean or filtered fields. For all their merits, it is also fair to recall that basic stochastic models, such as the SLM, live up to their names and remain, in fact, crude pictures. Indeed, looking back at the general structure of a Langevin model, as in Eq. (6.11b), it is seen that it involves three terms: a deterministic one, a purely ordered term (since all fluctuations are driven back to the local mean value in the same manner) and a purely disordered term manifested by the White-noise increments. Thus, it may not be an altogether surprise if such a formulation remains limited. In that respect, new ideas are called for and suggestions are made in Sect. 8.4.

## References

1. J.P. Minier, E. Peirano, Phys. Rep. **352**(1–3), 1 (2001). https://doi.org/10.1016/S0370-1573(01)00011-4
2. J.P. Minier, E. Peirano, S. Chibbaro, Phys. Fluids **16**(7), 2419 (2004). https://doi.org/10.1063/1.1718972
3. A. Monin, A. Yaglom, *Statistical Fluid Mechanics, Volume II: Mechanics of Turbulence*. Dover Books on Physics, vol. 2 (Dover Publications, Mineola, 2013). https://store.doverpublications.com/products/9780486458915
4. J. Eaton, J. Fessler, Int. J. Multiphase Flow **20**, 169 (1994). https://doi.org/10.1016/0301-9322(94)90072-8

5. S. Balachandar, J.K. Eaton, Ann. Rev. Fluid Mech. **42**, 111 (2010). https://doi.org/10.1146/annurev.fluid.010908.165243

6. R. Monchaux, M. Bourgoin, A. Cartellier, Int. J. Multiphase Flow **40**, 1 (2012). https://doi.org/10.1016/j.ijmultiphaseflow.2011.12.001

7. J. Pozorski, S.V. Apte, Int. J. Multiphase Flow **35**(2), 118 (2009). https://doi.org/10.1016/j.ijmultiphaseflow.2008.10.005

8. S. Pope, *Turbulent Flows* (Cambridge University Press, Cambridge, 2000). https://doi.org/10.1017/CBO9780511840531

9. S.B. Pope, Progress Energy Combustion Sci. **11**(2), 119 (1985). https://doi.org/10.1016/0360-1285(85)90002-4

10. S.B. Pope, Ann. Rev. Fluid Mech. **26**(1), 23 (1994). https://doi.org/10.1146/annurev.fl.26.010194.000323

11. D.C. Haworth, Progress Energy Combustion Sci. **36**(2), 168 (2010). https://doi.org/10.1016/j.pecs.2009.09.003

12. D.C. Haworth, S.B. Pope, Phys. Fluids **29**(2), 387 (1986). https://doi.org/10.1063/1.865723

13. S.B. Pope, Phys. Fluids **6**(2), 973 (1994). https://doi.org/10.1063/1.868329

14. J.P. Minier, J. Pozorski, Phys. Fluids **9**(6), 1748 (1997). https://doi.org/10.1063/1.869291

15. C. Gardiner, *Stochastic Methods*, 4th edn. (Springer, Berlin, 2009). https://link.springer.com/book/9783540707127

16. J.P. Minier, Progress Energy Combustion Sci. **50**, 1 (2015). https://doi.org/10.1016/j.pecs.2015.02.003

17. S.R. De Groot, P. Mazur, *Non-equilibrium Thermodynamics* (Dover, Mineola, 1984). https://store.doverpublications.com/products/9780486647418

18. H.C. Öttinger, *Beyond Equilibrium Thermodynamics* (John Wiley & Sons, Hoboken, 2005). https://doi.org/10.1002/0471727903

19. D.C. Venerus, H.C. Öttinger, *A Modern Course in Transport Phenomena* (Cambridge University Press, Cambridge, 2018). https://www.cambridge.org/us/universitypress/subjects/engineering/chemical-engineering/modern-course-transport-phenomena

20. J.P. Minier, S. Chibbaro, S.B. Pope, Phys. Fluids **26**(11), 113303 (2014). https://doi.org/10.1063/1.4901315

21. L.Y.M. Gicquel, P. Givi, F.A. Jaberi, S.B. Pope, Phys. Fluids **14**(3), 1196 (2002). https://doi.org/10.1063/1.1436496

22. M.R.H. Sheikhi, T.G. Drozda, P. Givi, S.B. Pope, Phys. Fluids **15**(8), 2321 (2003). https://doi.org/10.1063/1.1584678

23. M.R.H. Sheikhi, P. Givi, S.B. Pope, Phys. Fluids **19**(9) (2007). https://doi.org/10.1063/1.2768953

24. M.R.H. Sheikhi, P. Givi, S.B. Pope, Phys. Fluids **21**(7) (2009). https://doi.org/10.1063/1.3153907

25. T.D. Dreeben, S.B. Pope, Phys. Fluids **9**(1), 154 (1997). https://doi.org//10.1063/1.869157

26. T.D. Dreeben, S.B. Pope, J. Fluid Mech. **357**, 141 (1998). https://doi.org/10.1017/S0022112097008008

27. M. Waclawczyk, J. Pozorski, J.P. Minier, Phys. Fluids **16**(5), 1410 (2004). https://doi.org/10.1063/1.1683189

28. A.J. Chorin, *Vorticity and Turbulence*, vol. 103 (Springer Science & Business Media, Berlin, 2013). https://doi.org/10.1007/978-1-4419-8728-0

29. S. Pope, Phys. Fluids A Fluid Dyn. **3**(8), 1947 (1991). https://doi.org/10.1063/1.857925

30. P.L. Johnson, C. Meneveau, J. Fluid Mech. **837**, 80 (2018). https://doi.org/10.1017/jfm.2017.838

31. S. Chibbaro, J.P. Minier, Int. J. Multiphase Flow **37**(3), 293 (2011). https://doi.org/10.1016/j.ijmultiphaseflow.2010.10.010

32. C. Marchioli, Acta Mech. **228**(3), 741 (2017). https://doi.org/10.1007/s00707-017-1803-x

# Numerical Investigations

# 7

**Abstract**

The title of this chapter was chosen to indicate that its purpose is to complement the analysis through numerical simulations. In a way, we are pursuing our investigations by other means. This is a different objective from software development where the emphasis is on computational science, and from code validation where the emphasis is on numerical methods. In the following, we discuss several situations in the tracer-particle limit before addressing the case of discrete particles. There are reasons to do that. First, we have seen that present formulations of the stochastic model for the velocity of the fluid seen are built as extensions of the ones for fluid particles, and retrieving the fluid limit needs to be assessed in numerical simulations. Second, considering the fluid-limit case allows to discuss key physical phenomena, such as the convective or diffusive regimes in point-source dispersion, in a simplified setting. It also allows to reveal similarities or differences between fluid and discrete particles, e.g. the role of the pressure gradient and the impact of the incompressibility constraint. Only then can we assume that we rely on solid foundations and turn to disperse two-phase flow situations.

**Chapter Content** The numerical approach chosen for these applications is outlined in Sect. 7.1. Based on this formulation, we discuss point-source dispersion in Sect. 7.2. The constraints resulting from the incompressibility of the fluid and the consistency issues are investigated in Sect. 7.3 through both free shear and wall-bounded flows. We then turn our attention towards a situation involving discrete particles in Sect. 7.4 before stating some recommendations in Sect. 7.5.

© The Author(s) 2025
J.-P. Minier et al., *Understanding Turbulent Systems*,
Lecture Notes in Physics 1039, https://doi.org/10.1007/978-3-031-84466-9_7

## 7.1    The Numerical Context

Not only a detailed presentation but even an overview of the numerical methods involved in the simulations of disperse two-phase flows are outside the scope of this book. Therefore, we limit ourselves to providing a few indications about the numerical context.

When handling particle stochastic systems, one possibility is to use these stochastic particles to extract all the mean fields entering the particle evolution equations. For obvious reasons, this is called a stand-alone method. Such a self-sufficient formulation is attractive but brings in specific requirements. Indeed, it can only be applied if the particle state vector contains the relevant particle-attached variables in the sense that all the needed mean fields, or statistics, can be extracted from the particle set without requiring an external source of information. If this is the case, note that we are using McKean type of SDEs, cf. Sect. 4.3.3, in which statistical noise due to the Monte Carlo estimations of the statistics is fed back in the evolution equations making them potentially sensitive to numerical bias [1]. In our situation, the physical system comprises the fluid and the particle phases. To apply a stand-alone method implies therefore that we must use two sets of stochastic particles, one for the fluid and one for the disperse particles. Such a complete fluid-particle stochastic model was proposed some time ago [2, section 8], including the various terms accounting for the exchanges between these two sets of particles, but has not been implemented (to the best of the authors' knowledge). In the context of polymer solutions (discussed later in Chap. 9), recent works have started to consider formulations based on Brownian dynamics coupled to smooth particle hydrodynamics (SPH) [3], thus with one set of particles to model the polymer blobs and another one to model fluid particles. This shows that interest is growing and that progress is being made but stand-alone numerical formulations are still at an early stage of development.

For the numerical applications presented in this book, we rely on a coupled moment/PDF approach, sketched in Fig. 7.1. During each step of the simulation, the fluid mean fields are first updated to their values at the next time step $t_{n+1} = t_n + \Delta t$ by the moment solver and provided to the Lagrangian solver in which particle variables are advanced and particle statistics extracted. Source terms accounting for

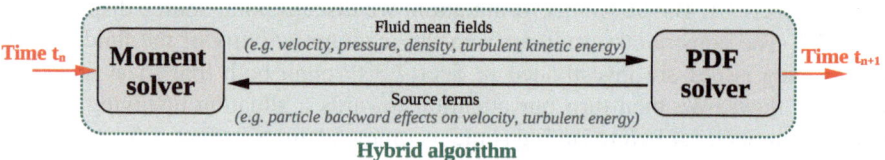

**Fig. 7.1**  Representation of the hybrid algorithm: during each time step, the fluid mean fields are calculated by the moment solver and fed to the PDF solver to advance particle variables and calculate source terms that are fed back to the moment solver

the exchange of momentum and kinetic energy are then fed back into the moment solver (these source terms are studied in Chap. 11).

One of the interests of this hybrid formulation is to benefit from the considerable body of work dedicated to the development of mesh-based finite-volume type of algorithms for the solution of the PDEs representing second-moment turbulence closures. Another interesting aspect is that these fluid mean fields are free of statistical noise. Nevertheless, the SDEs handled in the PDF solver are still of McKean type since all the particle statistics are obtained from the set of stochastic particles by Monte Carlo estimations, like the particle mean velocity $\langle U_p \rangle$ used in the Langevin equation of the two-phase SLM. Advancing the particle-attached variables implies steps such as integrating the governing SDEs while accounting for boundary conditions, tracking each particle in the mesh used for the fluid calculation and extracting statistics (which is the inverse operation compared to interpolating mean fields at particle positions). The literature is very large on some aspects, for instance on interpolation methods in mesh/particle, or more limited on others, such as on numerical schemes to integrate present Langevin type of SDEs. More details can be found in reference textbooks addressing one such topic [4, 5] or in specialized articles on single-phase PDF methods [1, 6–9] or disperse two-phase flows [10–13]. For all its usefulness as a general-purpose numerical method, this hybrid formulation introduces also specific features, in particular due to the fact that we are handling two descriptions of turbulence in the moment and the PDF approaches. This leads to a consistency issue which is addressed in Sect. 7.3.

## 7.2  Point-source Turbulent Dispersion

Point-source dispersion corresponds to the case where either discrete particles or fluid elements are released from a point-wise source into a surrounding turbulent flow. This is a situation that we can meet on a daily basis, as already indicated in Chap. 2 with the spread of a plume shown in Fig. 2.6d, and an interesting configuration to discuss the convective and diffusive regimes. It is illustrated in Fig. 7.2 by a snapshot of a numerical simulation of particles injected continuously from a point-source and spreading as they are carried downstream by a turbulent flow.

To analyze the convective and diffusive regimes, we consider fluid particles which simplifies the discussion but without modifying what happens also for inertial particles (see the analysis in [14]). This can represent a heated fluid stream (provided that the temperature difference with the surrounding fluid is small enough not to induce thermal modifications of the fluid properties) or some passive fluid markers introduced in the flow. To release a heated stream, a hot wire is usually placed in the flow so that it corresponds to a line-dispersion problem rather than a point-dispersion one. We can simplify the context by considering that the mean fields of the surrounding flow (velocity, temperature, turbulent kinetic energy and its dissipation rate) are homogeneous, stationary and even isotropic. The problem becomes then a one-dimensional problem or can be addressed as such. Indeed,

**Fig. 7.2** Simulation of a
particle point-source
dispersion: particles are
released from a point source
in a surrounding turbulent
flow and spread as they move
downstream with the velocity
of the surrounding flow. In
this plot of particle positions
at a given time, particles are
colored as a function of their
residence time in the domain

whether we follow the spread of the surface of a circle (line-dispersion) or of a
spherical cloud (point-source) only modifies some constants in front of the formulas
presented below and can, therefore, be ignored for our present purpose.

With these simplifications, the dynamics of the released fluid particles is reduced
to a simple form for $\mathbf{Z}_f = (X_f, U_f)$. Indeed, if we place ourselves in the inertial
reference frame traveling at the constant velocity $U$ of the surrounding flow (so that
we have $\langle U_f \rangle = 0$ and $u_f = U_f$ in this frame), the time evolution of $\mathbf{Z}_f$ is given by

$$dX_f = U_f \, dt \, , \tag{7.1a}$$

$$dU_f = -\frac{U_f}{T_L} \, dt + \sigma \, dW \, , \tag{7.1b}$$

where $T_L$ and $\sigma$ are constants. Since turbulence is at equilibrium, $T_L$, $\sigma$ and $\langle (u_f)^2 \rangle$
are related through the fluctuation-dissipation relation, i.e. $\langle (u_f)^2 \rangle = 1/2(\sigma^2 T_L)$.

There are several possible ways to derive the expression of the second-order
moments of $\mathbf{Z}_f = (X_f, U_f)$ from the model given in Eq. (7.1). A relatively straightforward method is to derive the set of ODEs satisfied by $\langle (U_f)^2 \rangle(t)$, $\langle X_f U_f \rangle(t)$ and
$\langle (X_f)^2 \rangle(t)$ and integrate from given initial conditions. We use directly the results
provided in [10, section 4.1] when the initial velocity has a fixed value $U_{f,0}$, which
gives for the variance of particle positions

$$\langle (X_f)^2(t) \mid U_{f,0} \rangle = U_{f,0}^2 \, T_L^2 \left[ 1 - \exp(-t/T_L) \right]^2$$
$$+ \sigma^2 T_L^2 \left\{ t - \frac{T_L}{2} \left[ 1 - \exp(-t/T_L) \right] \left[ 3 - \exp(-t/T_L) \right] \right\} \tag{7.2}$$

Note that similar considerations appear later when considering two-particle relative
dispersion, cf. Sect. 8.3.2.2, and for the source terms representing particle back
effects on the fluid in Sect. 11.4.4, albeit for different set of variables.

When the initial velocity $U_{f,0}$ is taken as a random variable with a variance given
by the equilibrium value, i.e. $1/2(\sigma^2 T_L)$, we obtain the expression of $\langle (X_f)^2 \rangle(t)$ by

integrating over all values of $U_{f,0}$ in Eq. (7.2). This yields

$$\langle (X_f)^2 \rangle (t) = \sigma^2 T_L^2 t$$
$$+ \frac{\sigma^2 T_L^3}{2} \left[ 1 - \exp(-t/T_L) \right] \left\{ 1 - \exp(-t/T_L) - \left[ 3 - \exp(-t/T_L) \right] \right\} , \qquad (7.3)$$

which is immediately rearranged to give

$$\langle (X_f)^2 \rangle (t) = \sigma^2 T_L^2 \left[ t - T_L \left( 1 - \exp(-t/T_L) \right) \right] , \qquad (7.4)$$

or alternatively

$$\langle (X_f)^2 \rangle (t) = 2 \langle (u_f)^2 \rangle T_L \left[ t - T_L \left( 1 - \exp(-t/T_L) \right) \right] . \qquad (7.5)$$

which is the formula given in [15, chapter 12]. The short- and long-time limits of the particle position variance $\sigma_{X_f}^2 = \langle (X_f)^2 \rangle (t)$ are easily obtained from Eq. (7.5) to bring out the convective and diffusive regimes

$$\frac{\langle (X_f)^2 \rangle (t)}{\langle (u_f)^2 \rangle T_L^2} = \begin{cases} (t/T_L)^2 & \text{if } t \ll T_L \text{ (convective regime)}, \\ 2\, t/T_L & \text{if } t \gg T_L \text{ (diffusive regime)}. \end{cases} \qquad (7.6)$$

These results are presented here as a function of time by considering that we are traveling with the center of the released cloud and observing the time-evolution of the cloud spread from a non-moving position. They can be transformed into a space-dependent formulation from a fixed-point perspective by making the simple transformation $x = U t$ with $x$ the streamwise coordinate. In that case, the short- and long-time regimes are mapped into the near- and far-field regions. This is how numerical results are presented in Figs. 7.3 and 7.4. In Fig. 7.3, the downstream evolution of the particle position standard deviation is plotted and the convective and diffusive regimes are also indicated to show their respective range of validity. In the same figure, we can see different profiles of the particle concentration, which is proportional to the particle position PDF. These concentrations are dependent upon the spatial coordinate corresponding to the variable handled in Eq. (7.1a), taken here as the vertical or cross-stream variable $z$, and upon time $t$ or, conversely, the downstream location $x$ from the source. We can observe how the shape of the concentration profile evolves from the convective to the diffusive regime where a Gaussian form prevails and that this evolution is nicely reproduced by the Lagrangian stochastic simulation. This is further illustrated in Fig. 7.4 which displays the evolution of the particle concentration at the center of the released cloud, emphasizing again the two regimes and the transition between them.

Although apparently simple, these results help to clarify the range of validity of classical modeling approximations. For instance, from Eq. (7.5), we see that the dispersion coefficient $\mathcal{D}_f = 1/2 \, d\langle (X_f)^2 \rangle (t)/dt$ is only constant in the diffusive

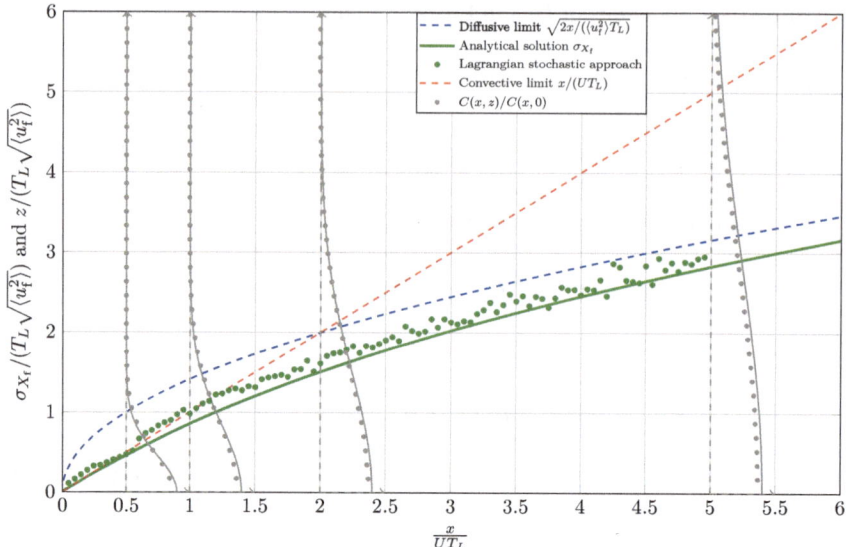

**Fig. 7.3** Study of particle dispersion from a source: downstream evolution of the standard deviation of particle positions as well as three profiles of the particle concentration as a function of the cross-stream coordinate and taken at three positions downstream of the source. The short-time or near-field convective regime and the long-time or far-field diffusive regime are indicated to reveal their range of validity

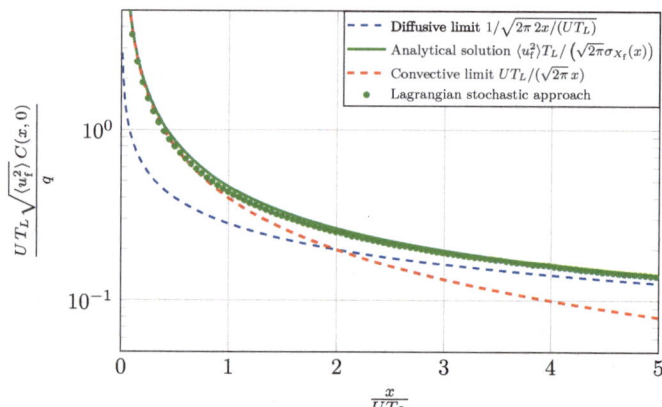

**Fig. 7.4** Study of particle dispersion from a source: evolution of the particle concentration at the center of the released cloud as a function of the downstream location. In both figures, the short-time or near-field convective regime and the long-time or far-field diffusive regime are indicated to reveal their range of validity

regime (thus called so for this very reason). This explains that models based on a gradient hypothesis (such as the Newton, Fick and Fourier laws introduced in Eqs. (3.3)–(3.5) in Sect. 3.2) and assuming the existence of a particle diffusivity are only acceptable in the long-time limit or in the far-field region but severely misrepresent the short-time limit or the near-field region. It is worth pointing out that this is not a question of whether we should use a moment or a PDF approach (though, obviously, the PDF approach does an excellent job for this problem). What matters is to treat properly convection, if possible without approximation. This is precisely what is achieved by the Lagrangian approach due to its particle formulation but it would also be captured (albeit with more difficulties) by second-moment closures which handle convective terms with transport equations contrary to models based on local closures and diffusion coefficients.

If we wish to pursue this argument one step further, we can note from Eq. (7.5) that we would always be in the diffusive regime if we take the limit $\langle (u_f)^2 \rangle \rightarrow +\infty$ while $T_L \rightarrow 0$ so that $\langle (u_f)^2 \rangle T_L \rightarrow \mathcal{D}_f$, where the manner with which the limit is taken is still to be determined. This is, therefore, an echo of the notion of fast-variable elimination already introduced in Chap. 5 and, in particular, in the discussion of the Markovian limit in Sect. 5.2.2 with the toy model. This topic is of the main undercurrents in this book and resurfaces several times before being addressed more directly and with additional details in Chap. 10.

## 7.3   The Incompressibility and Consistency Constraints in the Fluid Limit

In the point-source dispersion case, we only used particles to represent tracer elements released at the source in the surrounding fluid flow. We could also have simulated particles in the whole domain to describe the entire flow problem. In this stationary homogeneous situation, this was not needed since the mean fields have constant values which were assumed beforehand. However, for general non-homogeneous flows described by a particle stochastic approach, this leads us to ask: what are the conditions that must be met in order for a Lagrangian stochastic model to be a valid description of turbulent flows? The answer to that question was provided in Pope's seminal 1985 paper [16, section 4.7] in which it was demonstrated that the realizability and consistency conditions, which guarantee the Lagrangian and Eulerian MDFs to be true ones, are satisfied when the mean continuity equation is respected which, in turn, is equivalent to saying that the mean pressure-gradient must be calculated so as to ensure local mass conservation. In the present work, we consider incompressible flows and the mass continuity equation takes the form of a divergence-free constraint for the fluid mean velocity field.

A straightforward consequence is that, when formulated in terms of fluid particle instantaneous velocities, a Lagrangian stochastic model provides a valid description of turbulent flows if the model is such that we have

$$\frac{1}{dt} \left\langle dU_{f,i} \mid \mathbf{X}_f(t) = \mathbf{x} \right\rangle = -\frac{1}{\rho_f} \frac{\partial \langle P_f \rangle}{\partial x_i} (t, \mathbf{x}) \,. \tag{7.7}$$

This requirement is obviously met by the GLMs introduced in Sect. 6.2 and stems from the high Reynolds-number form of the NS equations written in a Lagrangian form as already indicated in Sect. 6.2.1. Rather surprisingly, this point has been surrounded with some confusion in spite of repeated clarifications [2, 16–18], for instance through misleading associations of this constraint to specific forms of the Langevin part of the stochastic model. When proposing general guidelines to the formulation of stochastic models for single- and disperse two-phase flows in [17], this issue was revisited and it was demonstrated that a stochastic model written as (using the same notations as in [17])

$$dX_{f,i} = U_{f,i} \, dt \,, \tag{7.8a}$$

$$dU_{f,i} = -\frac{1}{\rho_f} \frac{\partial \langle P_f \rangle}{\partial x_i} \, dt + dM_i(t; \mathbf{X}_f, \mathbf{U}_f) \,, \tag{7.8b}$$

is a valid description as long as the mean pressure-gradient is obtained from the divergence-free constraint for the fluid mean velocity field. In Eq. (7.8b), d$\mathbf{M}$ stands for the increments of a stochastic model $\mathbf{M}$ written in terms of the particle state vector for fluid particles, $\mathbf{Z}_f = (\mathbf{X}_f, \mathbf{U}_f)$, and is only requested to satisfy the two conditions that $\langle dM_i \rangle(t, \mathbf{x}) = 0$ and $\langle u_{f,i} \circ dM_i \rangle(t, \mathbf{x}) = -2 \langle \epsilon_f \rangle(t, \mathbf{x})$ to account for the mean dissipation rate of the turbulent kinetic energy (the notation $\circ$ refers to Stratonovich stochastic calculus which is used for the sake of allowing a more compact expression, as explained in Sect. 4.3.1). This general formulation emphasizes that the important point is the proper account of the mean pressure-gradient term and not the specific details entering the Langevin model, or any such stochastic model for that concern.

The troubles encountered around that issue could be related to the double aspect of the incompressibility condition in Lagrangian formulations (already hinted to at the beginning of Sect. 6.1.3). Indeed, in field-based methods, the density of the fluid at each point is simply taken as a constant, leaving only the divergence-free as a consequence of the mass continuity equation. On the other hand, in particle-based approaches, particle position becomes a variable whose related statistic is the particle concentration field while we also obtain the mean velocity field as one statistic extracted from the simulated fluid particle instantaneous velocities (through conditional Monte Carlo estimations [2, 15, 17]). Since each stochastic particle represents the same amount of mass, it follows that the particle concentration field (or the particle position PDF) must be uniform within the domain, as any depletion or accumulation would translate a local loss or a gain of mass at variance with the condition of an incompressible flow. In other words, the incompressibility requirement is translated into a double constraint in particle-based methods: (1) the particle concentration must remain uniform, and (2) the mean velocity field must be of zero divergence. Actually, these two constraints represent two sides of the same coin. Indeed, it can be shown that a uniform particle position distribution is preserved when these particles are transported by a divergence-free velocity field. In practical applications, due to numerical errors in time-splitting and fractional-step algorithms or to the presence of statistical noise in simulations performed with a necessarily-finite number of particles, these two conditions need to be addressed at the same time (see discussions developed for stochastic as well as non-stochastic particle-based methods in [19, 20]).

The first conclusion to draw is that we must always check that a theoretical model and its numerical implementation are such that, when we start from a uniform concentration for fluid particles, this concentration remains uniform in the course of the simulations to accept this description of turbulent flows as valid. The second side of the incompressibility coin, namely the divergence-free condition for the mean velocity field, takes different forms depending on whether we use a stand-alone or an hybrid numerical formulation. In stand-alone methods, the mean velocity field is directly obtained from the simulated stochastic particles and the mean-pressure gradient is then obtained to enforce the divergence-free condition on this velocity field, using for example a particle/mesh approach to do so (see details on the corresponding algorithm in [19]). In hybrid formulations, things take a slightly different expression and shift the emphasis on a related aspect. In these methods, the mean velocity field and the mean pressure-gradient are first calculated using a mesh-based moment approach and are provided to the PDF solver. Given that the GLMs involve a return-to-equilibrium term towards this externally-provided and divergence-free mean velocity field, we can expect that the mean velocity field extracted from the particle instantaneous velocities is then driven, by construction of the hybrid method, towards a divergence-free field and is nearly equal to the mean velocity field coming from the moment solver. In fact, there are always some subtleties induced by the different formulations (such as the different ways with which the velocity triple correlation are handled [15]) as well as details in the numerical implementations (such as how mean fields are interpolated at particle positions) which induce small discrepancies between these two mean velocity fields. Although these numerical specifics are outside the scope of the present work, the relevant outcome of such hybrid formulations is that we end up with a double prediction for the fluid mean velocity field. Actually, this is not limited to the velocity first-order moment but concerns all velocity statistics computed by both methods and includes therefore not only the mean velocity field but also its second-order moments or, conversely, the components of the Reynolds stress tensor. In hybrid moment/PDF formulations, the velocity first- and second-order moments are duplicate fields. This raises immediately the consistency issue that both numerical predictions should be identical since they correspond to the same physical entity. Note that this consistency issue appears therefore as the manifestation at the level of numerical simulations of the theoretical consistency requirement mentioned in Sect. 6.5.

**To Summarize** a mandatory step is to assess that the numerical implementation of a Lagrangian stochastic model preserves a uniform fluid particle distribution in general non-homogeneous flows. When using an hybrid moment/PDF method, the next step is to check that the mean velocity field as well as the Reynolds stress tensor provided by the moment and the PDF solvers are identical. To investigate whether these foundations are solid enough before turning our attention towards disperse two-phase situations, we consider two typical situations: a free shear-flow and a wall-bounded turbulent flow.

### 7.3.1 Free Shear Flows and Mixing Layers

The name free shear flows designates a class of flows in which streams having different velocities are mixed (thus creating shear effects through mean-velocity differences) in unbounded environments (free in the sense of being unconstrained by the presence of bounding walls). The typical examples of such free shear flows are jets, mixing layers and also wakes behind objects. Over the years, they have been studied in considerable details both from an experimental and a theoretical point of view so that their properties are well known (a detailed presentation can be found, for instance, in [15, chapter 5]).

In the following, we consider a plane mixing layer which forms when two uniform and parallel streams of different velocities, which were first separated by a splitter plate, start to mix once they reach the end of this splitter plate. This situation is sketched in Fig. 7.5 In that figure, the dominant direction of flow is $x$, the cross-stream coordinate is $y$ while statistics are independent of the spanwise coordinate $z$.

Compared to other free shear flows, a specific characteristic of the mixing layer is that the two stream velocities, $U_{max}$ and $U_{min}$, are imposed by the configuration

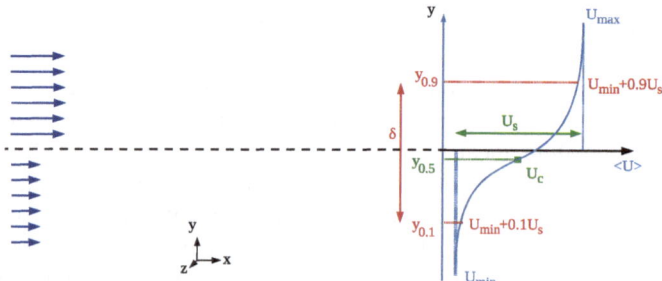

**Fig. 7.5** Sketch of a plane mixing layer indicating the low- and high-speed stream velocities $U_{min}$ and $U_{max}$, respectively; the characteristic convective velocity $U_c$; the velocity difference $U_s$, the width of the flow $\delta(x)$ and the lateral position $y_{0.5}$

and remain constant downstream of the splitter plate as the turbulent flow develops. From the configuration displayed in Fig. 7.5, a number of characteristics can be expressed. This includes the convective velocity, $U_c = 1/2\,(U_{min} + U_{max})$, and the velocity difference $U_s = U_{max} - U_{min}$ which is the main driving force for the development of this turbulent flow. A key variable is the characteristic width of the flow, noted $\delta(x)$, which is defined here as the extent of the region where the fluid velocity ranges from $U_{min} + 0.1\,U_s$ to $U_{min} + 0.9\,U_s$, so that we have

$$\delta(x) = y_{0.9}(x) - y_{0.1}(x) \,. \tag{7.9}$$

It is also interesting to define the characteristic reference lateral (or centerline) position, $y_{0.5}(x) = 1/2\,(y_{0.1}(x) + y_{0.9}(x))$, which can differ from zero since the flow tends to be shifted towards the low-speed stream with an effect more or less marked depending on the ratio $U_s/U_c$. As for other free shear flows, a fundamental property is that the mixing layer is self-similar. With the above definitions, this means that, for example, the scaled mean streamwise velocity $(\langle U_f \rangle - U_c)/U_s$ depends only on the scaled cross-stream coordinate $\xi$ which is

$$\xi = \frac{y - y_{0.5}(x)}{\delta(x)} \,, \tag{7.10}$$

so that $(\langle U_f \rangle - U_c)/U_s = f(\xi)$, with similar relations holding for all velocity statistics. Experimental data have also shown that mixing layers spread linearly and that the spreading parameter S defined as

$$S = \frac{U_c}{U_s}\frac{d\delta(x)}{dx} \tag{7.11}$$

is independent of the velocity ratio and constant, with reported values ranging from $S \simeq 0.06$ to $S \simeq 0.11$. For a given mixing layer, since $U_{max}$ and $U_{min}$ are constant, we obtain indeed that $\delta(x)$ is a linear function of the streamwise coordinate $x$.

For the results presented below, we considered air at normal pressure and temperature and we chose $U_{max} = 3\,\mathrm{m.s}^{-1}$ and $U_{min} = 2\,\mathrm{m.s}^{-1}$. This entails that the ratio $U_s/U_c = 0.4$ is small enough for the mixing layer to remain nearly symmetric even if a small shift towards the low-speed stream is noticeable. This configuration is therefore slightly different from experiments which are often carried out with $U_{min} = 0\,\mathrm{m.s}^{-1}$. The length of the domain was $L_{domain} = 10\,\mathrm{m}$, its height $H_{domain} = 0.8\,\mathrm{m}$ while its width was $l_{domain} = 0.004\,\mathrm{m}$ (the spanwise dimension is irrelevant due to the symmetry in that direction).

Even if comparisons with experimental data are provided in some of the results shown below, it must be remembered that our objective is mainly to discuss the consistency issues set forth at the beginning of Sect. 7.3. For this reason, we selected a simple turbulence model, i.e. the SLM for the PDF approach and its corresponding second-moment closure in the moment approach (which is the Rotta model). In the simulations, we used $C_0 = 3.5$ and the consistent value of the Rotta constant is then $C_R = 1 + 3/2C_0 = 6.25$, cf. Eq. (6.27).

A two-dimensional mesh made up by $100 \times 100$ cells was used (only one cell was considered in the spanwise direction) and calculations were performed with the open-source software code_saturne, using a time step of $\Delta t = 10^{-2}$ s and about 4000 iterations to ensure that the stationary regime is reached. For the particle simulation, since each particle represents the same amount of mass and has therefore the same statistical weight in the estimations, the number of particles injected per time step must correspond to the ratio of the low- and high-speed velocities to respect the inlet mass fluxes. For these computations, 80 fluid particles were injected during each iteration in the low-speed stream and 120 in the high-speed stream. These fluid particles are simulated by assigning a very small diameter to them so that their inertia becomes completely negligible and they act as fluid tracers. The same time step was used in the PDF solver and statistics are extracted from the variables attached to each particle present in a given cell by Monte Carlo estimations. Since we are considering a stationary flow, these Monte Carlo estimations obtained at each time step are further averaged in time (once a stationary number of particles inside the domain is reached) to reduce statistical noise to a very small level.

An iso-color plot of the mean streamwise velocity obtained from the moment approach is shown in Fig. 7.6, where it is seen that the mixing layer starts to develop immediately after the end of the splitter plate and spreads in a near-linear growth with a small shift towards the low-speed stream. The three vertical lines (indicated in green in this figure) are used to plot the rescaled variables using the parameters defined above to assess self-similarity.

When the mean streamwise velocity component is plotted as a function of $\xi$ at various downstream locations, it is clear from Fig. 7.7a that all the results collapse on a single profile, showing that self-similarity is satisfied. The numerical prediction of the value of the spreading parameter is $S = 0.068$, thus within the range of values measured experimentally.

Having validated the mean fields provided to the PDF solver, we analyze the consistency requirements set forth in the introduction of Sect. 7.3, starting with the particle concentration field. Two profiles along the crosswise direction of this particle concentration taken at two downstream locations are shown in Fig. 7.7b. It is obvious that, with a proper account of the mean pressure-gradient calculated from the fluid continuity equation, we obtain a near-perfect respect of the uniformity

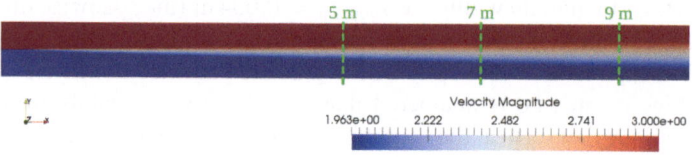

**Fig. 7.6** Snapshot of the fluid velocity obtained from the fluid-flow simulation inside the domain showing the formation and growth of the mixing layer (the vertical dashed lines indicated at $x = 5$, 7 and 9 m are used to plot data to assess self-similarity)

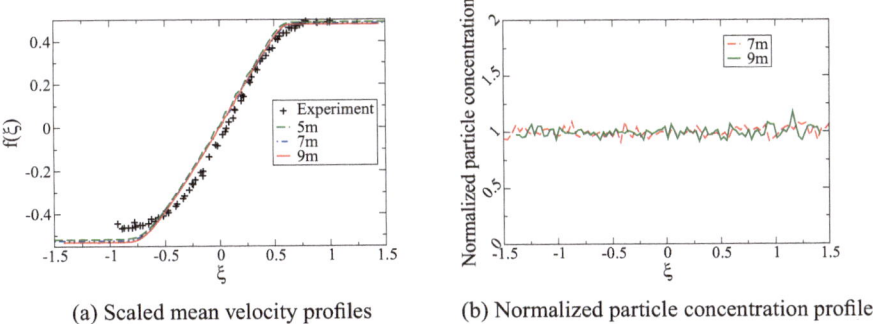

(a) Scaled mean velocity profiles          (b) Normalized particle concentration profiles

**Fig. 7.7** Scaled mean velocity profiles at various distances from the inlet (lines) compared to experimental data [21,22] (symbols) (**a**); Concentration profiles at two downstream distances from the inlet showing that a uniform concentration is maintained in the domain (**b**). Reprinted with permission from [18]. © 2019 Springer Nature B.V. All rights reserved

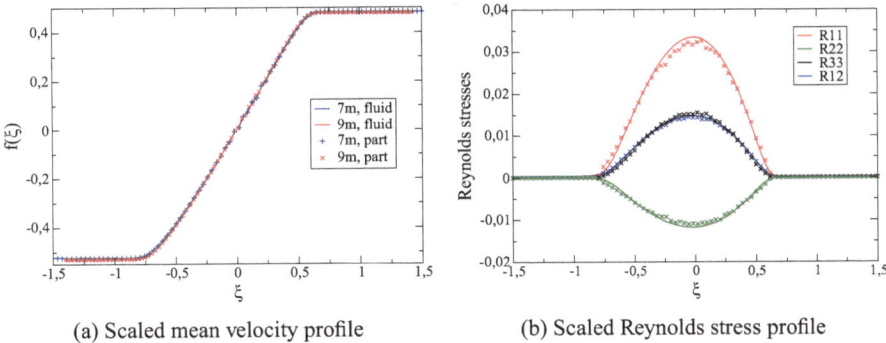

(a) Scaled mean velocity profile          (b) Scaled Reynolds stress profile

**Fig. 7.8** Profiles of the scaled mean velocity (**a**) and of the normalized non-zero components of the Reynolds stress tensor $R_{f,ij}/U_s^2$ (**b**) showing consistency between the numerical results from the moment approach (lines) and ones from the stochastic particle or PDF approach (symbols). Several results sampled at various distances from the inlet are plotted for the scaled mean velocity to indicate self-similarity but are not shown for the Reynolds stress profiles in (**b**) to bring out the comparison between the predictions of the moment and PDF approaches. Reprinted with permission from [18]. © 2019 Springer Nature B.V. All rights reserved

constraint as we go from the low-speed to the high-speed nearly laminar streams throughout the turbulent region (located approximately between $-0.7 \leq \xi \leq 0.7$).

The second step in the consistency analysis consists in collating the first- and second-order velocity moments coming from the moment and the PDF approaches. This is presented in Fig. 7.8, where it is seen that excellent agreement is obtained between the predictions from the two formulations. Actually, these results do not have the same status. Indeed, the mean velocity from the moment solver is provided to the PDF solver and is used directly in the return-to-equilibrium term of the SLM, cf. Eq. (6.20) (where $\mathbf{G}^a = 0$) so that the good agreement appearing in Fig. 7.8a was perhaps to be expected. On the other hand, the values of the Reynolds stress tensor

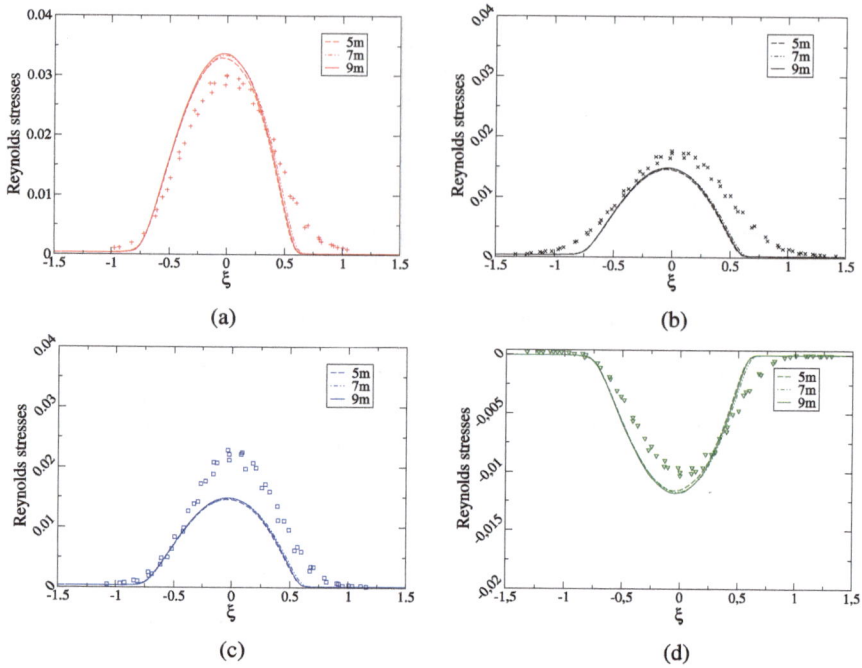

**Fig. 7.9** Profiles of the scaled Reynolds stresses $R_{f,ij}/U_s^2$: comparison between experimental data [21, 22] (symbols) and numerical results sampled at various distances from the inlet (lines). Reprinted with permission from [18]. © 2019 Springer Nature B.V. All rights reserved. (**a**) $R_{f,11} = R_{f,xx}$. (**b**) $R_{f,22} = R_{f,yy}$. (**c**) $R_{f,33} = R_{f,zz}$. (**d**) $R_{f,12} = R_{f,xy}$

are not used explicitly in the expression of the SLM (apart from its trace $k_f$ which appears in the timescale $T_L$, cf. Eqs. (6.18) and (6.33)), so that the very good accord displayed in Fig. 7.8b is a significant result.

Even though this was not our main objective, we can give a better feeling of where we stand by comparing numerical results to experimental data [21, 22]. For the mean streamwise velocity, such a comparison was already given in Fig. 7.7a, showing that numerical results are in line with the experimental data. The characteristic features of the turbulence models (the GLMs) appear essentially at the level of the second-order velocity moments and we must keep in mind that we have used the simplest candidate (the SLM). Nevertheless, it is seen in Fig. 7.9 that a reasonable, and sometimes satisfactory, agreement is obtained. Note that a property of the SLM in the boundary-layer approximation is that the kinetic energies in the cross-stream and spanwise direction ($R_{f,33}$ and $R_{f,22}$, respectively) are equal whereas there is a marked difference in the experiments. For more accurate predictions, other models than the SLM would probably do a better job.

Finally, we present in Fig. 7.10 a comparison between numerical results and experimental data for the skewness and flatness factors of the streamwise and crosswise velocity components. The numerical results were obtained with a different

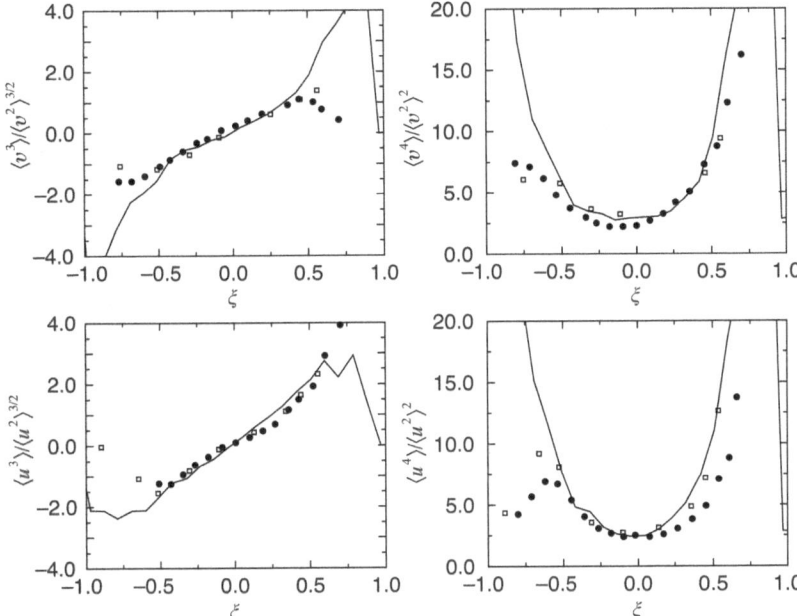

**Fig. 7.10** Skewness and flatness profiles of the axial (u) and lateral (v) velocities in the self-similar plane mixing layer. Lines, calculations by Minier and Pozorski [23] based on the lognormal/refined Langevin model of Pope [24]; symbols, experimental data of Wygnanski and Fiedler (filled circle) [22] and of Champagne et al. (open square) [21]. Reprinted with permission from [23]. © 1996 EDF. All rights reserved

model [23], called the refined Langevin model (already mentioned in Sect. 6.5), but still in the same spirit as the present SLM (see [24]). Not only are the numerical results in very good agreement with the experimental data but this comparison is also useful for two purposes. First, it is worth recalling that a PDF approach contains more information than a second-moment closure since we have immediately access to the infinite set of one-point velocity moments, represented in Fig. 7.10 by the third- and fourth-order velocity moments, respectively. A second, and more important, purpose is to illustrate the physical interpretation given in Sect. 4.3.2 about the Gaussian and non-Gaussian nature of present stochastic diffusion models such as the GLMs. Since for a Gaussian random variable the skewness and flatness factors are 0 and 3, respectively, it is clear from Fig. 7.10 that the model, although based on a Gaussian assumption for the conditional increments, is quite capable of capturing marked deviations from Gaussianity. This is manifest in the non-homogeneous parts of turbulent core of the mixing layer (roughly speaking, for $0.1 \leq |\xi| \leq 0.5$) while the model retrieves a near-Gaussian behavior at the center (around $\xi \simeq 0$) where the flow is locally homogeneous.

## 7.3.2    Wall-bounded flows and Wall-function Boundary Conditions

Another category of major importance is constituted by turbulent flows bounded by solid surfaces. The no-slip condition at a solid surface leads to the formation of a thin region near the bounding walls, called the boundary layer, characterized by a sharp decrease of turbulent quantities to zero at the wall and by strong anisotropies. In the situation of fully developed turbulent flows, the mean velocity is parallel to the wall and homogeneous in the streamwise direction. The search is then to obtain the profiles of the mean velocity and other statistics with respect to the wall-normal coordinate through universal functions written in terms of non-dimensional variables. To that end, a central issue is to determine the shear stress exerted by the fluid on the wall, which is expressed as $\tau_{\text{wall}} = \rho_f u_*^2$. This introduces the important notion of the friction velocity $u_*$ which, along with the fluid viscosity $\nu_f$, are used to work out the expression of what is referred to as the law of the wall.

Although detailed presentations (see [15, chapter 7]) are needed to provide a more precise description, we can divide a boundary layer into, at least, two zones. In the immediate vicinity of the wall, viscous effects dominate in the so-called viscous sub-layer while, further away, turbulent motions take over in what is called the log-law region since the streamwise mean velocity follows a logarithmic profile in this region. This has direct impact on turbulence models which have to be complemented with appropriate boundary conditions. From the above account, it seems natural to apply the no-slip condition at the boundary (meaning that there is no relative velocity between the fluid and the surface boundaries). However, this also implies to extend the turbulence models with a formulation accounting for the effect of viscous terms occurring in the viscous sub-layer. This puts a heavy burden on the numerical implementation at high Reynolds-number flows. In fact, we need to capture the stiff variations of the velocity moments in this exceedingly thin region, whose width diminishes when the Reynolds number increases. Another possibility is to apply the boundary conditions at an interface located within the turbulent part of the boundary layer, that is in the log-law region in which the shear stress in approximately constant. The interest of this second option is that we can then retain the same high-Reynolds form of the turbulence models under consideration. On the other hand, we must now develop what seems an artificial condition at a fluid interface. This is referred to as the wall-function treatment [15]. When doing so, we give up the idea to describe the details of the turbulent boundary layer and only aim to bridge it while still providing the correct flux of momentum which integrates the local exchanges between the flow and the wall to account for the net flux towards the bulk of the flow.

For PDF models, it appears that only a few works have studied how to describe turbulent boundary layers [25]. The down-to-the-wall formulation with an extended form of the Langevin model for fluid particle velocities was developed in [26, 27] while another proposal based on a different expression of the viscous terms entering the stochastic model was proposed later [28]. The wall-function approach was first put forward in [29] and soon after in [19]. It was revisited in [18] with a view towards discussing the incompressibility constraint along the lines developed in the introduction of Sect. 7.3 and, in an updated and comprehensive work focused on atmospheric surface boundary layers, in [13]. In this section, we also follow the wall-function treatment and take up the same concern to assess the incompressibility and consistency issues. In the course of the discussion, we rely on a few numerical results obtained in these two studies [13, 18] but also propose a new approach for the formulation of the wall-function type of boundary conditions. This new description is meant to complement the above-mentioned developments by suggesting a more physics-based formulation in relation with the theoretical considerations put forward in previous chapters.

### 7.3.2.1 Macroscopic Balance for Longitudinal Momentum Exchange

Insights into the formulation of the wall-function boundary conditions for fluid particles can be obtained by expressing the momentum exchange between two adjacent volumes. To that effect, we consider two fluid volumes located one above the other in the vertical direction, see Fig. 7.11, and analyze the mean longitudinal momentum exchange between these two fluid volumes through their interface at $z = z_{pl}$. This momentum exchange is due to instantaneous vertical fluid motions which carry different longitudinal velocities. Since the interface marks the limit of the domain, vertical fluid motions starting from the top volume and crossing that interface are leaving the domain. For that reason, the top volume (or cell) is referred to as the 'out-cell' while the fluid volume below the interface, from which vertical fluid motions crossing the interface are entering the domain, is labeled as the 'in-cell'.

At the continuum level, the mean longitudinal momentum exchange experienced by the out-cell due to actions of the in-cell stems from stresses working at the interface $z = z_{pl}$. These mean stresses include the mean viscous and pressure terms and the Reynolds stresses $R_{f,13} = \langle u_f w_f \rangle$. While pressure and viscous forces arise from short-range interactions between molecules on either side of the interface (in the case of pressure, this is essentially molecular transport and collision in a classical kinetic theory picture), the Reynolds stresses represent momentum fluxes induced by the fluctuating velocity field. As indicated by their name, they are often interpreted as stresses, though they basically arise from convective transport [15, 17].

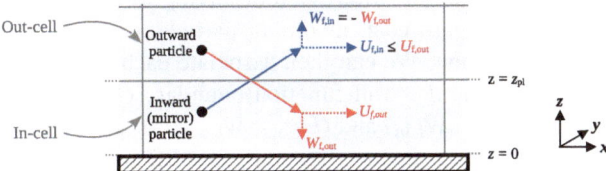

**Fig. 7.11** The wall-function boundary conditions are effective at an interface at a distance $z_{pl}$ from the wall which divides the out-cell (above $z_{pl}$) from the in-cell (below $z_{pl}$). Particles leaving the out-cell through the interface are replaced by in-coming ones from the in-cell whose instantaneous velocities have to be determined

In the present context, this suggests that these surface forces are best regarded as momentum fluxes. In the logarithmic region of turbulent boundary layers, viscous effects are negligible and the mean pressure is approximately constant in the vertical direction leaving no net effect at the interface $z = z_{pl}$, while the Reynolds shear stress is given by $R_{f,13} = -u_*^2$. This means that the net longitudinal momentum flux going into the out-cell due to the actions of the in-cell through the mean shear stress at the interface $z = z_{pl}$ is $-\rho_f u_*^2$ (note that the surface unit vector is pointing inward the in-cell).

The concept of stresses as momentum fluxes is naturally embodied in the PDF approach due to its particle formulation. Indeed, in the Monte Carlo formulation of the PDF description of turbulent flows, fluid statistics at a given location are extracted from the set of particles located in a small volume around that location. If each particle represents an identical amount of mass $\Delta m_f$ and the number of particles per unit volume is $N$ (m$^{-3}$), the fluid density is thus equal to $\rho_f = N \, \Delta m_f$, showing that $N$ must be constant for constant-density flows. The longitudinal momentum exchange between the two fluid volumes is due to the difference between the longitudinal velocities of incoming and outgoing particles in one of these two fluid volumes. Noting $N_{in}$ the number of particles entering the out-cell from the in-cell through the interface and $N_{out}$ the number of particles leaving the out-cell through the same interface ($N_{in}$ and $N_{out}$ stand for particle number fluxes and have units in m$^{-2}$.s$^{-1}$), the resulting gain of longitudinal momentum $\Delta Q_{in \to out}$ for the out-cell writes

$$\Delta Q_{in \to out} = \sum_{l=1}^{N_{in}} \Delta m_f \, U_{f,in}^{[l]} - \sum_{m=1}^{N_{out}} \Delta m_f \, U_{f,out}^{[m]}, \tag{7.12}$$

where the suffixes $[l]$ and $[m]$ are particle labels.

From the description of fluid vertical fluctuating motions outlined above, a natural boundary condition for $W_f$ is a purely reflective one at $z_{pl}$. It ensures that there is no net mass flux through the interface at $z_{pl}$ and a zero mean vertical velocity (i.e., $\langle W_f \rangle = 0$) in line with fluid statistics in a fully-developed channel flow and, in particular, with the divergence-free condition. A direct consequence of this boundary condition is that the particle number fluxes through the interface at $z_{pl}$ are identical, i.e., that $N_{in} = N_{out} = N_b$ with $N_b$ the number of particles crossing the interface in one direction without having to specify which one. This also means that we can regard each incoming particle as the mirror, or reflected, particle of each outgoing one. We can then associate each incoming particle to an outgoing one and formulate the wall-function boundary conditions by developing the relations between ($U_{f,in}$, $W_{f,in}$) and ($U_{f,out}$, $W_{f,out}$), as sketched in Fig. 7.11. For the wall-normal velocity, the reflective boundary condition is simply expressed by

$$W_{f,in} = -W_{f,out}, \tag{7.13}$$

while we can express the gain of longitudinal mean momentum of the out-cell as

$$\Delta Q_{\text{in}\rightarrow\text{out}} = \sum_{l=1}^{N_b} \Delta m_{\text{f}} \left( U_{\text{f,in}}^{[l]} - U_{\text{f,out}}^{[l]} \right) , \qquad (7.14)$$

which gives

$$\Delta Q_{\text{in}\rightarrow\text{out}} = N_b \, \Delta m_{\text{f}} \left\{ \frac{1}{N_b} \sum_{l=1}^{N_b} \left( U_{\text{f,in}}^{[l]} - U_{\text{f,out}}^{[l]} \right) \right\} \simeq N_b \, \Delta m_{\text{f}} \left\langle \left( U_{\text{f,in}}^{[1]} - U_{\text{f,out}}^{[1]} \right) \right\rangle .$$
$$(7.15)$$

Since particles are statistically identical, we can choose any one of them on the right-hand side term of Eq. (7.15), for instance $l = 1$, but, for the sake of simplicity, this suffix is omitted from now on. The issue is then to work out a closed relation for the longitudinal velocity difference for each pair of outgoing and incoming particles, i.e. $U_{\text{f,in}} - U_{\text{f,out}}$, to derive the resulting mean longitudinal momentum flux through the interface located at $z_{\text{pl}}$.

*Linear Force-flux Relations*  In a manner consistent with the modeling steps presented in Sect. 6.2.1 leading to the formulation of GLMs, we can rely on the linear response theory and the notion of force-flux relations. From the description of equilibrium turbulent wall-boundary layers, it appears that vertical fluctuating motions are the driving phenomenon for macroscopic transport and mixing. Using the terminology introduced in Sect. 6.2.1, we can then consider that $X = W_{\text{f}}$ and that $F(W_{\text{f}}) = -W_{\text{f}}/\langle (W_{\text{f}})^2 \rangle$ is the 'phenomenological force' $\mathbf{F(X)}$.

In the spirit of linear irreversible thermodynamics and force-flux relations [30, 31], the next step consists in proposing a linear relation between the longitudinal velocity difference of each reflected particle (involved in the resulting momentum flux, cf. Eq. (7.15) above) and their vertical velocity (the generalized force), which writes

$$U_{\text{f,in}} - U_{\text{f,out}} = 2C \frac{W_{\text{f,out}}}{\langle (W_{\text{f}})^2 \rangle} , \qquad (7.16)$$

where $C$ is a coefficient, or a function of local mean fluid properties, to be determined but which does not depend on the particle instantaneous velocities (the factor 2 is used for convenience for later manipulations, as shown below).

*Wall-boundary Conditions for a Symmetrical Wall-normal Velocity PDF*  Based on the relation in Eq. (7.16), we can now work out the complete expression of the longitudinal mean momentum exchange. Indeed, inserting Eq. (7.16) in Eq. (7.15), we have

$$\Delta Q_{\text{in}\rightarrow\text{out}} = N_b \, \Delta m_{\text{f}} \frac{2C}{\langle (W_{\text{f}})^2 \rangle} \langle W_{\text{f,out}} \rangle . \qquad (7.17)$$

The particle number flux $N_b$ is obtained from the distribution function $f_f = N\,p_f$ as

$$N_b = \int_0^{+\infty} W_f\, f_f(t, z_{pl}; W_f)\, dW_f = N \int_0^{+\infty} W_f\, p_f(t, z_{pl}; W_f)\, dW_f\,, \qquad (7.18)$$

while $\langle W_{f,out}\rangle$, which is an example of surface statistics to be discussed below, writes

$$\langle W_{f,out}\rangle = -\frac{1}{N_b} \int_0^{+\infty} (W_f)^2\, f_f(t, z_{pl}; W_f)\, dW_f$$

$$= -\frac{N}{N_b} \int_0^{+\infty} (W_f)^2\, p_f(t, z_{pl}; W_f)\, dW_f\,. \qquad (7.19)$$

This yields for the mean longitudinal momentum exchange term $\Delta Q_{in \to out}$

$$\Delta Q_{in \to out} = -N \Delta m_f\, \frac{2C}{\langle (W_f)^2 \rangle} \int_0^{+\infty} (W_f)^2\, p_f(t, z_{pl}; W_f)\, dW_f\,. \qquad (7.20)$$

With the purely reflective boundary condition applied for $W_f$ at $z_{pl}$, see Eq. (7.13), it follows that the PDF for fluid particle wall-normal velocities is symmetrical and we have then

$$\int_0^{+\infty} (W_f)^2\, p_f(t, z_{pl}; W_f)\, dW_f = \frac{1}{2} \int_{-\infty}^{+\infty} (W_f)^2\, p_f(t, z_{pl}; W_f)\, dW_f$$

$$= \frac{1}{2} \langle (W_f)^2 \rangle (z_{pl})\,, \qquad (7.21)$$

showing that

$$\Delta Q_{in \to out} = -N \Delta m_f\, C = -\rho_f\, C\,. \qquad (7.22)$$

As indicated above, the theoretical value of this longitudinal mean momentum flux is $-\rho_f\, u_*^2$, proving that the coefficient $C$ in Eq. (7.16) is $C = u_*^2$. Note that we have only used the symmetry of the wall-normal velocity PDF without having to assume that this PDF is Gaussian.

The complete form of the boundary conditions for fluid particles in a wall-function approach of equilibrium turbulent boundary layers can be summed up as follows:

(a) Each fluid particle crossing the interface at $z_{pl}$ is reflected back into the domain;
(b) The wall-normal position, longitudinal and wall-normal velocities ($Z_f$, $U_f$, $W_f$) of each incoming particle are deduced from the values of each outgoing one

through the expressions

$$Z_{f,in} = 2\,z_{pl} - Z_{f,out} \, , \tag{7.23a}$$

$$W_{f,in} = -W_{f,out} \, , \tag{7.23b}$$

$$U_{f,in} = U_{f,out} + \frac{2\,u_*^2}{\langle (W_f)^2 \rangle}\,W_{f,out} \, , \tag{7.23c}$$

with $\langle (W_f)^2 \rangle$ the value sampled at $z_{pl}$ by averaging over all particles present locally (regardless of whether they cross the interface or not). Other particle variables are left unchanged.

*Volume and Surface Statistics*   More light can be shed on the wall-function boundary conditions in Eq. (7.23) by manipulating the shear-stress $\langle U_f\,W_f \rangle$. This requires however to distinguish carefully between what are volumetric or surface statistics in order to obtain consistent results.

To derive explicit relations and illustrate our argument, we consider here that the PDF of the fluid particle wall-normal velocity is a Gaussian of zero mean and with a variance $\langle (W_f)^2 \rangle_{out} = \langle (W_f)^2 \rangle(z_{pl})$. From Eqs. (7.18)–(7.19), the particle number flux is then (using $N_{in} = N_{out} = N_b$)

$$N_b = N\,\frac{1}{\sqrt{2\,\pi}}\,\langle (W_f)^2 \rangle_{out}^{1/2} \, , \tag{7.24}$$

while the expression for $\langle W_{f,out} \rangle$ is

$$\langle W_{f,out} \rangle = -\frac{1}{2}\sqrt{2\,\pi}\,\langle (W_f)^2 \rangle_{out}^{1/2} \, . \tag{7.25}$$

The notation used for $\langle (W_f)^2 \rangle_{out}$ and $\langle W_{f,out} \rangle$ is helpful to reveal a significant difference in the nature of the statistics involved. In the Monte Carlo expression of the PDF description, $\langle (W_f)^2 \rangle_{out}$ is an example of statistics evaluated from particles located in a small volume around $z_{pl}$ or at the center of the out-cell, and can be referred to as a volumetric statistic. On the other hand, $\langle W_{f,out} \rangle$ is an example of statistics evaluated from particles crossing the interface at $z_{pl}$ from the out-cell in a unit time interval, and can be referred to as a surface statistic. These statistics are derived from the same set of particles but conditioned on different events and are therefore not to be confused (for instance, we have $\langle W_f \rangle_{out} = 0$ while $\langle W_{f,out} \rangle < 0$).

Note then that a quantity such as $\langle U_f\,W_f \rangle(z_{pl})$ is obtained from particles located near the interface at $z_{pl}$. This is different from the same correlation but conditioned on the subset of particles crossing the interface in a unit time. We denote the former as $\phi^{vol}$ (with $\phi^{vol} = \langle U_f\,W_f \rangle(z_{int})$) while the latter is written as $\phi^{surf}$. This surface statistics is due to the contribution of incoming and outgoing particles and writes

$$\phi^{surf} = \frac{1}{2} \left[ \langle U_{f,in} \, W_{f,in} \rangle + \langle U_{f,out} \, W_{f,out} \rangle \right] , \qquad (7.26)$$

where the factor $1/2$ comes from the relation $N_{in} = N_{out}$, meaning that half of the particles crossing the interface are incoming ones and the other half is made up by outgoing ones.

By decomposing the velocities of incoming and outgoing particles with respect to local mean values (or volumetric averages), noted $\langle U_f \rangle (z_{f,in})$ and $\langle U_f \rangle (z_{f,out})$, respectively, and associated to the in- and out-cells in a manner to be precised below, we have

$$U_{f,in} = u_{f,in} + \langle U_f \rangle (z_{f,in}) , \qquad (7.27a)$$

$$U_{f,out} = u_{f,out} + \langle U_f \rangle (z_{f,out}) . \qquad (7.27b)$$

Note that vertical velocities are left unchanged since $\langle W_f \rangle = 0$. This gives

$$\phi^{surf} = \frac{1}{2} \left[ \langle u_{f,in} \, W_{f,in} \rangle + \langle u_{f,out} \, W_{f,out} \rangle + \langle W_{f,in} \rangle \left\{ \langle U_f \rangle (z_{f,in}) - \langle U_f \rangle (z_{f,out}) \right\} \right] . \qquad (7.28)$$

By considering the fluxes of incoming and outgoing particles in a small-enough time increment, the two points $z_{f,in}$ and $z_{f,out}$ at which the mean values $\langle U_f \rangle (z_{f,in})$ and $\langle U_f \rangle (z_{f,out})$ are taken can be assumed to be in the immediate vicinity of $z_{pl}$. The continuity of the mean longitudinal velocity field $\langle U_f \rangle (\mathbf{x})$ implies then that $\langle U_f \rangle (z_{f,in}) \simeq \langle U_f \rangle (z_{f,out}) \simeq \langle U_f \rangle (z_{pl})$. The expression of the surface statistic $\phi^{surf}$ in Eq. (7.28) is thus reduced to

$$\phi^{surf} = \frac{1}{2} \left[ \langle u_{f,in} \, W_{f,in} \rangle + \langle u_{f,out} \, W_{f,out} \rangle \right] . \qquad (7.29)$$

If we consider the equilibrium statistical laws for the in- and out-cells within the constant-stress region, we know that $\langle u_f \, W_f \rangle (z_{pl}) = \langle u_f \, W_f \rangle (z_{f,out}) = -u_*^2$. However, the two correlations entering the right-hand side of Eq. (7.29) are surface statistics. By assuming a Gaussian form for the joint PDF of $(U_f, W_f)$, we can derive that

$$\langle u_{f,in} \, W_{f,in} \rangle = 2 \langle u_f \, W_f \rangle (z_{f,in}) = -2 \, u_*^2 , \qquad (7.30a)$$

$$\langle u_{f,out} \, W_{f,out} \rangle = 2 \langle u_f \, W_f \rangle (z_{f,out}) = -2 \, u_*^2 , \qquad (7.30b)$$

which yields $\phi^{surf} = -2 \, u_*^2$ , whereas a naive estimation would give $\phi^{surf} = -u_*^2$ .

On the other hand, by introducing the boundary conditions formulated in Eq. (7.23) into the definition of $\phi^{surf}$ in Eq. (7.26), we obtain the alternative estimation

$$\widetilde{\phi}^{surf} = -\frac{1}{2} \frac{2C}{\langle (W_f)^2 \rangle} \langle W_{f,out}^2 \rangle , \qquad (7.31)$$

which, using the result for the surface statistic $\langle W_{f,out}^2 \rangle = 2 \langle (W_f)^2 \rangle$ gives that $\widetilde{\phi}^{surf} = -2C$. Equating the two estimations of $\phi^{surf}$ retrieves the correct expression that $C = u_*^2$. The important outcome of these developments is that the correct result is only obtained provided that a proper distinction between volumetric and surface statistics is made.

*General Formulation and Equilibrium Second-order Moments* For fluid particle velocities, the boundary conditions can be put into the general form

$$\mathbb{U}_{f,in} = \mathbb{U}_{f,out} - \frac{2 \langle W_f \mathbb{U}_f \rangle}{\langle (W_f)^2 \rangle} W_{f,out} \qquad (7.32)$$

with $\mathbb{U}_f$ any component of the fluid particle velocity $(U_f, V_f, W_f)$, as can be checked by comparing to the complete formulation in Eqs. (7.23b)–(7.23c) for $\mathbb{U}_f = U_f$ and $\mathbb{U}_f = W_f$. In the transverse direction, particle velocity are left unchanged since $\langle V_f W_f \rangle = 0$, which is to be expected because this direction is unaffected by the wall-function boundary conditions and other boundary conditions, such as specular reflection, are due to symmetry arguments.

The interest of the compact form in Eq. (7.32) is to bring out that there are no special requirements on the mean second-order values entering the boundary conditions. Indeed, the value of $\langle (W_f)^2 \rangle$ should be the local one, $\langle (W_f)^2 \rangle (z_{pl})$, corresponding to the wall-normal kinetic energy of particles near the interface (so that, for instance, we have that $\langle W_{f,out}^2 \rangle = 2 \langle (W_f)^2 \rangle (z_{pl})$ as discussed earlier). The only requirement is that the shear-stress $\langle U_f W_f \rangle$ be equal to the theoretical value, i.e. that we have $\langle U_f W_f \rangle = -u_*^2$. Since this is the key physical variable to retrieve known results of the logarithmic region, it appears therefore as the minimal requirement to put on model performance. In other words, the wall-function boundary conditions in Eq. (7.23) are applicable whatever the stochastic model retained to describe the dynamics of particles.

To appreciate this point, it is worth emphasizing that, in contrast, second-order velocity moments in the logarithmic region depend upon the characteristics of each stochastic model. As an illustration, we consider the reference model which is the GLM (for the sake of simplified manipulations, we switch here to tensor notations for particle positions and velocities, which means that $U_{f,1} = U_f$, $U_{f,2} = V_f$ and $U_{f,3} = W_f$ and the same for positions), which writes

$$dX_{f,i} = U_{f,i}\, dt \,, \tag{7.33a}$$

$$dU_{f,i} = -\frac{1}{\rho_f}\frac{\partial\langle P_f\rangle}{\partial x_i}\, dt + G_{ij}\left(U_{f,j} - \langle U_{f,j}\rangle\right) dt + \sqrt{C_0\,\langle \epsilon_f\rangle}\, dW_i \,, \tag{7.33b}$$

where the matrix $\mathbf{G}$ is subject to the constraint (cf. Sect. 6.2.2)

$$\left(1 + \frac{3}{2}C_0\right)\langle \epsilon_f\rangle + G_{ij}\, R_{f,ij} = 0 \,. \tag{7.34}$$

Following the method introduced in earlier studies [19], particle fluctuating velocities, $u_{f,i} = U_{f,i} - \langle U_{f,i}\rangle$, are regarded as stationary processes within the logarithmic region of turbulent boundary layers. By considering the corresponding stochastic equations for these fluctuating velocities [2, 15, 17], this leads directly to the following matrix equation

$$\mathcal{P}_{R_f} + \mathbf{G}\mathbf{R}_f + \mathbf{R}_f\mathbf{G}^{\perp} + C_0\,\langle \epsilon_f\rangle\,\mathbb{1} = 0 \tag{7.35}$$

where $\mathbb{1}$ is the identity matrix and $\mathcal{P}_{R_f} = (\mathcal{P}_{R_{f,ij}})$ the production tensor given in Eq. (6.24). In the boundary-layer approximation, the only non-zero mean velocity gradient is $\partial\langle U_{f,1}\rangle/\partial x_3$ and the production tensor reduces to

$$\mathcal{P}_{R_{f,ij}} = -\langle u_{f,3}\, u_{f,j}\rangle\frac{\partial\langle U_{f,1}\rangle}{\partial x_3}\delta_{i1} - \langle u_{f,3}\, u_{f,i}\rangle\frac{\partial\langle U_{f,1}\rangle}{\partial x_3}\delta_{j1} \,, \tag{7.36}$$

which leaves only two non-zero components, which are

$$\mathcal{P}_{R_{f,11}} = -2\langle u_{f,1}\, u_{f,3}\rangle\frac{\partial\langle U_{f,1}\rangle}{\partial x_3} \,, \quad \mathcal{P}_{R_{f,13}} = -\langle (u_{f,3})^2\rangle\frac{\partial\langle U_{f,1}\rangle}{\partial x_3} \,. \tag{7.37}$$

Using these results together with known statistics, such as $\mathcal{P}_{R_{f,11}} = 2\langle \epsilon_f\rangle$ and $\langle u_{f,1}\, u_{f,3}\rangle = -u_*^2$, the Reynolds tensor $\mathbf{R}_f$ can be obtained from Eq. (7.35). It is thus clear that the non-zero values and, especially, the normal stresses $\langle (u_{f,i})^2\rangle$ depend on the closure selected for the matrix $\mathbf{G}$. For example, if we retain the SLM in which the matrix $\mathbf{G}$ is isotropic and given by Eq. (6.18), the normal stresses are easily worked out and are [19, 29]

$$\langle u_f\, w_f\rangle_{\log} = -u_*^2 \,, \quad \langle (u_f)^2\rangle_{\log} = u_*^2\frac{C_0 + 2}{\sqrt{C_0}} \,, \quad \langle (v_f)^2\rangle_{\log} = \langle (w_f)^2\rangle_{\log} = u_*^2\sqrt{C_0} \,. \tag{7.38}$$

The conclusion of these considerations is that, for each model, potentially different values of the resulting Reynolds tensor $\mathbf{R}_f$ are to be expected, apart from the shear-stress which must always be equal to $-u_*^2$. Nevertheless, regardless of the choice of different stochastic models corresponding to different matrices $\mathbf{G}$ in Eq. (7.33b), the same form of the wall-function boundary conditions in Eq. (7.23) still apply.

To bring out the significance, first, of the mean pressure gradient in relation with particle concentration and, second, of the present wall-boundary conditions to achieve consistency for the first- and second-order velocity moments, we consider in the following well-established turbulent flow cases: one in a channel (see Sect. 7.3.2.2) and one in a surface boundary layer (see Sect. 7.3.2.3).

### 7.3.2.2 Application to a Channel Flow

The role of the mean pressure gradient is best analyzed in the case of an infinitely long channel flow which, by reason of symmetry at the channel half-height, can be reduced to a one-dimensional flow where variables depend only on the wall-normal coordinate $z$. In this configuration, the boundary-layer approximation gives for the wall-normal mean momentum equation

$$\frac{1}{\rho_f}\frac{d\langle P_f\rangle}{dz} + \frac{d\langle (w_f)^2\rangle}{dz} = 0 , \qquad (7.39)$$

which corresponds to a Bernoulli-like result since $\langle P_f\rangle + \rho_f\langle (w_f)^2\rangle = C_{P_f}$ where $C_{P_f}$ is a constant (we used normal derivative since functions depend only on $z$).

Due to the symmetry condition at the channel half-height, the flow is non-homogeneous in the vertical, or wall-normal, direction so that both $\langle (w_f)^2\rangle$ and $\langle P_f\rangle$ are space dependent. The impact of different values assigned to the mean pressure gradient onto the particle concentration profile was addressed in detail in [18] whose results are used for the present discussion. When the mean pressure gradient is such that Eq. (7.39) is satisfied, a uniform particle concentration is indeed obtained, as can be seen in Fig. 7.12a.

Further support is given in Fig. 7.12a since it is clear that the correct uniform profile is maintained in time as the simulation proceeds over a time lapse representing several times the typical timescales [18]. On the contrary, when the mean pressure gradient is poorly estimated, here represented by a test case where it was put to zero, the particle concentration profile shown in Fig. 7.12b reveals that a marked deviation from uniformity builds up. In that case, particles are driven away from the near-wall zone and tend to concentrate near the center half-height (with an error of nearly 50%). This simply reflects that $\langle (w_f)^2\rangle$ is stronger at the wall and decreases monotonically as we move towards the center of the channel, thereby inducing a net particle flux away from the wall (the region of high-intense wall-normal kinetic energy) towards the center (the region of low-intense wall-normal kinetic energy). This is a basic example of a force-flux relation and we have, in fact, created a turbophoresis effect. This is, however, a spurious one, leading to what is then called particle spurious drifts. What is missing is to account for the mean pressure-gradient whose role, as manifested by the equilibrium relation in Eq. (7.39), is precisely to

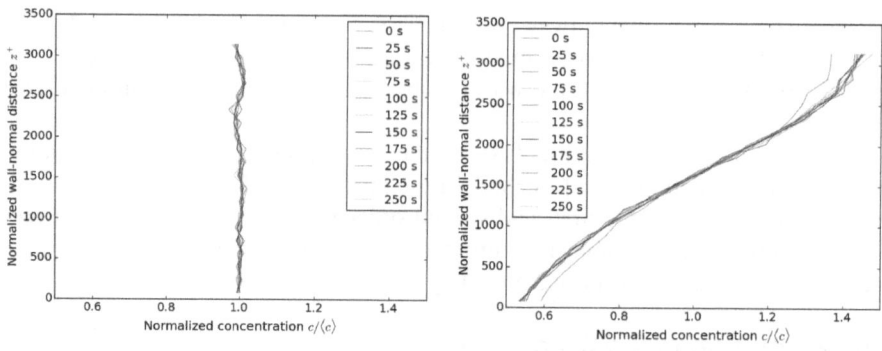

(a) Particle concentration with $\nabla\langle P_f\rangle$        (b) Particle concentration without $\nabla\langle P_f\rangle$

**Fig. 7.12** Evolution of the particle concentration as a function of time: when the mean pressure gradient is included in the particle velocity model maintaining a uniform distribution (**a**) and when the mean pressure gradient is not included leading to a turbophoresis effect (**b**). In both figures, the wall-normal coordinate is normalized by the wall units, $z^+ = z\,u_*/\nu_f$, so that its value at the channel half-height corresponds to the Reynolds number which is 3330. Reprinted with permission from [18]. © 2019 Springer Nature B.V. All rights reserved

counter-balance such 'turbophoresis motions' and ensure that mass conservation is respected. This is not the case for inertial particles for which the pressure-gradient term in the extended particle momentum equation, Eq. (3.23), becomes very small as soon as the density ratio $\rho_p/\rho_f$ increases. There is no pressure-like effects reacting to a discrete-particle concentration buildup and discrete particles can accumulate in some regions (up to a limit when back effects on the fluid or direct collisions start to play a role) or avoid some others, making the particle volume fraction variable in time and space. As discussed elsewhere [11, section 2.3], the phenomenon of turbo-phoresis for discrete particles is already accounted for in present formulations through the actions of the velocity of the fluid seen $\mathbf{U}_s$ and does not require extra terms to be added to the particle momentum equation.

### 7.3.2.3 Application to Surface Boundary Layers

As shown in Fig. 7.13a, we now consider a surface boundary layer which is similar to the case of the half-height infinitely-long channel flow but with a different boundary condition applied at the top (i.e. opposite of the wall). Contrary to the channel flow where a zero-stress condition was used, we now apply a constant shear stress at the top location. Although simplified, this configuration is nevertheless representative of neutral and near-ground atmospheric flows when thermal stratification or stabilizing effects are ignored, and is thus of direct interest for atmospheric applications. The constant shear-stress condition at the top of the surface layer with the bottom condition away from the viscous layer creates a region where the second-order velocity moments have constant values given as the solutions of the equilibrium relations, cf. Eq. (7.35), while the streamwise

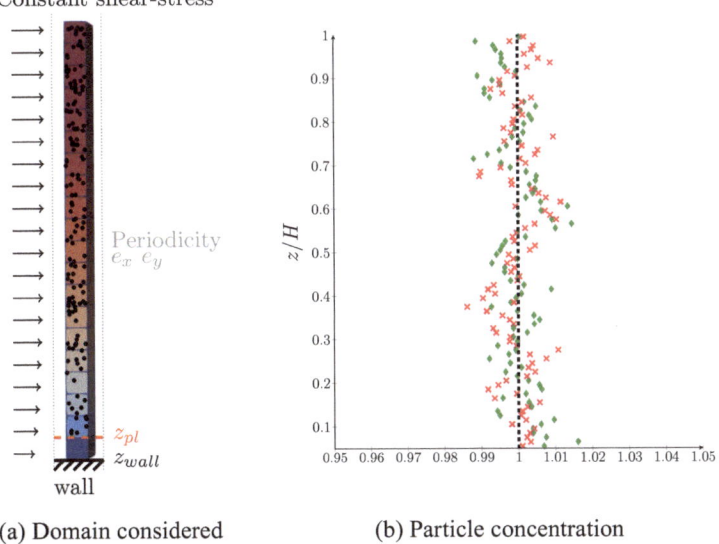

(a) Domain considered                    (b) Particle concentration

**Fig. 7.13** Sketch of the surface boundary layer domain (**a**); and profiles of the particle concentration for the elastic (symbols red cross) and the an-elastic boundary conditions (symbols green triangle) (**b**). In both figures, the wall-normal coordinate $z$ is normalized by the height of the surface boundary layer $H$

component of the mean velocity follows a logarithmic law which, for smooth walls, is

$$\langle U_{\mathrm{f}}\rangle(z) = u_* \left[ \frac{1}{\kappa} \ln\left( \frac{z\, u_*}{\nu_{\mathrm{f}}} \right) + C_{\log} \right] , \qquad (7.40)$$

where $\kappa$ is the von Karman constant ($\kappa = 0.42$) and $C_{\log}$ a constant taken as 5.2 [15]. For the SLM, the equilibrium values of the Reynolds stress tensor are the ones given in Eq. (7.38).

In this constant-stress layer, application of Eq. (7.39) yields that the mean pressure is constant and plays no role in the distribution of particle locations which is only governed by the constant-value wall-normal kinetic energy. Since this kinetic energy is maintained by the rebound elastic condition, cf. Eq. (7.23b), we expect particle concentration to be maintained whatever the condition used for the velocity streamwise component, as observed in Fig. 7.13b. The picture is however quite different for the velocity first- and second-order moments presented in Figs. 7.14 and 7.15, respectively. It is seen that, while the wall-normal and transverse kinetic energies are unaffected as explained above, the statistics involving the velocity streamwise component show serious discrepancies compared to the theoretical profiles when the elastic condition is used. Even the mean velocity deviates from the logarithmic profile near the bottom part of the layer, in spite of the fact that

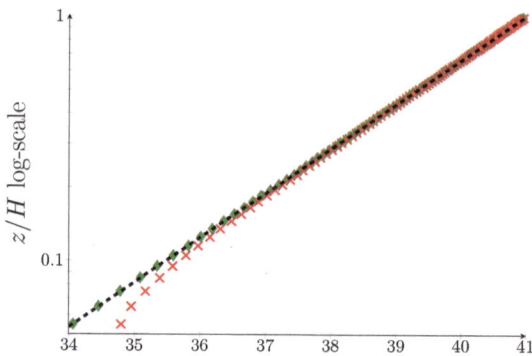

**Fig. 7.14** Profile of the streamwise mean velocity in the surface boundary layer. When the an-elastic wall-boundary condition is used, the logarithmic law is retrieved (symbols green triangle) whereas deviations near the wall are induced by the elastic wall-boundary condition (symbols red cross)

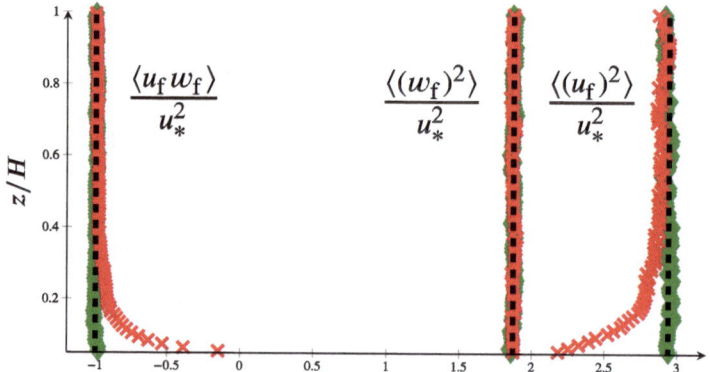

**Fig. 7.15** Profiles of the non-zero components of the Reynolds stress tensor in the surface boundary layer. The spanwise component $\langle (v_f)^2 \rangle$ is identical to the wall-normal component $\langle (w_f)^2 \rangle$ with the SLM and is not shown. Satisfactory results are obtained with the an-elastic wall-boundary condition (symbols green triangle) whereas deviations are created with the elastic wall-boundary condition (symbols red cross)

the logarithmic law in Eq. (7.40), which is retrieved by the moment approach, was provided to the PDF solver in the particle velocity Langevin equation. Furthermore, the zero stress that this elastic condition implies is clearly apparent in Fig. 7.15 in the near-wall profile of $\langle u_f w_f \rangle$, as well as in the decrease of $\langle (u_f)^2 \rangle$. These results reveal the serious inconsistencies induced by the application of the elastic boundary condition in the overall hybrid formulation. On the contrary, it is evident from Figs. 7.14 and 7.15 that a good agreement is obtained when the correct an-elastic condition, cf. Eq. (7.23c), is used, which demonstrates the importance of ensuring consistency in the hybrid formulation not only in terms of the models being used but also in terms of the boundary conditions at the wall.

### 7.3.2.4 An Open Issue: Wall-function Boundary Conditions for Inertial Particles

To the best of the authors' knowledge, this issue has rarely been addressed with the risk that it could even go unnoticed. Indeed, in many applications, a boundary condition such as elastic rebound (or a similar one) is applied for discrete particles whereas a wall-function condition is used for the fluid phase. In hybrid numerical approaches, this confusion is compounded by the fact that the fluid mean fields are handled by the moment approach in its current mesh-based finite-volume formulation whereas the discrete particles are treated by the PDF approach in its present particle stochastic expression. Going back to the sketch in Fig. 7.11, this means that, on the one hand, the wall-boundary conditions for the fluid are considered at the position of the interface $z_{pl}$ (as detailed above), whereas the chosen wall-boundary conditions for the discrete particles are written for particles hitting the 'real wall' at $z = 0$. This is not just a matter of approximate locations but also of the variables involved in each model approach. Once a wall-function treatment is chosen and applied for the fluid phase, then all the particle-attached variables entering the discrete particle state vector $\mathbf{Z}_p = (\mathbf{X}_p, \mathbf{U}_p, \mathbf{U}_s)$ must be handled simultaneously and at the same interface at $z_{pl}$. This is revealed by considering the tracer-particle limit where $\mathbf{U}_p \simeq \mathbf{U}_s$ for which we have seen that the specific an-elastic wall-boundary conditions given in Eq. (7.23) must be used (since here $\mathbf{U}_s = \mathbf{U}_f$, as further demonstrated in Sect. 8.1.3). More specifically, for small-inertia discrete particles, it is inconsistent to rely on a wall-function treatment for the fluid phase and not to apply the related wall-boundary for the velocity of the fluid seen which, for these particles, is the main particle dynamics driving force. If, for instance, no condition is considered for $\mathbf{U}_s$ which is left unchanged by the boundary conditions, this means that we are actually applying a zero-stress type of condition which, as shown above, is at variance with the physics of boundary layers and in contradiction with the treatment enforced in the fluid phase calculations. In the following, we suggest new possibilities with the underlying purpose to raise attention on this open issue.

*Boundary Conditions for Discrete Particle Mass Fluxes and Positions*  For discrete particles with non-zero inertia, the situation is different than for fluid particles for which we know, from the incompressibility constraint, that there is no net wall-normal flux at a given interface within the boundary layer. This is not necessarily so for discrete particles which, like a compressible phase, can accumulate near the wall or leave that zone depleted. Therefore, we cannot consider that each discrete particle crossing the interface at $z_{pl}$ is to be replaced by an incoming one and, even if this is the case, that the location of the incoming particle is the reflection of the location of the outgoing one across the interface. To provide a general formulation, we propose to consider that we know either the values of the two particle volume fractions (or concentrations) in the in- and out-cells or that we know the net mass flux towards the wall. Based on this information, we can derive a probability $\mathbb{P}$ for each outgoing discrete particle to be reflected at the interface. If $\mathbb{P} = 1$ there is no net mass flux at the interface at $z_{pl}$ whereas, as soon as $\mathbb{P} < 1$, a net mass flux towards the wall is induced. For the discrete particles reflected at the interface, we can assume a reflected position or sample a new one within the out-cell according to a profile of the particle volume fraction (if available). Should there exist a net mass flux away from the wall and towards the bulk of the flow, the situation is different as we need then not only to reflect each outgoing particles but to generate new ones through the interface at $z_{pl}$. Alternatively, we can generate new discrete particles to respect the given inward mass flux (as for an inlet surface) whose locations are sampled within the out-cell regardless of the locations of the outgoing discrete particles crossing the interface. Once this condition is established, we need to consider how the discrete particle velocity and the velocity of the fluid seen are expressed.

*Boundary Conditions for the Velocity of the Fluid Seen* We propose to retain for $\mathbf{U}_s$ the same boundary conditions as for fluid particles, cf. Eq. (7.23), simply substituting $\mathbf{U}_s$ for $\mathbf{U}_f$.

*Boundary Conditions for Discrete Particle Velocities* The proposed wall-function boundary conditions for the discrete particle velocities are:

$$W_{p,in} = W_{p,out} - \frac{2 \langle W_s W_p \rangle_{log}}{\langle (W_s)^2 \rangle} W_{s,out} + (\delta W_p)_{wall} \exp(-\tau/\tau_p) , \qquad (7.41a)$$

$$U_{p,in} = U_{p,out} - \frac{2 \langle W_s U_p \rangle_{log}}{\langle (W_s)^2 \rangle} W_{s,out} + (\delta U_p)_{wall} \exp(-\tau/\tau_p) . \qquad (7.41b)$$

In these equations, $\tau$ stands for the characteristic time for discrete particles to travel from the interface at $z = z_{pl}$ down to the wall at $z = 0$ while $(\delta U_p)_{wall}$ and $(\delta W_p)_{wall}$ represent the boundary conditions that apply at the wall for such discrete particles in the longitudinal and wall-normal directions, respectively. This characteristic travel time can be estimated as the minimal time between a free-flight or diffusive typical time lapses, which gives

$$\tau \simeq \min \left( \frac{z_{pl}}{W_p} ; \frac{z_{pl}^2}{(T_L + \tau_p) \langle (W_p)^2 \rangle} \right) . \qquad (7.42)$$

If we consider typical particle boundary conditions, for instance elastic rebound at the wall, we have $(\delta W_p)_{wall} = -2 W_{p,out}$ and $(\delta U_p)_{wall} = 0$. Furthermore, the fluid-particle correlations entering Eq. (7.41) are obtained by assuming equilibrium expressions in the logarithmic region taken at the vicinity of $z = z_{pl}$ (cf. the Tchen's formulas, detailed in [2, section 7.5.5])

$$\langle W_s W_p \rangle = \langle (W_s)^2 \rangle \frac{1}{1 + \tau_p/T_L} , \qquad (7.43a)$$

$$\langle W_s U_p \rangle = \langle U_s W_s \rangle \frac{1}{1 + \tau_p/T_L} . \qquad (7.43b)$$

It is seen that these relations involve the statistics of the velocity of the seen, which can differ from those of fluid particles due to particle inertia and crossing trajectory effects. In principle, the values of $\langle (W_s)^2 \rangle$ and $\langle U_s W_s \rangle$ in Eq. (7.43) can be deduced from the particle simulation itself. However, in the spirit of the equilibrium derivation leading to the expressions in Eq. (7.43) and to provide well-monitored driving forces towards the fluid limit case for small inertia particles, it may be preferable to retain fluid statistics. This means that in Eq. (7.41) the fluid-particle correlations are estimated by the following relations

$$\langle W_s W_p \rangle_{\log} = \langle (W_f)^2 \rangle_{\log} \frac{1}{1 + \tau_p / T_L} , \qquad (7.44a)$$

$$\langle W_s U_p \rangle_{\log} = \langle U_f W_f \rangle_{\log} \frac{1}{1 + \tau_p / T_L} , \qquad (7.44b)$$

where the fluid velocity second-order moments are typically given by Eq. (7.38).

From a physical standpoint, it is seen that the difference between incoming and outgoing particle velocities in Eq. (7.41), such as $W_{p,in} - W_{p,out}$, is due to the contribution of two terms which reflect the two main physical phenomena involved in particle dynamics, i.e. fluid-induced motions and particle inertia. The second terms on the right-hand side of Eqs. (7.41a)–(7.41b) are similar to the ones appearing in the fluid particle boundary conditions in Eq. (7.23) (see also the compact form in Eq. (7.32)) and reflect particle velocities induced by the underlying instantaneous driving force, which is $W_s$, through the respective fluid-particle correlations. The last terms on the right-hand side of Eq. (7.41) account for particle inertia and potential interactions with the wall itself. Further light on the physics captured by these boundary conditions can be shed by considering the two limits of high- and small-inertia particles.

High-inertia particles are not sensitive to the fluid turbulence and tend to be bullet-like while crossing fluid regions. In that sense, they are not correlated with the fluid and we can assume that $\langle W_s W_p \rangle$ as well as $\langle W_s U_p \rangle$ are negligible. Note that this is what the relations in Eq. (7.43) predict. Furthermore, the time $\tau$ to travel through the near-wall region must be small while $\tau_p$ is large. In other words, we have $\tau / \tau_p \ll 1$ and $\exp(-\tau / \tau_p) \simeq 1$. Using these approximations in Eq. (7.41), we obtain for an elastic rebound at wall

$$W_{p,in} \simeq W_{p,out} + \left( \delta W_p \right)_{wall} = -W_{p,out} , \qquad (7.45a)$$

$$U_{p,in} \simeq U_{p,out} + \left( \delta U_p \right)_{wall} = U_{p,out} , \qquad (7.45b)$$

which are, indeed, the expected boundary conditions to apply. Note that they are applied at the interface $z_{pl}$ rather than at the wall itself but the difference is not significant for such inertial particles. In short, the boundary conditions in Eq. (7.41) retrieve the correct limit in the case of large Stokes numbers. This is also valid if we consider non-elastic rebound at the wall.

On the other hand, small inertia particles tend to be well correlated with the fluid. In that limit, the travel time $\tau$ to the wall can be large while $\tau_p$ is small, showing that we have now $\tau/\tau_p \gg 1$ and $\exp(-\tau/\tau_p) \simeq 0$. This gives the boundary conditions as

$$W_{p,in} = W_{p,out} - \frac{2 \langle W_s W_p \rangle_{\log}}{\langle (W_s)^2 \rangle} W_{s,out} , \tag{7.46a}$$

$$U_{p,in} = U_{p,out} - \frac{2 \langle W_s U_p \rangle_{\log}}{\langle (W_s)^2 \rangle} W_{s,out} . \tag{7.46b}$$

in which to the first order in $St = \tau_p/T_L$, the fluid-particle correlations become

$$\langle W_s W_p \rangle_{\log} \simeq \langle (W_f)^2 \rangle_{\log} , \tag{7.47a}$$

$$\langle W_s U_p \rangle_{\log} \simeq \langle U_f W_f \rangle_{\log} , \tag{7.47b}$$

showing that we tend indeed towards the boundary conditions for fluid particles, cf. Eq. (7.23), when small inertia discrete particle become fluid tracers. Since the same fluid boundary conditions have been retained for the velocity of the fluid seen, this shows that the wall-function treatment for $U_p$ and $U_s$ becomes identical, which is now fully consistent with the fact that $U_p$ reverts to $U_s$ in the tracer-particle limit.

At the moment, these wall-function boundary conditions for discrete particles are mere proposals. They remain to be tested to outline their interest and assess their range of validity. It is however hoped that the above discussion will encourage research of this topic.

*To Be Well-mixed or Not* In several studies, the incompressibility constraint is referred to as the 'well-mixed criterion'. This expression could appear vague to a reader discovering this issue since it could be believed that we are talking about a two-component system with one solute mixed down to its molecular details in a solvent. This is nothing misleading in that suggestive image, as long as we do not take it at face value. It is relevant if it is understood as meaning pure disorder given the constraint of having fluid particles everywhere in the domain under study (i.e. uniformly distributed). Interestingly, this terminology carries also the notion that statistics derived from a subset of marked particles are representative of the fluid ones, thus free of statistical bias. This is useful when comparing the statistics of the fluid seen by discrete particles with respect to the fluid ones. On the other hand, one should be careful not to apply this criterion to a sub-stream injected into a non-homogeneous turbulent flow.

## 7.4   A Particle-laden Jet Flow

In Chap. 1, we pointed out that, in contrast to the development of PDF models for single-phase flows which follows a top-down approach, the PDF models for disperse two-phase flows are built according to a bottom-up approach. This implies that, once we have assessed theoretical models with respect to a set of criteria (see the discussion in Sect. 6.3.1 in reference to [17]), there are less constraints to respect for their numerical formulations compared to the ones in the fluid limit, as we saw in Sects. 7.2 and 7.3. Since the objectives put forward at the beginning of this chapter are oriented more towards investigating the physical issues involved than going into a detailed validation process, this explains also the apparent puzzle that there are less cases dedicated to discrete inertial particles than for tracer ones.

An interesting and challenging test case is investigated in details in Sect. 11.6, once we have formulated how to account for the two-way exchanges of momentum and energy between the fluid and particle phases. In the present section, we propose to consider a first example which serves to illustrate that it is useful to give beforehand enough consideration to simple questions such as: Which physical phenomenon is prevalent? And, what are we actually testing?

### 7.4.1   A Study of Particle Dispersion in a Round Jet

The experimental set-up is sketched in Fig. 7.16. It consists in a downward turbulent round jet of air at normal pressure and temperature conditions being injected in a co-current air stream having a lower speed. The velocity of the air flow in the injector is around 25–29 m.s$^{-1}$ while the co-current stream has a velocity of 15 m.s$^{-1}$. The diameter of the injector is 13 mm and the length of the domain is 300 mm. In this experiment, two types of particles are used. In the first case, particles are mono-

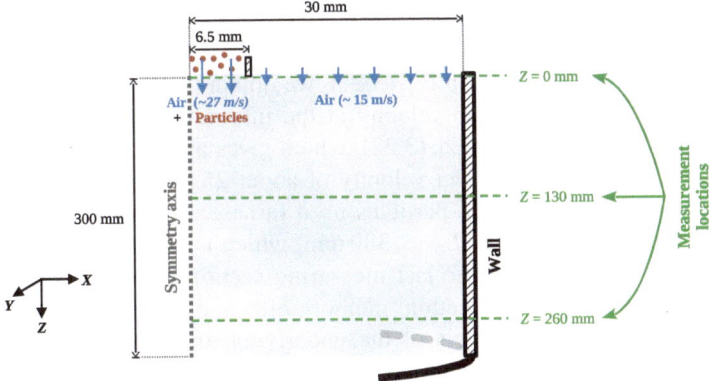

**Fig. 7.16** Sketch of the particle dispersion case: discrete particles are injected in a high-speed air round jet in a co-current unladen low-speed stream

disperse with a diameter of $d_p = 64.4\,\mu m$ and a density $\rho_p = 2590\,kg.m^{-3}$, whereas, in the second case, particles are still mono-disperse but with a diameter $d_p = 80.1\,\mu m$ and a density $\rho_p = 280\,kg.m^{-3}$. In both cases, the particles are only injected in the centered high-speed jet while the co-current low-speed air-jet is unladen and conditions are such that we are still in a dilute regime which means that the statistics of the air flow are unaffected by the particle phase. Inlet conditions are slightly different between case 1 and case 2 with particles in case 1 injected at a velocity of around 23–24 m.s$^{-1}$ while the inlet particle velocity for case 2 is around 27–28 m.s$^{-1}$. Measurements were taken at four downstream locations corresponding to $z = 0, 65, 130$ and 260 mm, respectively. The first one at $z = 0$ is the inlet section and no data are available at the first section at $z = 65$ mm for case 2, which means that we are essentially interested in the results at the last two sections, at $z = 130$ mm and $z = 260$ mm.

Since this is a statistically stationary flow with no back effects of the particle phase on the turbulent air jet, calculations are carried out in a sequential way: the moment solver is first run until stationary results are obtained for the fluid flow mean fields, which are then provided to the PDF solver. In both cases 1 and 2, the number of particles injected is calculated so as to respect the given experimental profiles of the particle mass flow rates at the injector.

### 7.4.2 What Is Actually Tested?

When running such simulations, it is hard not to rush to see how the numerical predictions at the last two sections compare to the experimental data. It is, however, best to wonder first: what aspect do we wish to test? If we want to test the particle turbulence model, which means here a Langevin model for the velocity of the fluid seen, are we sure that the measurements against which we assess numerical predictions reflect the effects of the turbulence of the air flow?

Insight into this question is provided by considering the particle relaxation timescale, cf. Eq. (3.20), and the time (or distance) needed for a particle to loose the memory of its inlet conditions. For case 1, if we use the expression corresponding to the Stokes expression given in Eq. (3.21), we obtain that $\tau_p \simeq 3.4 \times 10^{-2}$ s but we need to account for the slip velocity at the inlet (around 5 m.s$^{-1}$ for case 1) through the correction term in Eq. (3.22), which gives a more accurate estimate of $\tau_p \simeq 1.5 \times 10^{-2}$ s. With an inlet velocity of about 25 m.s$^{-1}$, this means that the typical distance traveled by the particles used in case 1 over a time of the order of their relaxation timescale is $L_p \simeq 380$ mm, which is noticeably larger than the distance from the injector of the last measuring section ($z = 260$ mm). In other words, at this last measuring section, memory effects of the inlet conditions have not been lost whereas the effects of the underlying turbulence of the air jet are still marginal. We are therefore basically testing how initial conditions are being transported, an aspect for which particle-based models are very well-suited since they treat convection without approximation. The relevant point is that this is not a good situation to test a dispersion model. In case 2, particles have a higher diameter but a much smaller density so that the particle relaxation timescale can be estimated from the Stokes expression (there is a negligible slip velocity at the inlet), which yields that $\tau_p \simeq 5.6 \times 10^{-3}$ s. With an inlet velocity of about 27–28 m.s$^{-1}$, this means that the typical distance covered by particles over a time of the order of their memory time is around $L_p \simeq 150$ mm, which is now appreciably smaller than the distance of the last measuring section while being of the order of the last-but-one section at $z = 130$ mm. The conditions used in case 2 are therefore better if we wish the turbulence of the air jet to have an effect on particle statistics.

### 7.4.3   Numerical Results

For these reasons, we only show in Figs. 7.17 and 7.18 simulation results obtained for case 2 with the two-phase SLM presented in Sect. 6.3. Without going into a comprehensive validation analysis, these outcomes call for some comments. The results for the particle mass flow rate are probably the most significant ones since they refer to the actual transport of material elements. In the experiment, no specific measurements were made for the particle volume fraction but we can get an indirect estimation by combining the particle mass flow rate and the particle mean axial velocity. Indeed, by retaining the convective mass flow rate as the main contribution (neglecting the diffusive mass flow rate) to the particle mass flow rate $\dot{m}_p$, we get that $\dot{m}_p = \alpha_p \rho_p \langle W_p \rangle$ from which we can evaluate the particle volume fraction $\alpha_p$. The results shown in Fig. 7.17a indicate then that the particle volume fraction becomes negligible for $r \geq 10$ mm at $z = 130$ mm and $r \geq 15$ mm at $z = 260$ mm. In the predictions for the mean velocity in Fig. 7.17b as well as for the fluctuating velocity components in Fig. 7.18, it appears that the profiles of the numerical results

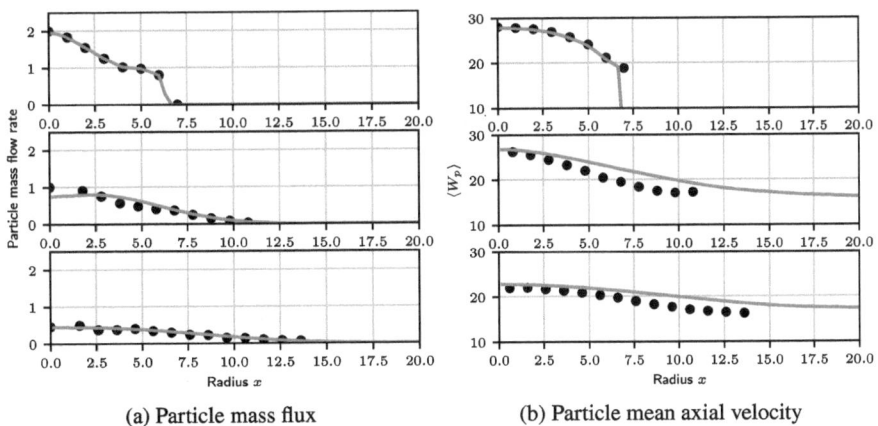

(a) Particle mass flux              (b) Particle mean axial velocity

**Fig. 7.17** Radial profiles for case 2 of the jet experiment at the inlet and at locations $z = 130$ mm and $z = 260$ mm: the particle mass flow rate (**a**) and the particle mean axial velocity (**b**)

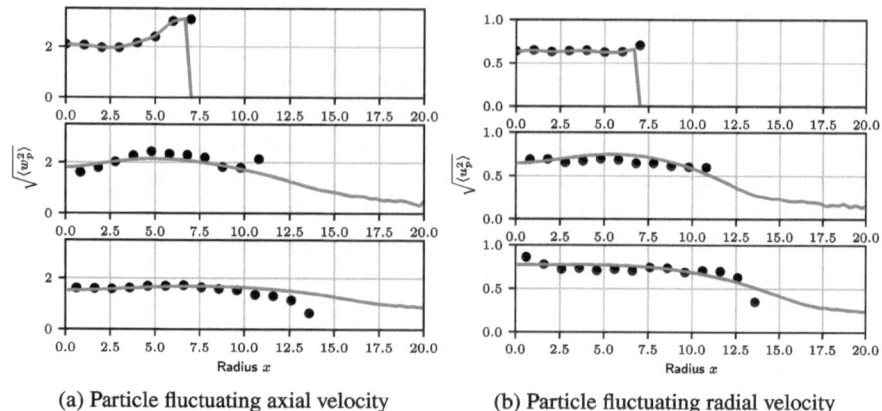

(a) Particle fluctuating axial velocity              (b) Particle fluctuating radial velocity

**Fig. 7.18** Radial profiles for case 2 of the jet experiment at the inlet and at locations $z = 130\,\mathrm{mm}$ and $z = 260\,\mathrm{mm}$: the particle fluctuating axial velocity (**a**) and the particle fluctuating radial velocity (**b**)

extend beyond these limits. This is in fact a statistical artifact due to values obtained from ensemble averaging but over very few particles, as demonstrated by the very low value of the particle volume fraction deduced from the near-zero value of the mass flow rate in these outer regions of the jet. In that sense, these results are not significant (they can be eliminated by using a cut-off value for the number of particles over which ensemble averaging is carried out but this amounts also to introducing an artificial statistical parameter). There is a small tendency to over-predict the mean axial velocity but the anisotropy of the fluctuating velocities is correctly captured and, all in all, the simulation results are in line with the experimental data even with the simple formulation of the two-phase SLM.

## 7.5  If You Are Given a Dispersed Two-phase Code

Imagine that you are provided with a code to simulate turbulent dispersed two-phase flows or have implemented a model of your choice. What is the best way to proceed to evaluate how such a numerical formulation perform?

Drawing on the examples discussed above, it appears that a sound methodology is to assess first the fluid limit by considering tracer particles. Capturing the ballistic and diffusive regimes in point-source dispersion should not be too difficult. A more stringent test case consists in selecting a non-homogeneous flow (either wall-bounded or free shear) and checking that an initially uniform distribution of fluid particles remains uniform and, thus, that no unphysical buildup or depletion in the particle concentration appears in the domain when the simulation is run over a long-enough time interval (longer than the fluid characteristic timescales). Any deviation of the fluid particle position distribution from a constant would manifest spurious drifts induced by flaws in the formulations of the model and/or the numerical schemes. Once the particle position PDF is validated, we can consider the velocity moments extracted from the particle sets to see if they match the ones provided by the fluid phase solver, as done for the mixing layer and boundary surface layer cases. Both steps need to be carefully addressed, and failure to satisfy one of these constraints should rule out a numerical formulation from being further considered for turbulent dispersed two-phase flow situations.

An example of interest is the near-wall region for discrete particle deposition problems (see also the discussion in Sect. 8.4). We are simulating the dynamics of small-inertia particles, say in an infinitely long channel flow, to determine the flux of particles going to the wall where, for instance, adhesive forces prevail and induce deposition, corresponding to a sticking wall-boundary condition for particles. The deposition rate is then directly function of the flux of discrete particles moving towards the wall. Since the main driving force for this particle flux is the velocity of the fluid seen, it is of paramount importance to check that, for fluid particles, there is no resulting mean flux going either towards or away from the wall at any location within the boundary layer. Should such a flux exist for fluid particles, it would imply that there is a spurious force either pushing the discrete particles towards the wall or pulling them away from it. The existence of this artifact would then question any validation of the deposition rate of inertia particles. For this problem, validation consists in demonstrating that, first, there is no mean drift for fluid particles, and, second, that the deposition rate profile is correctly captured as a function of particle inertia. Without the first step, the second one is pointless.

# References

1. J. Xu, S. Pope, J. Comput. Phys. **152**(1), 192 (1999). https://doi.org/10.1006/jcph.1999.6241
2. J.P. Minier, E. Peirano, Phys. Rep. **352**(1–3), 1 (2001). https://doi.org/10.1016/S0370-1573(01)00011-4
3. V. Ahuja, J. Van Der Gucht, W. Briels, J. Chem. Phys. **148**(3) (2018). https://doi.org/10.1063/1.5006627
4. R.W. Hockney, J.W. Eastwood, *Computer Simulation Using Particles* (CRC Press, Boca Raton, 2021). https://doi.org/10.1201/9780367806934
5. P.E. Kloeden, E. Platen, P.E. Kloeden, E. Platen, *Stochastic Differential Equations* (Springer, Berlin,1992). https://doi.org/10.1007/978-3-662-12616-5_4
6. P. Jenny, S.B. Pope, M. Muradoglu, D.A. Caughey, J. Comput. Phys. **166**(2), 218 (2001). https://doi.org/10.1006/jcph.2000.6646
7. M. Muradoglu, S.B. Pope, D.A. Caughey, J. Comput. Phys. **172**(2), 841 (2001). https://doi.org/10.1006/jcph.2001.6861
8. M. Muradoglu, P. Jenny, S.B. Pope, D.A. Caughey, J. Comput. Phys. **154**(2), 342 (1999). https://doi.org/10.1006/jcph.1999.6316
9. P.P. Popov, R. McDermott, S.B. Pope, J. Comput. Phys. **227**(20), 8792 (2008). https://doi.org/10.1016/j.jcp.2008.06.021
10. E. Peirano, S. Chibbaro, J. Pozorski, J.P. Minier, Progress Energy Combustion Sci. **32**(3), 315 (2006). https://doi.org/10.1016/j.pecs.2005.07.002
11. J.P. Minier, Progress Energy Combustion Sci. **50**, 1 (2015). https://doi.org/10.1016/j.pecs.2015.02.003

12. G. Balvet, J.P. Minier, C. Henry, Y. Roustan, M. Ferrand, Monte Carlo Methods Appl. **29**(2), 95 (2023). https://doi.org/10.1515/mcma-2023-2002
13. G. Balvet, J.P. Minier, Y. Roustan, M. Ferrand, Monte Carlo Methods Appl. **29**(4), 275 (2023). https://doi.org/10.1515/mcma-2023-2017
14. J.P. Minier, E. Peirano, S. Chibbaro, Monte Carlo Methods Appl. **9**(2), 93 (2003). https://doi.org/10.1515/156939603322663312
15. S. Pope, *Turbulent Flows* (Cambridge University Press, Cambridge, 2000). https://doi.org/10.1017/CBO9780511840531
16. S.B. Pope, Progress Energy Combustion Sci. **11**(2), 119 (1985). https://doi.org/10.1016/0360-1285(85)90002-4
17. J.P. Minier, S. Chibbaro, S.B. Pope, Phys. Fluids **26**(11), 113303 (2014). https://doi.org/10.1063/1.4901315
18. M.L. Bahlali, C. Henry, B. Carissimo, Boundary-Layer Meteorol. **174**, 275 (2020). https://doi.org/10.1007/s10546-019-00486-9
19. J.P. Minier, J. Pozorski, Phys. Fluids **11**(9), 2632 (1999). https://doi.org/10.1063/1.870125
20. K. Szewc, J. Pozorski, J.P. Minier, Int. J. Numer. Methods Eng. **92**(4), 343 (2012). https://doi.org/10.1002/nme.4339
21. F. Champagne, Y. Pao, I.J. Wygnanski, J. Fluid Mech. **74**(2), 209 (1976)
22. I. Wygnanski, H.E. Fiedler, J. Fluid Mech. **41**(2), 327 (1970)
23. J.P. Minier, J. Pozorski, Analysis of a PDF model in a mixing layer case. Technical Report, Electricite de France (EDF), 1996. https://inis.iaea.org/search/search.aspx?orig_q=RN:28030614. EDF–96–NB–00130
24. S. Pope, Phys. Fluids A Fluid Dyn. **3**(8), 1947 (1991). https://doi.org/10.1063/1.857925
25. D.C. Haworth, Progress Energy Combustion Sci. **36**(2), 168 (2010). https://doi.org/10.1016/j.pecs.2009.09.003
26. T.D. Dreeben, S.B. Pope, Phys. Fluids **9**(1), 154 (1997). https://doi.org/10.1063/1.869157
27. T.D. Dreeben, S.B. Pope, J. Fluid Mech. **357**, 141 (1998). https://doi.org/10.1017/S0022112097008008
28. M. Waclawczyk, J. Pozorski, J.P. Minier, Phys. Fluids **16**(5), 1410 (2004). https://doi.org/10.1063/1.1683189
29. T.D. Dreeben, S.B. Pope, Phys. Fluids **9**(9), 2692 (1997). https://doi.org/10.1063/1.869381
30. H.C. Öttinger, *Beyond Equilibrium Thermodynamics* (John Wiley & Sons, Hoboken, 2005). https://doi.org/10.1002/0471727903
31. D.C. Venerus, H.C. Öttinger, *A Modern Course in Transport Phenomena* (Cambridge University Press, Cambridge, 2018). https://www.cambridge.org/us/universitypress/subjects/engineering/chemical-engineering/modern-course-transport-phenomena

# Modeling the Velocity of the Fluid Seen: New Propositions

**8**

**Abstract**

For all its merits, the two-phase SLM presented in Chap. 6 suffers from limitations. Indeed, its construction is tailor-made on one fluid turbulence model (the SLM) so that it is unclear how to proceed from another Langevin model. It seems also to be applicable only for the reduced particle momentum equation and requires closed expressions for the timescale of the velocity of the fluid seen as inputs, even if these drawbacks are related to a class of Langevin formulations. Clearly, new ideas are called for to remove these bottlenecks. In that sense, the objective of this chapter is threefold. It is, first, to report on recent advances allowing to build two-phase GLMs in a consistent manner. By using an alternative formulation, a second objective is to bring out that present two-phase Langevin models contain enough information to be used with the extended particle momentum equation. Finally, the third objective is to develop new approaches to predict the timescales of the velocity of the fluid seen or to derive them from stochastic models based on a two-step formulation, as well as to suggest structure-based models.

**Chapter Content** The recently-proposed methodology to construct two-phase GLMs is developed in Sect. 8.1. By carefully revisiting their expressions, it is shown in Sect. 8.2 that an alternative formulation of the stochastic models for the velocity of the fluid seen allows to treat also the extended particle momentum equation. A central part of this chapter is made up by Sect. 8.3 where new ideas are set forth in connection with suggestions on space-time transformations of the stochastic models. To complement these new research directions, other formulations based on structure-based models are expressed in Sect. 8.4.

© The Author(s) 2025
J.-P. Minier et al., *Understanding Turbulent Systems*,
Lecture Notes in Physics 1039, https://doi.org/10.1007/978-3-031-84466-9_8

## 8.1      Proposals for General Two-phase GLM

In RANS approaches, viscous terms can be disregarded for high Reynolds-number turbulent flows with the exception of the remaining finite value of the mean turbulent kinetic energy dissipation rate $\langle \epsilon_f \rangle$. On the other hand, retaining only the SLM to account for turbulent fluctuations is clearly limited and, at least, direct responses to mean fluid velocity gradients should be included (see discussions in terms of slow and rapid pressure in [1]). This was recognized very early in turbulence modeling and is the reason behind the formulation of a wide range of models, as manifested in our context by the development of GLMs. Though the LRR-IP model is often considered as the basic model [1], no particular proposition can be singled out and we need to consider various formulations. This has direct bearing on two-phase flow modeling, indicating that what is needed is a general methodology to go from single-phase GLMs to two-phase flow ones without relying on the specific characteristics of a given GLM (as was done for the SLM).

### 8.1.1    The Extended Local Linear Response Theory

Such a methodology was proposed recently [2] and is sketched in Fig. 8.1:

(i) A general operator, represented by a matrix $\mathbf{H}$, is introduced and transforms fluid particle response functions to those of the fluid seen. Conditioned on a given location $\mathbf{x}$, the response function is defined as the derivative of the mean conditional increments of the velocity over a small time interval $\Delta t$ with respect to the velocity at time $t$. As already given in Sect. 6.2.1, this is translated for a fluid particle located at $\mathbf{x}$ at time $t$ by

$$\frac{1}{\Delta t} \frac{\delta}{\delta \mathbf{U}_f} \langle \Delta \mathbf{U}_f[\mathbf{U}_f] | \mathbf{X}_f = \mathbf{x} \rangle \, , \tag{8.1}$$

and it is therefore assumed that we have

$$\frac{1}{\Delta t} \frac{\delta}{\delta \mathbf{U}_s} \langle \Delta \mathbf{U}_s[\mathbf{U}_s] | \mathbf{X}_p = \mathbf{x} \rangle = \mathbf{H} \left( \frac{1}{\Delta t} \frac{\delta}{\delta \mathbf{U}_f} \langle \Delta \mathbf{U}_f[\mathbf{U}_f] | \mathbf{X}_f = \mathbf{x} \rangle \right) . \tag{8.2}$$

**Fig. 8.1** Representation in discrete time of the operator $\mathbf{H}$ that maps the velocity of a fluid particle issued from $\mathbf{X}_p(t)$ into the velocity of the fluid seen at time $t + \Delta t$

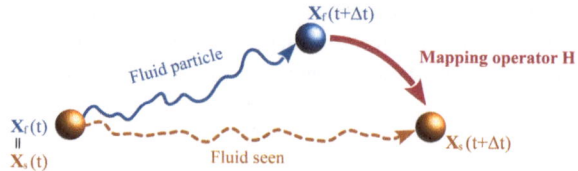

(ii) At the moment, the form given in Eq. (6.30) is retained but is not regarded as the definition of **H** but as one proposition among a class of possible expressions;

(iii) The two-phase GLM must be an extension of the fluid one built with minimum additions and so that it reverts to the form above in the tracer-particle limit;

## 8.1.2 Complete Formulations

Based on these guidelines, the general form of the proposed two-phase GLM is:

$$dU_{s,i} = -\frac{1}{\rho_f}\frac{\partial\langle P_f\rangle}{\partial x_i}\,dt + \langle U_{r,j}\rangle\frac{\partial\langle U_{f,i}\rangle}{\partial x_j}dt + G^*_{ij}\left(U_{s,j} - \langle U_{f,j}\rangle\right)dt + B_{s,ij}\,dW_j\ .$$

(8.3)

With the fluid GLM given in Eq. (6.11b), the local value of the fluid response function is the matrix **G**, while Eq. (8.3) indicates that the response function of the velocity of the fluid seen is **G***. From the first principle stated above, it follows therefore that the matrix $G^*_{ij}$ is built from $G_{ij}$ as

$$G^*_{ij} = (\mathbf{H\,G})_{ij} = H_{ik}\,G_{kj}\ .$$

(8.4)

In that sense, we are evolving in the frame of linear response theories. Note that the operator transformation is applied only on the 'relaxation timescale' characterizing the return-to-equilibrium term (the matrix **G**) and not on the mean part of the drift coefficients (the first two terms on the right-hand side of Eq. (8.3)) which is transformed as in the two-phase SLM presented in Sect. 6.3.1. In Eq. (8.3), the diffusion matrix $B_{s,ij}$ is still obtained as the square root of the matrix $L_{ij}$ (i.e. $\mathbf{B}_s\,\mathbf{B}_s^T = \mathbf{L}$) expressed by Eq. (6.35) but where the coefficients $L_{\|}$ and $L_\perp$ are now given by the following relations

$$L_{\|} = \langle\epsilon_f\rangle\left[-\frac{4}{3}\frac{\mathrm{Tr}(\mathbf{H\tilde{G}R_f})}{\mathrm{Tr}(\mathbf{HR_f})}\frac{\tilde{k}_f}{k_f}b_{\|} - \frac{2}{3}\right],$$

(8.5a)

$$L_\perp = \langle\epsilon_f\rangle\left[-\frac{4}{3}\frac{\mathrm{Tr}(\mathbf{H\tilde{G}R_f})}{\mathrm{Tr}(\mathbf{HR_f})}\frac{\tilde{k}_f}{k_f}b_\perp - \frac{2}{3}\right],$$

(8.5b)

in which $\mathbf{\tilde{G}}$ is the normalized matrix **G** defined by

$$\tilde{G}_{ij} = \frac{k_f}{\langle\epsilon_f\rangle}\,G_{ij}\ .$$

(8.6)

As in Eq. (6.38), these expressions for the components of the matrix $\mathbf{L}$ rely on the same decomposition as $\mathbf{G}^*$ and involve the same transformation operator $\mathbf{H}$ as in Eq. (8.4) since we have

$$L_{ij} = -2\frac{\langle \epsilon_f \rangle}{k_f} \frac{\mathrm{Tr}(\mathbf{H}\widetilde{\mathbf{G}}\mathbf{R}_f)}{\mathrm{Tr}(\mathbf{H})} H_{ij} - \frac{2}{3}\langle \epsilon_f \rangle \delta_{ij} \ . \tag{8.7}$$

To carry out a few simple checks on the new formulation, it is useful to use the same decomposition of $\mathbf{G}$ given in Eq. (6.19) from which we have

$$\widetilde{G}_{ij} = -\left(\frac{1}{2} + \frac{3}{4}C_0\right)\delta_{ij} + \frac{k_f}{\langle \epsilon_f \rangle} G_{ij}^{\mathrm{a}} = -\left(\frac{1}{2} + \frac{3}{4}C_0\right)\delta_{ij} + \widetilde{G_{ij}^{\mathrm{a}}} \ . \tag{8.8}$$

Introducing this decomposition into Eq. (8.5), we get

$$L_{\|} = \langle \epsilon_f \rangle \left[\left(C_0 + \frac{2}{3}\right)\frac{\tilde{k}_f}{k_f}b_{\|} - \frac{2}{3}\right] - \langle \epsilon_f \rangle \left[\frac{4}{3}\frac{\mathrm{Tr}(\mathbf{H}\widetilde{\mathbf{G}}^{\mathrm{a}}\mathbf{R}_f)}{\mathrm{Tr}(\mathbf{H}\mathbf{R}_f)}\frac{\tilde{k}_f}{k_f}b_{\|}\right], \tag{8.9a}$$

$$L_{\perp} = \langle \epsilon_f \rangle \left[\left(C_0 + \frac{2}{3}\right)\frac{\tilde{k}_f}{k_f}b_{\perp} - \frac{2}{3}\right] - \langle \epsilon_f \rangle \left[\frac{4}{3}\frac{\mathrm{Tr}(\mathbf{H}\widetilde{\mathbf{G}}^{\mathrm{a}}\mathbf{R}_f)}{\mathrm{Tr}(\mathbf{H}\mathbf{R}_f)}\frac{\tilde{k}_f}{k_f}b_{\perp}\right]. \tag{8.9b}$$

Note that the clipping condition by zero applies to $L_{\|}$ and $L_{\perp}$ in Eqs. (8.5) or (8.9).

The formulation of the diffusion coefficients in Eq. (8.9) is convenient to study the fluid limit case, that is when discrete particles become fluid ones in the absence of any remaining drift velocity ($\mathbf{U}_r = 0$). In this tracer-particle limit, $H_{ij} = \delta_{ij}$ while $b_{\|} = b_{\perp} = \tilde{k}_f/k_f = 1$. From Eq. (8.9), it is then seen that $\mathbf{L}$ is isotropic so that we retrieve also an isotropic formulation for the diffusion coefficients $B_{ij}$, since

$$L_{ij} = \left(C_0 \langle \epsilon_f \rangle - \frac{2}{3}\mathrm{Tr}(\mathbf{G}^{\mathrm{a}}\mathbf{R}_f)\right)\delta_{ij} \tag{8.10}$$

which, with $\mathbf{G}^* = \mathbf{G}$, is indeed the general GLM formulation for single-phase turbulence, cf. Eq. (6.21).

As a second verification, we can check that the two-phase GLM reverts to the two-phase SLM when $\mathbf{G}^{\mathrm{a}} = 0$. Indeed, Eq. (8.4) is then the same as Eq. (6.29) and since the first terms on the right-hand side of Eq. (8.9) are identical to the ones of Eq. (6.36), this proves that, when $\mathbf{G}^{\mathrm{a}} = 0$, we retrieve the current form of the two-phase SLM. To emphasize that the two-phase GLM in Eqs. (8.3)–(8.6) is actually an extension of the current two-phase SLM, we can express $\mathbf{G}^*$ as the sum of two components. From Eq. (8.4), we have

$$G_{ij}^* = G_{ij}^{\mathrm{slm},*} + G_{ij}^{\mathrm{a},*} \tag{8.11}$$

with $G_{ij}^{\text{slm},*}$ as in Eq. (6.29) and $G_{ij}^{\text{a},*} = H_{ik} G_{kj}^{\text{a}}$. The decomposition of the diffusion coefficients $L_{ij}$ given in Eq. (8.9) shows that we can rewrite Eq. (8.3) as the sum of two Langevin models

$$dU_{\text{s},i} = -\frac{1}{\rho_{\text{f}}} \frac{\partial \langle P_{\text{f}} \rangle}{\partial x_i} \, dt + \langle U_{\text{r},j} \rangle \frac{\partial \langle U_{\text{f},i} \rangle}{\partial x_j} \, dt + G_{ij}^{\text{slm},*} \left( U_{\text{s},j} - \langle U_{\text{f},j} \rangle \right) dt + B_{\text{s},ij}^{\text{slm}} \, dW_j^{\text{slm}}$$

$$+ G_{ij}^{\text{a},*} \left( U_{\text{s},j} - \langle U_{\text{f},j} \rangle \right) dt + B_{\text{s},ij}^{\text{a}} \, dW_j^{\text{a}} , \qquad (8.12)$$

with $\mathbf{W}^{\text{slm}}$ and $\mathbf{W}^{\text{a}}$ two independent Wiener vector processes. In Eq. (8.12), the first line corresponds to the current formulation with $B_{\text{s},ij}^{\text{slm}}$ obtained as the square root of the matrix $L_{ij}^{\text{slm}}$ given by Eq. (6.36), while the second line accounts for new effects of the two-phase GLM with $G_{ij}^{\text{a},*}$ defined as above and $B_{\text{s},ij}^{\text{a}}$ obtained as the square root of the matrix $L_{ij}^{\text{a}}$ defined from Eq. (6.35) based on the coefficients $L_{\parallel}^{\text{a}}$ and $L_{\perp}^{\text{a}}$

$$L_{\parallel}^{\text{a}} = -\langle \epsilon_{\text{f}} \rangle \left[ \frac{4}{3} \frac{\text{Tr}(\mathbf{H}\widetilde{\mathbf{G}^{\text{a}}}\mathbf{R}_{\text{f}})}{\text{Tr}(\mathbf{H}\mathbf{R}_{\text{f}})} \frac{\widetilde{k}_{\text{f}}}{k_{\text{f}}} b_{\parallel} \right] , \qquad (8.13\text{a})$$

$$L_{\perp}^{\text{a}} = -\langle \epsilon_{\text{f}} \rangle \left[ \frac{4}{3} \frac{\text{Tr}(\mathbf{H}\widetilde{\mathbf{G}^{\text{a}}}\mathbf{R}_{\text{f}})}{\text{Tr}(\mathbf{H}\mathbf{R}_{\text{f}})} \frac{\widetilde{k}_{\text{f}}}{k_{\text{f}}} b_{\perp} \right] . \qquad (8.13\text{b})$$

It is then seen that, since $\text{Tr}(\mathbf{L}^{\text{a}}) = -2\text{Tr}(\mathbf{H}\mathbf{G}^{\text{a}}\mathbf{R}_{\text{f}}) = -2\text{Tr}(\mathbf{G}^{\text{a},*}\mathbf{R}_{\text{f}})$, the deviatoric Langevin model from the SLM one, represented by the second line in Eq. (8.12), does not yield any contribution to the kinetic energy budget. This is the counterpart of what happens in the fluid case and is in line with the criteria set forth in [3].

### 8.1.3 Study of the Tracer-particle Limit

As embodied by the formulation in Eq. (8.12), the main feature of the new class of stochastic models for the velocity of the fluid seen consists in the extension of the standard two-phase SLM model to general two-phase GLM ones. There is also a second difference which concerns the expression of the mean drift term. Indeed, by comparing the first two terms on the right-hand side of Eq. (6.28) to those of Eq. (8.12), it is seen that the mean part of the drift vector which was written as

$$-\frac{1}{\rho_{\text{f}}} \frac{\partial \langle P_{\text{f}} \rangle}{\partial x_i} \, dt + \left( \langle U_{\text{p},j} \rangle - \langle U_{\text{f},j} \rangle \right) \frac{\partial \langle U_{\text{f},i} \rangle}{\partial x_j} \, dt , \qquad (8.14)$$

in the standard two-phase SLM model in Sect. 6.3.1 is now expressed as

$$-\frac{1}{\rho_{\text{f}}} \frac{\partial \langle P_{\text{f}} \rangle}{\partial x_i} \, dt + \langle U_{\text{r},j} \rangle \frac{\partial \langle U_{\text{f},i} \rangle}{\partial x_j} \, dt \qquad (8.15)$$

The difference between the two expressions is therefore equal to

$$\left(\langle U_{s,j}\rangle - \langle U_{f,j}\rangle\right) \frac{\partial \langle U_{f,i}\rangle}{\partial x_j} dt = U_{d,j} \frac{\partial \langle U_{f,i}\rangle}{\partial x_j} dt \,, \tag{8.16}$$

where $\mathbf{U}_d$ is the drift velocity already introduced in Sect. 5.1. To bring out the significance of this term, it is useful to consider the tracer-particle limit obtained by taking $\tau_p \to 0$. In that case, we have $\mathbf{U}_r \to 0$ which, of course, entails that $\langle \mathbf{U}_r\rangle \to 0$, so that the formulation in Eq. (8.12) retrieves exactly the GLM expression for fluid particle velocities as in Eq. (6.11b) in Sect. 6.2. On the other hand, the standard two-phase SLM in Eq. (6.28) yields for tracer particles

$$dU_{s,i} = -\frac{1}{\rho_f} \frac{\partial \langle P_f\rangle}{\partial x_i} dt + U_{d,j} \frac{\partial \langle U_{f,i}\rangle}{\partial x_j} dt$$

$$+ G_{ij}\left(U_{s,j} - \langle U_{f,j}\rangle\right) dt + \sqrt{C_0 \langle \epsilon_f\rangle} \, dW_i \,. \tag{8.17}$$

The significance of the extra mean drift term corresponds to the possible differences between fluid statistics extracted from the subset of tracer particles compared to the complete fluid ones extracted from the whole set of fluid particles (regardless of whether they are tracer particles or not). This issue is therefore related to the notion of well-mixed particles and of a potential bias if these tracer particles make up a subclass with associated statistics which differ from the complete local fluid ones. If these tracer particles are sufficiently well-mixed, then $\mathbf{U}_d = 0$ and both formulations are identical. Note that this is, implicitly, the situation which is considered when analyzing the macroscopic behavior of stochastic models in order to derive the corresponding mean field equations. If, however, we consider that tracer particles are not well-mixed, as when they are injected only at specific locations in the domain and with surrounding fluid streams, then $\mathbf{U}_d \neq 0$ and the second term on the right-hand side of Eq. (8.17) does not vanish.

It is worth pointing out that the appearance of the extra term involving the drift velocity (i.e., the second term on the right-hand side of Eq. (8.17)) does not necessarily point to a flaw. Indeed, for not well-mixed tracer particles, it can be noted that the return-to-equilibrium term (i.e., the third term on the right-hand side of Eq. (8.17)) is not anymore a fluctuating term around a zero mean value since its average is $G_{ij} U_{d,j}$, so that all these terms on the right-hand side of Eq. (8.17) are actually similar. Nevertheless, for consistency reasons (for example, with the use of the same mean relative velocity $\langle \mathbf{U}_r\rangle$ in the closure of the Csanady's factors in Eq. (6.32) in Sect. 6.3.1) and to obtain exactly the same fluid model in the limit of tracer particles when $\tau_p \to 0$, the new formulation as in Eqs. (8.3) and (8.12) appears preferable and is retained from now on.

### 8.1.4    Remarks on the Transformation Operator

A few remarks can be made. First, this new methodology rests upon the proposition of a mapping operator $\mathbf{H}$. The form given in Eq. (6.30) has support in locally isotropic turbulence but, since we are mapping turbulent fluctuations rather than small-scale components, this is not necessarily the only possibility. New ideas based on underlying first principles are called for to obtain extended formulations. Second, if a direct mapping through the operator $\mathbf{H}$ turns out to be difficult to express, it is also possible to build two-step models with an explicit step dedicated to the CTE. Third, a general observation is that, in all the previously-discussed expressions, the timescales of the velocity of the fluid seen rely only on the Csanady's formulas. Given the central role played by these timescales, it is surprising that so little attention has been devoted to clarifying their physical justification. These open issues are addressed in Sect. 8.3. In line with the second objective indicated in the introduction of this chapter, we first propose an alternative formulation of two-phase GLMs to reveal that we can already handle the extended particle momentum equation.

## 8.2    Alternative Formulation for the Extended Particle Momentum Equation

With the formulation of the general Langevin model in Eqs. (8.3)–(8.7), we have a complete stochastic model for discrete particle dynamics in turbulent flows. This is evident for the reduced particle momentum equation in which only the drag force is retained, cf. Eq. (3.19), since $\mathbf{U_s}$ can be directly plugged in the drag force formula. However, this is less obvious for the extended particle momentum equation, cf. Eq. (3.23), because the pressure-gradient term requires to simulate $D\mathbf{U_s}/Dt$ which is the acceleration of the fluid particle located at the discrete particle one or, more rigorously in the present stochastic framework, the velocity increments of that fluid particle along its own trajectory. In other words, we need to simulate not only $d\mathbf{U_s}$ but also $D\mathbf{U_s}$ (or, loosely speaking, the two accelerations $d\mathbf{U_s}/dt$ and $D\mathbf{U_s}/Dt$) which are likely to be correlated but represent nevertheless different velocity increments. On the other hand, we have seen that present stochastic models for $\mathbf{U_s}$ are extensions of a corresponding GLM for fluid particles and revert to them when particle inertia becomes negligible. This hints to the fact that, even when $\tau_p \neq 0$, Langevin models such as the ones in Eqs. (8.3)–(8.7) contain a sub-model for the correlated velocity increments of the associated fluid particle along its trajectory.

To bring out this sub-model, we propose an alternative formulation of the general two-phase Langevin model in Eqs. (8.3)–(8.7), making use of an important characteristic of the timescale of the velocity of the fluid seen. Following the analysis developed in Sect. 6.1.1, we have indeed concluded that $T_{\mathrm{L}}^* \leq T_{\mathrm{L}}$, which allows us to write $1/T_{\mathrm{L}}^*$ as

$$\frac{1}{T_{\mathrm{L}}^*} = \frac{1}{T_{\mathrm{L}}} \left(1 + \delta b\right), \text{ with } \delta b \geq 0 . \tag{8.18}$$

This is exemplified by the Csanady factors in Eq. (6.31), but constitutes a general result regardless of the specific closures retained for the coefficients entering the mapping operator $\mathbf{H}$. This feature plays a key role to work out the correlated models for $\mathrm{d}\mathbf{U}_{\mathrm{s}}$ and $\mathrm{D}\mathbf{U}_{\mathrm{s}}$ which are now formulated following a step-by-step approach.

As a first step, we do not consider the anisotropy of the matrix $\mathbf{H}$ which is then written as $H_{ij} = (1 + \delta b)\, \delta_{ij}$. This allows us to concentrate on the main physical ideas while keeping simple tensor notations, without loss of generality as demonstrated further below. In that case, the diffusion matrix $\mathbf{L}$ in Eq. (8.7) is isotropic, i.e. $L_{ij} = L^{\mathrm{iso}}\, \delta_{ij}$ with

$$L^{\mathrm{iso}} = -\frac{2}{3} \frac{\langle \epsilon_{\mathrm{f}} \rangle}{k_{\mathrm{f}}} \, \mathrm{Tr}(\widetilde{\mathbf{G}} \mathbf{R}_{\mathrm{f}}) \, (1 + \delta b) - \frac{2}{3} \langle \epsilon_{\mathrm{f}} \rangle . \tag{8.19}$$

Using the decomposition of $\mathbf{G}$ given in Eq. (6.19) or of $\widetilde{\mathbf{G}}$ in Eq. (8.8) leads to

$$L^{\mathrm{iso}} = C_0 \langle \epsilon_{\mathrm{f}} \rangle - \frac{2}{3} \mathrm{Tr}(\mathbf{G}^{\mathrm{a}} \mathbf{R}_{\mathrm{f}}) + \delta b \left[ \left( \frac{2}{3} + C_0 \right) \langle \epsilon_{\mathrm{f}} \rangle - \frac{2}{3} \mathrm{Tr}(\mathbf{G}^{\mathrm{a}} \mathbf{R}_{\mathrm{f}}) \right] , \tag{8.20}$$

and the general two-phase GLM in Eq. (8.3) can then be written as

$$\mathrm{d}U_{\mathrm{s},i} = -\frac{1}{\rho_{\mathrm{f}}} \frac{\partial \langle P_{\mathrm{f}} \rangle}{\partial x_i} \, \mathrm{d}t + G_{ij} \left( U_{\mathrm{s},j} - \langle U_{\mathrm{f},j} \rangle \right) \mathrm{d}t$$

$$+ \langle U_{\mathrm{r},j} \rangle \frac{\partial \langle U_{\mathrm{f},i} \rangle}{\partial x_j} \mathrm{d}t + \delta b\, G_{ij} \left( U_{\mathrm{s},j} - \langle U_{\mathrm{f},j} \rangle \right) \mathrm{d}t + \sqrt{L^{\mathrm{iso}}} \, \mathrm{d}W_i . \tag{8.21}$$

The diffusion coefficient $L^{\mathrm{iso}}$ in Eq. (8.20) has the form $L^{\mathrm{iso}} = L^{\mathrm{s-p}} + \delta b\, L^{\mathrm{t-p}}$ where $L^{\mathrm{s-p}}$ is the remaining coefficient in the fluid limit case when $\delta b = 0$ (as labeled by the superscript s-p for single-phase) while $L^{\mathrm{t-p}}$ corresponds to the term in the bracket which appears in the two-phase flow situation (thus labeled by the superscript t-p for two-phase). Note that since we assume that $L^{\mathrm{s-p}}$ is positive to have a meaningful GLM in the fluid limit, $L^{\mathrm{t-p}}$ is even more likely to be positive and, furthermore, we have that $\delta b \geq 0$. The diffusion coefficient $L^{\mathrm{iso}}$ is therefore the sum of two positive terms. In a weak formulation, the white-noise term $\sqrt{L^{\mathrm{iso}}} \, \mathrm{d}W_i$ in Eq. (8.21) can then be expressed as the sum of two independent Wiener processes, noted $W^{s-p}$ and $W^{t-p}$ respectively, i.e. $\sqrt{L^{iso}}\,dW_i = \sqrt{L^{s-p}}\,dW_i^{s-p} + \sqrt{\delta b\,L^{t-p}}\,dW_i^{t-p}$. This yields the following expression for the two-phase GLM

$$
dU_{s,i} = -\frac{1}{\rho_f}\frac{\partial\langle P_f\rangle}{\partial x_i}\,dt + G_{ij}\left(U_{s,j} - \langle U_{f,j}\rangle\right)dt + \sqrt{\left(C_0\langle\epsilon_f\rangle - \frac{2}{3}\mathrm{Tr}(\mathbf{G^a R_f})\right)}\,dW_i^{s-p}
$$

$$
+ \langle U_{r,j}\rangle\frac{\partial\langle U_{f,i}\rangle}{\partial x_j}\,dt + \delta b\,G_{ij}\left(U_{s,j} - \langle U_{f,j}\rangle\right)dt + \sqrt{\delta b\,L^{t-p}}\,dW_i^{t-p}\,, \qquad (8.22)
$$

where the complete expression of $L^{s-p}$ has been kept to reveal that the first line of Eq. (8.22) corresponds to the GLM for fluid particles, cf. Eq. (6.21). Said otherwise, the first line on the right-hand side of Eq. (8.22) is the associated GLM for fluid particles and can be written as $DU_{s,i}$, which leads to the alternative formulation

$$
dU_{s,i} = DU_{s,i}
$$

$$
+ \langle U_{r,j}\rangle\frac{\partial\langle U_{f,i}\rangle}{\partial x_j}\,dt + \delta b\,G_{ij}\left(U_{s,j} - \langle U_{f,j}\rangle\right)dt + \sqrt{\delta b\,L^{t-p}}\,dW_i^{t-p}\,. \qquad (8.23)
$$

Given that $\delta b$ is a function of $\langle \mathbf{U_r}\rangle$, as in the Csanady factors in Eq. (6.31), note that the last three terms on the right-hand side of Eq. (8.23) vanish when the mean relative velocity goes to zero. They correspond to the additional part of the two-phase GLM compared to its associated single-phase one. In that sense, this formulation clearly manifests one of the great benefits of having devised two-phase Langevin models as consistent extensions of single-phase GLMs.

In the general case where the anisotropy of the mapping operator is accounted for, similar developments can be pursued. To that effect, the key properties of the timescales of the velocity of the fluid seen still allow us to write the matrix $\mathbf{H}$ as $H_{ij} = (1 + \delta b)\,\delta_{ij} + \delta H_{ij}$ with $\mathrm{Tr}(\delta\mathbf{H}) = 0$ and where each component of the matrix $\delta\mathbf{H}$ scales as $\delta b$ (i.e., $\delta H_{ij} \sim \delta b$). In the two-phase GLM, the diffusion matrix $\mathbf{L}$ in Eq. (8.7) is no longer isotropic but has the form

$$
L_{ij} = \left(C_0\langle\epsilon_f\rangle - \frac{2}{3}\mathrm{Tr}(\mathbf{G^a R_f})\right)\delta_{ij}
$$

$$
+ \delta b\left[\left(\frac{2}{3} + C_0\right)\langle\epsilon_f\rangle - \frac{2}{3}\mathrm{Tr}(\mathbf{G^a R_f})\right]\delta_{ij} - \frac{2}{3}\frac{\mathrm{Tr}((\delta\mathbf{H})\mathbf{G R_f})}{(1 + \delta b)}\delta H_{ij}\,. \qquad (8.24)
$$

Applying the same methodology as above for the white-noise terms in the GLM for $\mathbf{U_s}$ leads to a formulation akin to the one in Eq. (8.23) which reads

$$
dU_{s,i} = DU_{s,i} + \langle U_{r,j}\rangle\frac{\partial\langle U_{f,i}\rangle}{\partial x_j}\,dt
$$

$$
+ \left[\delta b\,G_{ij} + ((\delta\mathbf{H})\mathbf{G})_{ij}\right]\left(U_{s,j} - \langle U_{f,j}\rangle\right)dt + B_{ij}^{t-p}\,dW_j^{t-p}\,, \qquad (8.25)
$$

where the diffusion matrix $\mathbf{B}^{t-p}$ is obtained as the square root of matrix $\mathbf{L}^{t-p}$ which corresponds to the second line of Eq. (8.24). Therefore, though the tensor notations are more involved, we reach the same conclusion that the general two-phase Langevin model for $d\mathbf{U}_s$ contains a consistent sub-model for $D\mathbf{U}_s$ to which it reverts when $\delta b$ (and, thus, $\delta H_{ij}$) vanishes.

From a physical standpoint, the outcome of these formulations is that general two-phase Langevin models, such as the ones expressed in Eqs. (8.3)–(8.7), provide also a consistent model for the pressure-gradient term entering the extended particle momentum equation. This model for $D\mathbf{U}_s$ is the very single-phase GLM on which each two-phase GLM is built. Furthermore, it is also seen from the decomposition of the Wiener processes that the correlation between the white-noise terms entering the correlated models for $d\mathbf{U}_s$ and $D\mathbf{U}_s$ is naturally obtained.

Finally, it can be noted that the model for the extended particle momentum equation involves now correlated stochastic models for the discrete particle velocity $\mathbf{U}_p$ and the velocity of the fluid seen $\mathbf{U}_s$. Indeed, with the decompositions introduced above and still relying on the expression of the matrix $\mathbf{G}$ as in Eq. (6.19), the complete stochastic model for $\mathbf{Z}_p = (\mathbf{X}_p, \mathbf{U}_p, \mathbf{U}_s)$ takes the following form (retaining only the drag and pressure-gradient forces in the particle momentum equation)

$$dX_{p,i} = U_{p,i}\, dt \, , \tag{8.26a}$$

$$dU_{p,i} = \frac{U_{s,i} - U_{p,i}}{\tau_p}\, dt + \frac{\rho_f}{\rho_p}\, DU_{s,i} \, , \tag{8.26b}$$

$$DU_{s,i} = -\frac{1}{\rho_f}\frac{\partial\langle P_f\rangle}{\partial x_i}\, dt$$

$$+ G_{ij}\left(U_{s,j} - \langle U_{f,j}\rangle\right)dt + \sqrt{\left(C_0\langle\epsilon_f\rangle - \frac{2}{3}\mathrm{Tr}(\mathbf{G}^a\mathbf{R}_f)\right)}\,dW_i^{s-p} \, , \tag{8.26c}$$

$$dU_{s,i} = DU_{s,i} + \langle U_{r,j}\rangle\frac{\partial\langle U_{f,i}\rangle}{\partial x_j}dt$$

$$+ \left(\delta b\, G_{ij} + ((\delta\mathbf{H})\mathbf{G})_{ij}\right)\left(U_{s,j} - \langle U_{f,j}\rangle\right)dt + B_{ij}^{t-p}\,dW_j^{t-p} \, , \tag{8.26d}$$

where the diffusion matrix $\mathbf{B}^{t-p}$ is defined as above, and with $\mathbf{W}^{s-p}$ and $\mathbf{W}^{t-p}$ still two independent Wiener processes. This formulation turns out to be useful when analyzing how source terms representing particle back effects on the fluid behave in the fluid-particle limit when particle inertia vanishes, as addressed in Chap. 11.

## 8.3 New Macroscopic Approaches and Microscopic Models

### 8.3.1 Extended Kolmogorov Hypotheses for Lagrangian Timescales

As recalled in Sect. 5.3, the classical Kolmogorov similarity hypotheses are developed from a Lagrangian standpoint by considering space-time fluid correlations in a small domain described as locally isotropic around fluid particles (which form an equivalence class). In the following, we propose to extend this hypothesis to discrete particles having a mean velocity drift $\langle \mathbf{U}_r \rangle$ with respect to the fluid. This means that we are observing similar space and time fluid correlations but from the standpoint of discrete particles (which form also an equivalence class, though different from the one made up by fluid particles). Said otherwise, we have two different points of view to describe the same fluid statistical events, as pictured in Fig. 8.2.

To formulate the relations between these two different reference systems, it is useful to introduce the notion of (fluid) statistical events: we consider fluid velocity statistical events $(\tau, l)$ having a typical timescale $\tau$ and a typical length scale $l$. In this time-length statistical space, the two equivalent standpoints (i.e., astride a fluid particle or a discrete one) imply that the corresponding coordinates are different but related. To illustrate that point, let us consider a purely Lagrangian event, as sketched in Fig. 8.2, and limit ourselves to a one-dimensional spatial setting for the sake of simplicity. If we follow fluid particles, we have a purely time event and since we are interested in the characteristic timescale of fluid velocities, this statistical event can be written as $(T_L, 0)_{\mathcal{E}_f}$ in this observation frame. However, if we observe the same statistical event from the standpoint of discrete particles, we have a time-length statistical event which we can write as $(T_L^*, |\langle U_r \rangle| T_L^*)_{\mathcal{E}_p}$ where $\mathcal{E}_p$ is used to denote the discrete particle observation frame. Obviously, the time-component $T_L^*$ is the characteristic timescale of the velocity of the fluid seen and we have the equivalence

$$(T_L, 0)_{\mathcal{E}_f} \equiv (T_L^*, |\langle U_r \rangle| T_L^*)_{\mathcal{E}_p} . \tag{8.27}$$

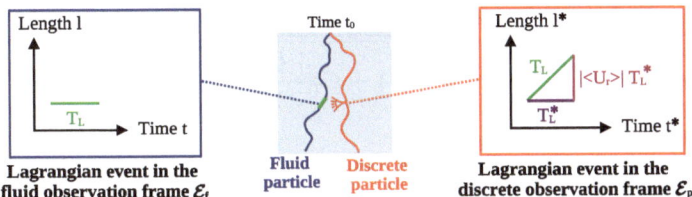

**Fig. 8.2** Graphical representation of a fluid Lagrangian event $T_L$ as seen from fluid particles, where it appears as $(T_L, 0)_{\mathcal{E}_f}$ in the observation frame $\mathcal{E}_f$ (left side) and as seen from discrete particles, where it appears as a time-length event $(T_L^*, |\langle U_r \rangle| T_L^*)_{\mathcal{E}_p}$ in the observation frame $\mathcal{E}_p$ (right side)

Loosely speaking, the timescale of the velocity of the fluid seen appears as a projection of the fluid Lagrangian timescale onto discrete particle trajectories and we need to retain both the time and length components to properly describe the same statistical event (namely, the characteristic timescale of fluid particle velocities).

To work out explicit closure relations, the next step consists in introducing a measure of such statistical events and requiring that this measure be an invariant when switching from one observation standpoint to another. If we retain the classical measure using the time and length coordinates, we can write that the invariant measure $\Delta s$ of an event $(\tau, l)_{\mathcal{E}}$ in any observation frame $\mathcal{E}$ is

$$(\Delta s)^2 = \left(\frac{\tau}{T_{\mathrm{L}}}\right)^2 + \left(\frac{l}{L_{\mathrm{E}}}\right)^2 , \tag{8.28}$$

where time increments are normalized by the Lagrangian timescale $T_{\mathrm{L}}$ whereas length increments are normalized by the Eulerian length scale $L_{\mathrm{E}}$. In the example illustrated in Fig. 8.2, the statistical even $(T_{\mathrm{L}}, 0)_{\mathcal{E}_{\mathrm{f}}}$ is then of unit measure. The same remains valid when observed from discrete particles and this implies that the coordinates $(T_{\mathrm{L}}^*, |\langle U_{\mathrm{r}}\rangle| T_{\mathrm{L}}^*)_{\mathcal{E}_{\mathrm{p}}}$ are such that we have

$$\left(\frac{T_{\mathrm{L}}^*}{T_{\mathrm{L}}}\right)^2 + \left(\frac{|\langle U_{\mathrm{r}}\rangle| T_{\mathrm{L}}^*}{L_{\mathrm{E}}}\right)^2 = 1 , \tag{8.29}$$

from which we get that

$$T_{\mathrm{L}}^* = \frac{T_{\mathrm{L}}}{\sqrt{1 + \left(\frac{|\langle U_{\mathrm{r}}\rangle| T_{\mathrm{L}}}{L_{\mathrm{E}}}\right)^2}} . \tag{8.30}$$

By using $L_{\mathrm{E}} = u_{\mathrm{f}} T_{\mathrm{E}}$ (with $u_{\mathrm{f}} = 2/3 k_{\mathrm{f}}$, in the locally isotropic formulation), we retrieve therefore the Csanady expression for the Lagrangian timescale of the velocity of the fluid seen (cf. Eq. (6.31) in Sect. 6.3.1)

$$T_{\mathrm{L}}^* = \frac{T_{\mathrm{L}}}{\sqrt{1 + C_{\mathrm{T}}^2 \frac{|\langle U_{\mathrm{r}}\rangle|^2}{u_{\mathrm{f}}^2}}} . \tag{8.31}$$

An interesting outcome of such reasoning is to bring out that the typical dependence of the Csanady expression, which scales as $(1 + x^2)^{-1/2}$, with $x = |\langle U_{\mathrm{r}}\rangle|/u_{\mathrm{f}}$ the normalized mean velocity drift, results from the choice of the measure used to evaluate $\Delta s$ for each statistical event. Other measures would result in different relations for $T_{\mathrm{L}}^*$. For example, if we consider

$$\Delta s = \frac{\tau}{T_{\mathrm{L}}} + \frac{l}{L_{\mathrm{E}}} \tag{8.32}$$

where all quantities are positive, we would obtain for the timescale of the velocity of the fluid seen

$$T_L^* = \frac{T_L}{1 + C_T \dfrac{|\langle U_r \rangle|}{u_f}} \,. \tag{8.33}$$

By re-expressing Eq. (8.33), it is seen that we are directly adding the inverse of the timescales

$$\frac{1}{T_L^*} = \frac{1}{T_L} + \frac{(|\langle U_r \rangle|/u_f)}{T_E} = \frac{1}{T_L} + \frac{|\langle U_r \rangle|}{L_E} \,. \tag{8.34}$$

Such expressions are more in line with stochastic models based on Langevin formulations since the return-to-equilibrium term is written with a friction or resistance coefficient which is the inverse of a timescale. Then, when devising a two-step Langevin model by adding two successive independent steps (a Lagrangian one followed by an Eulerian one), we are adding resistances as in Ohm's law. In comparison, it is seen that the Csanady expressions consist in adding the square of the same resistance coefficients since we have from Eq. (8.29)

$$\left( \frac{1}{T_L^*} \right)^2 = \left( \frac{1}{T_L} \right)^2 + \left( \frac{|\langle U_r \rangle|}{L_E} \right)^2 \,. \tag{8.35}$$

The formulation of the present extended Kolmogorov hypothesis can be further developed by writing transformation rules for the time-length coordinates of the same fluid statistical event observed from two different standpoints, written as $\mathcal{E}'$ and $\mathcal{E}$, corresponding to two different mean slip velocities $\langle U_r' \rangle$ and $\langle U_r \rangle$ with respect to the fluid (the relative velocity between $\mathcal{E}'$ and $\mathcal{E}$ is therefore $\Delta \langle U_r \rangle = \langle U_r \rangle - \langle U_r' \rangle$). By assuming a linear mapping between coordinates in $\mathcal{E}'$ and $\mathcal{E}$, the relations between $(t', x')$ in $\mathcal{E}'$ and $(t, x)$ in $\mathcal{E}$ are obtained from the requirement that the measure in Eq. (8.28) remains invariant. This yields

$$x = \gamma \left[ x' + (\Delta \langle U_r \rangle) \, t' \right] \,, \tag{8.36a}$$

$$t = \gamma \left[ t' - C_T^2 (\Delta \langle U_r \rangle) \frac{x'}{u_f^2} \right] \,, \tag{8.36b}$$

where $\gamma$ is the Csanady factor

$$\gamma = \frac{1}{\sqrt{1 + C_T^2 \dfrac{|\Delta \langle U_r \rangle|^2}{u_f^2}}} \,. \tag{8.37}$$

Then, if we revisit the example introduced above, the first standpoint following fluid particles corresponds to $\langle U_r' \rangle = 0$ (i.e., $\mathcal{E}' = \mathcal{E}_f$) and the transformation of a time event $(\Delta t,\, 0)$ is now given by the equivalence

$$(\Delta t,\, 0)_{\mathcal{E}_f} \iff (\gamma \Delta t,\, \gamma \langle U_r \rangle \Delta t)_{\mathcal{E}_p}\,, \tag{8.38}$$

from which we retrieve $T_L^* = \gamma T_L$ by taking $\Delta t = T_L$ in the above relation. Interestingly, in the limit of high velocity slips, $|\langle U_r \rangle|/u_f \gg 1$, a time event $(\Delta t',\, 0)$ in $\mathcal{E}_f$ is transformed into a spatial one in $\mathcal{E}_p$, since

$$\Delta t \simeq 0\,, \tag{8.39a}$$

$$\Delta x \simeq \frac{1}{C_T} u_f \Delta t' = \frac{T_E}{T_L} u_f \Delta t'\,, \tag{8.39b}$$

which is the translation of the frozen-turbulence hypothesis [1]. Indeed, by taking $\Delta t' = T_L$, it is seen that, when $|\langle U_r \rangle|/u \gg 1$, the statistical event $(T_L,\, 0)$ in $\mathcal{E}_f$ is transformed into $(0,\, L_E)$ in $\mathcal{E}_p$.

As indicated above, the previous relations have been worked out in a one-dimensional spatial setting and we need to consider the more realistic three-dimensional spatial version. In the Kolmogorov picture, turbulence is described as isotropic in small time and space domains around fluid particles. This means that we can still consider an isotropic Lagrangian timescale tensor and retain $T_L$ as a scalar timescale that provides the necessary time information. However, this is not the case for fluid velocity increments over a separation increments $r$ since the length scale for longitudinal velocity components is twice the one for transverse components. In other words, we must now consider that $L_E$, as well as $T_E$ which is deduced from it, become second-order tensors which are, however, isotropic functions of the separation vector $r$. In the general three-dimensional spatial setting, the space effects are due to the mean velocity drift between discrete particles and the fluid and the relevant separation vector is $\mathbf{r} = \langle \mathbf{U}_r \rangle \Delta t$ over a time interval $\Delta t$. This means that the Eulerian time tensor can be expressed as

$$T_{E,ij} = T_{E,\perp}\delta_{ij} + \left(T_{E,\|} - T_{E,\perp}\right) \frac{r_i r_j}{|\mathbf{r}|}\,, \tag{8.40}$$

where $T_{E,\|} = 2T_{E,\perp}$. It is then appropriate to express directly the timescale of the velocity of the fluid seen in the local reference system aligned with the mean velocity drift and to follow the previous developments. This leads therefore to retrieve the Csanady expressions, cf. Eq. (6.31).

A few remarks are in order. First, the previous relations must not be regarded as a new kinematics but merely as a proposed formulation in terms of time and length scales to provide new insights into the closure issue of the timescale of the velocity of the fluid seen. Should they prove useful, the ideas introduced in the present section are meant to reveal what the functional form entering Csanady expressions represents and how we can improve them or suggest alternative closures (e.g., as in Eq. (8.33)). Second, these considerations, which follow a relativity-like principle, still constitute a top-down approach where the timescales of the velocity of the fluid seen are obtained from guiding rules and are input into Langevin models. Another, more microscopic, approach consists in formulating stochastic models from which the timescales $T_L^*$ would be an outcome instead of being an input. In that respect, some new modeling roads are now suggested.

## 8.3.2   Capturing Space and Time Correlations Through Time Changes

Early stochastic models for turbulent dispersion were often developed by considering a Lagrangian step followed by an Eulerian one. Although these models suffered from shortcomings, in particular due to the fact that they used instantaneous relative velocities and were not properly written as stochastic diffusion processes in continuous time, it is interesting to revisit such two-step formulations, however with the Eulerian step governed by the mean velocity drift as indicated in Sect. 6.1. The basic situation is represented in Fig. 8.3 and the purpose of this section is to suggest new modeling possibilities.

### 8.3.2.1  Basic Ideas and Current Limitations of Two-step Stochastic Models

For this discussion, we limit ourselves to a one-dimensional setting where $T_L$ and $T_E$ are constant scalars (this amounts to considering stationary isotropic turbulence). In this case, we retain two simple Langevin models, for the Lagrangian step

$$dU_L = -\frac{U_L}{T_L}\,dt + \sqrt{\frac{2\,u_f^2}{T_L}}\,dW_L\,, \qquad (8.41)$$

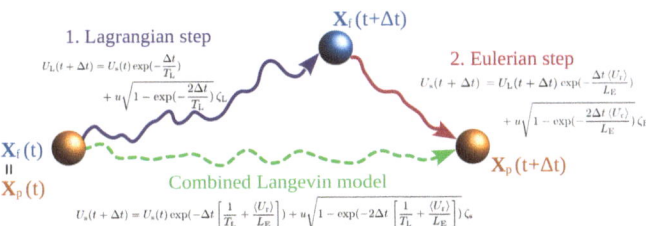

**Fig. 8.3** Graphical representation of a two-step formulation of the stochastic model for the velocity of the fluid seen with a Lagrangian step followed by an Eulerian one governed by the mean velocity drift

as well as for the Eulerian one

$$dU_E = -\frac{U_E}{L_E}\,dr + \sqrt{\frac{2\,u_f^2}{L_E}}\,dW_E'\,,\tag{8.42}$$

where $dW_L$ and $dW_E'$ are two independent Wiener processes (note that here $\langle(dW_E')^2\rangle = dr$). Since the Eulerian step is due to the mean velocity drift, we have $dr = \langle U_r\rangle\,dt$ and we can reformulate Eq. (8.42) as

$$dU_E = -\frac{U_E\,\langle U_r\rangle}{L_E}\,dt + \sqrt{\frac{2\,u_f^2\langle U_r\rangle}{L_E}}\,dW_E\,,\tag{8.43}$$

with $W_E$ a Wiener process in time (i.e., $\langle(dW_E)^2\rangle = dt$). For OU processes, exact integration of the trajectories is straightforward and we can proceed with the values of velocities at successive time intervals $\Delta t$, which corresponds to handling Markov chains. Starting from $U_s(t)$ at time $t$, we have

$$U_L(t + \Delta t) = U_s(t)\exp\left(-\frac{\Delta t}{T_L}\right) + u_f\sqrt{1 - \exp\left(-\frac{2\Delta t}{T_L}\right)}\,\zeta_L\,,\tag{8.44a}$$

$$U_E(t + \Delta t) = U_E(t)\exp\left(-\frac{\Delta t\,\langle U_r\rangle}{L_E}\right) + u_f\sqrt{1 - \exp\left(-\frac{2\Delta t\,\langle U_r\rangle}{L_E}\right)}\,\zeta_E\,,\tag{8.44b}$$

where $\zeta_L$ and $\zeta_E$ are sampled in two independent centered normalized Gaussian random variables, i.e., both $\zeta_L$ and $\zeta_E$ are $\mathcal{N}(0, 1)$. By construction of this two-step model, we have $U_s(t + \Delta t) = U_E(t + \Delta t)$ and $U_E(t) = U_L(t + \Delta t)$ (due to its transformation from $r$, time is actually fictitious in the Eulerian step). The Lagrangian and Eulerian steps can then be combined to yield

$$U_s(t + \Delta t) = U_s(t)\exp\left(-\Delta t\left[\frac{1}{T_L} + \frac{\langle U_r\rangle}{L_E}\right]\right)$$

$$+ u_f\sqrt{1 - \exp\left(-\frac{2\Delta t\,\langle U_r\rangle}{L_E}\right)}\,\zeta_E + u_f\exp\left(-\frac{\Delta t\,\langle U_r\rangle}{L_E}\right)\sqrt{1 - \exp\left(-\frac{2\Delta t}{T_L}\right)}\,\zeta_L\,.\tag{8.45}$$

The sum of the two independent Gaussian number can be lumped into a third one and straightforward calculation gives

$$U_s(t + \Delta t) = U_s(t) \exp\left(-\Delta t \left[\frac{1}{T_L} + \frac{\langle U_r \rangle}{L_E}\right]\right)$$

$$+ u_f \sqrt{1 - \exp\left(-2\Delta t \left[\frac{1}{T_L} + \frac{\langle U_r \rangle}{L_E}\right]\right)} \zeta_s \,, \qquad (8.46)$$

where $\zeta_s$ is another $\mathcal{N}(0, 1)$ random variable. In continuous time, this corresponds to the combined simple Langevin model for the velocity of the fluid seen $U_s$

$$dU_s = -\frac{U_s}{T_L^*} dt + \sqrt{\frac{2 u_f^2}{T_L^*}} dW \,, \qquad (8.47)$$

with $T_L^*$ given as in Eq. (8.33). From this simple two-step construction, we retrieve as an outcome the expression of the Lagrangian timescale of the fluid seen where the inverse of the timescales are added.

Even in this simplified case of turbulent flows, the weak point of the previous two-step model is clearly the formulation of the Eulerian step. Regardless of whether the resulting expression of the timescale of the fluid seen, cf. Eq. (8.33), is satisfactory or not, there are difficulties with respect to the predicted correlation in space. Indeed, from the assumption of a simple Langevin, as in Eq. (8.42), we have

$$\langle U_E(r_0 + r) U_E(r_0) \rangle = u_f^2 \exp\left(-\frac{r}{L_E}\right) \qquad (8.48)$$

which yields for the second-order moment of velocity differences over $r$, $\delta U_E(r) = U_E(r_0 + r) - U_E(r_0)$,

$$\langle (\delta U_E(r))^2 \rangle = 2 \left[u_f^2 - \langle U_E(r_0 + r) U_E(r_0) \rangle\right] = 2 u_f^2 \left[1 - \exp\left(-\frac{r}{L_E}\right)\right]. \qquad (8.49)$$

In the inertial range, where $r \ll L_E$, this implies $\langle (\delta U_E(r))^2 \rangle \simeq 2u_f^2 r/L_E$, whereas the proper Kolmogorov scaling should be $\langle (\delta U_E(r))^2 \rangle \simeq (\langle \epsilon \rangle r)^{2/3}$. To address this issue, it is instructive to consider first a simple model for two-particle relative motion.

### 8.3.2.2  A Reminder on Relative Dispersion

This situation is displayed in Fig. 8.4: at a time $t_0 = 0$, taken as an initial time, two fluid particles are located in a small region indicated as a zone of strong interaction (whose size is typically of the order of the Kolmogorov scale $\eta_K$). This means that

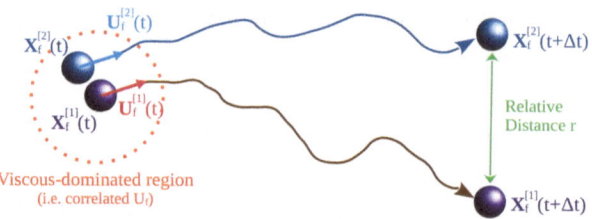

**Fig. 8.4** Representation of the two-fluid-particle relative dispersion situation. At an initial time $t$, two fluid particles are in a zone of strong interaction so that their velocities are identical. Over a time interval $\Delta t$, these two fluid particles leave this zone and evolve by independent simple Langevin models

viscous forces take over so that the two fluid particles have the same initial velocity, $U_f^{[1]}(0) = U_f^{[2]}(0) = U_0$. In terms of particle velocities, this amounts to a 'perfectly inelastic collision' (particles do not stick but their velocities relax immediately to the same value). Outside of this small viscous-dominated region, it is assumed that the two particle velocities are driven by two simple Langevin models

$$dU_f^{[i]} = -\frac{U_f^{[i]}}{T_L} \, dt + \sqrt{C_0 \langle \epsilon_f \rangle} \, dW^{[i]} \, , \qquad (8.50)$$

where the two Wiener processes, $W^{[i]}$, with $i = 1, 2$, are independent. The relative velocity, $\delta U_f^{[1-2]} = U_f^{[2]} - U_f^{[1]}$, follows therefore a similar Langevin model

$$d(\delta U_f^{[1-2]}) = -\frac{\delta U_f^{[1-2]}}{T_L} \, dt + \sqrt{2 \, C_0 \langle \epsilon_f \rangle} \, dW \, , \qquad (8.51)$$

where $W$ is another Wiener process. Given that the initial condition is $\delta U_f^{[1-2]}(0) = 0$, the exact solution is

$$\delta U_f^{[1-2]}(t) = \sqrt{2 \, C_0 \langle \epsilon_f \rangle} \exp\left(-\frac{t}{T_L}\right) \int_0^t \exp\left(\frac{s}{T_L}\right) \, dW(s) \, , \qquad (8.52)$$

from which we obtain

$$\langle (\delta U_f^{[1-2]})^2(t) \rangle = C_0 \langle \epsilon_f \rangle T_L \left[1 - \exp\left(-\frac{2t}{T_L}\right)\right] = 2u_f^2 \left[1 - \exp\left(-\frac{2t}{T_L}\right)\right] . \qquad (8.53)$$

The relative distance $r^{[1-2]}$ between the two fluid particles is obtained by integrating $\delta U_f^{[1-2]}(t)$, which gives

$$
r^{[1-2]}(t) = \int_0^t \delta U_f^{[1-2]}(s)\, ds
$$

$$
= \sqrt{2\, C_0 \langle \epsilon_f \rangle} \left( T_L W_t - T_L \exp\left(-\frac{t}{T_L}\right) \int_0^t \exp\left(\frac{s}{T_L}\right) dW_s \right), \qquad (8.54)
$$

from which a tedious but straightforward calculation shows that

$$
\langle (r^{[1-2]})^2(t) \rangle = 2\, C_0 \langle \epsilon_f \rangle T_L^2 \left\{ t - \frac{T_L}{2} \left[ 1 - \exp(-\frac{t}{T_L}) \right] \left[ 3 - \exp(-\frac{t}{T_L}) \right] \right\}.
$$
$$
(8.55)
$$

When the two fluid particles are outside the zone of strong interaction but still within the initial range, i.e., $t \ll T_L$, the interesting result is that

$$
\langle (\delta U_f^{[1-2]})^2(t) \rangle \simeq 2\, C_0 \langle \epsilon_f \rangle t \quad \text{and} \quad \langle (r^{[1-2]})^2(t) \rangle \simeq \frac{2}{3} C_0 \langle \epsilon_f \rangle t^3
$$

$$
\Longrightarrow \langle (\delta U_f^{[1-2]})^2 \rangle \sim \left( \langle \epsilon_f \rangle\, r^{[1-2]} \right)^{2/3}, \qquad (8.56)
$$

which shows that we retrieve the correct scaling for the Eulerian velocity statistics within the inertial range.

In physical terms, these results indicate that consistency with the expected Kolmogorov scaling both in time and space is obtained for a diffusive model in which fluid particles interact strongly when located nearby (within distances of the order of $\eta_K$) and evolve with a simple dynamical model involving a mean-field term (i.e., the return-to-equilibrium term of the Langevin models in Eq. (8.50)) and independent random forcing (i.e., the diffusive term driven by independent Wiener processes).

### 8.3.2.3 A New Two-step Stochastic Model

Coming back to the difficulty encountered with the correlated velocities generated by the Eulerian step with Eq. (8.42), it is seen that we are in fact trying to hold together two different regimes: on the one hand, a convective regime (where $r = \langle U_r \rangle t$) and, on the other hand, a diffusive one (where $r = (\langle \epsilon_f \rangle t^3)^{1/2}$). Since the Eulerian step is governed by the mean drift (see Sect. 6.1), the linear relation between $r$ and $t$ is relevant but the use of a Langevin model is more appropriate to reproduce velocity correlations in the diffusive regime. This suggests to consider a Langevin model formulated in terms of a rescaled time so that separation distances in the two regimes are statistically identical. More precisely, starting from a given initial time (to be discussed below), we introduce a rescaled time $\widehat{t}$ so that

$$r = \langle U_r \rangle t \simeq \left[ \langle \epsilon_f \rangle \widehat{t}^{\,3} \right]^{1/2} \Longrightarrow \widehat{t} = C \left( \frac{\langle U_r \rangle^2}{\langle \epsilon_f \rangle} \right)^{1/3} t^{2/3} \tag{8.57}$$

where $C$ is a constant (for example, we get from Eq. (8.56) that $C = (3/(2C_0))^{1/3}$).

In the discrete-time and Markov-chain setting, a modified two-step model can then be proposed. It consists in the same Lagrangian step, cf. Eq. (8.44a), and in an Eulerian step expressed by

$$U_E(t + \Delta t) = U_E(t) \exp \left( -\frac{\Delta \widehat{t}}{T_E} \right) + u_f \sqrt{1 - \exp \left( -\frac{2 \Delta \widehat{t}}{T_E} \right)} \, \zeta_E , \tag{8.58}$$

where $\Delta \widehat{t} = C \left( \langle U_r \rangle^{2/3} / \langle \epsilon_f \rangle^{1/3} \right) (\Delta t)^{2/3}$. As in Eq. (8.46), the two steps can be fused into one to yield

$$U_s(t + \Delta t) = U_s(t) \exp \left[ -\left( \frac{\Delta t}{T_L} + \frac{\Delta \widehat{t}}{T_E} \right) \right] + u_f \sqrt{1 - \exp \left[ -2 \left( \frac{\Delta t}{T_L} + \frac{\Delta \widehat{t}}{T_E} \right) \right]} \, \zeta_s , \tag{8.59}$$

whose behavior is more in line with the expected scaling laws in time and in space. Indeed, by replacing directly in the classical expression of the auto-correlation of first-order Markov chains, we get

$$\langle U_s(t) U_s(t + \Delta t) \rangle = u_f^2 \exp \left( -\frac{\Delta t}{T_L} - \frac{A \, (\Delta t)^{2/3}}{T_E} \right) , \tag{8.60}$$

where we have used $A = C \left( \langle U_r \rangle^{2/3} / \langle \epsilon_f \rangle^{1/3} \right)$. In the inertial range where $(\Delta t \ll T_L, T_E)$, this yields for the second-order moment of the velocity increments of $\delta U_s(\Delta t) = U_s(t + \Delta t) - U_s(t)$

$$\langle (\delta U_s(\Delta t))^2 \rangle \simeq 2u_f^2 \left( \frac{\Delta t}{T_L} + \frac{A \, (\Delta t)^{2/3}}{T_E} \right) \simeq C_0 \langle \epsilon_f \rangle \Delta t + C_E \, C \left( \langle \epsilon_f \rangle \langle U_r \rangle \, \Delta t \right)^{2/3} , \tag{8.61}$$

where $C_E$ is the constant entering the closure of $T_E$ (i.e., $T_E = 2u_f^2/(C_E \langle \epsilon_f \rangle)$), while it is recalled that $T_L = 2u_f^2/(C_0 \langle \epsilon_f \rangle)$). It is seen that we retrieve the linear relation in $\Delta t$ for the Lagrangian correlations and the 2/3 law for the Eulerian ones in terms of the separating distance $(\Delta r)_{\text{drift}} = \langle U_r \rangle \, \Delta t$ induced by the mean velocity drift over the time increment $\Delta t$. At this stage, it is tempting to rewrite Eq. (8.60), as

$$\langle U_s(t) \, U_s(t + \Delta t) \rangle = u_f^2 \exp(-\frac{\Delta t}{\widetilde{T}_L^*}) \quad \text{with} \quad \widetilde{T}_L^* = \frac{T_L}{1 + C_T \left( \dfrac{\langle U_r \rangle^2}{\langle \epsilon_f \rangle \Delta t} \right)^{1/3}} \cdot$$

(8.62)

However, though we are treating $\Delta t$ as a parameter in the time-discrete setting, it does not mean that $\widetilde{T}_L^*$ is the expression of a proper timescale since it is an explicit function of the time increment. Actually, a specific feature of this model is that, in contrast with previous formulations, the timescale of the velocity of the fluid seen is not an input but an output (see below). Another noteworthy point is that we have used each discrete time, $t_0 + k\Delta t$ (with $k \in \mathbb{N}$), as the initial time for the companion fluid-particle used in the reformulation of the Eulerian step, which means that the time step $\Delta t$ is the elapsed time from the moment of strong interaction used in the two-particle relative dispersion problem, (i.e. $t \equiv \Delta t$ as displayed in Fig. 8.5). This explains that we use the same relation between $\widehat{\Delta t}$ and $\Delta t$ as $\widehat{t}$ and $t$ in Eq. (8.57).

It appears therefore that the previous considerations have been developed somewhat loosely and need to be put on a more rigorous basis by formulating the corresponding stochastic differential equations in continuous time. Using the time rescaling given in Eq. (8.57) for the Eulerian step, the proposed equation for $U_s$ is obtained by applying the time change in Eq. (8.42) which, combined to Eq. (8.41), leads to

$$dU_s = -\frac{U_s}{T_L} \, dt - \frac{U_s}{T_E} \frac{2}{3} A(t - t_0)^{-1/3} \, dt + \sqrt{\langle \epsilon_f \rangle \left( C_0 + \frac{2C_E}{3} A \, (t - t_0)^{-1/3} \right)} \, dW \, .$$

(8.63)

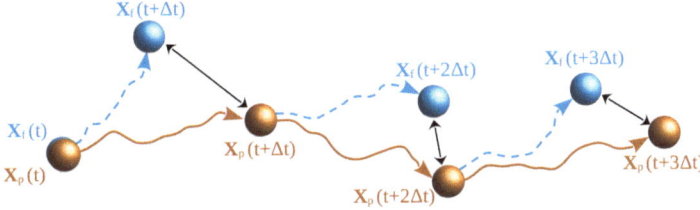

**Fig. 8.5** Representation of the two-step model in discrete time. As indicated by the black arrows, the correct spatial correlation is obtained at each time step between the velocity of the fluid seen $U_s(t + k\Delta t)$ at the particle locations $X_p(t + k\Delta t)$ and the velocity of its companion fluid-particle $U_f(t + k\Delta t)$ at the positions $X_f(t + k\Delta t)$ which are drifting away from the discrete particle trajectory due to the mean relative velocity slip. At each time step, the companion fluid-particle is re-initialized, as indicated by the dashed blue lines

However, the signification of the time(s) written as $t_0$ in this equation requires careful consideration.

From the two-particle relative dispersion study, it follows that the Eulerian step is devised to retrieve the correct correlation between the velocity of the fluid seen (the fluid velocity at the discrete particle location) and the velocity of a companion fluid particle whose location is moving away from the discrete particle according to the mean velocity drift. In Eq. (8.63), the time $t_0$ refers to the choice of the 'initial time' at which this companion fluid particle is released (this is the time at which these two fluid elements leave the 'strong interaction zone', see Fig. 8.4) and its value reflects a modeling choice for the size of the domain in which we wish to capture, at each time, the velocity correlations between these two fluid elements. Since the relative distance between the trajectory of a discrete particle and its companion fluid particle increases due to the mean velocity drift, it can grow unbounded and after a long-enough time lapse (i.e., $t - t_0 \gg L_f / \langle U_r \rangle$), these two fluid elements are so far apart that their velocities become independent. This is reflected in Eq. (8.63) by the fact that, if $t_0$ is a constant, then the drift and diffusion coefficients of the Eulerian step go to zero as $t - t_0$ increases. It is then more appropriate to consider that $t_0$ becomes a random variable whose value is somehow updated and attached to each discrete particle. For instance, in the discrete time setting presented above, we have implicitly assumed that this companion fluid particle is 're-initialized' at each time step, while $t - t_0 = \Delta t$. In the continuous time setting, this can be monitored by considering a fluid particle partner attached to the trajectory of each discrete particle. The evolution equation for the relative distance between the discrete particle and this fluid partner, say $\delta r_p^f$, is defined as

$$\frac{d\left(\delta r_p^f(t)\right)}{dt} = \langle U_r \rangle - k_{\text{spring}} \, \delta r_p^f(t) \,, \tag{8.64}$$

where $k_{\text{spring}}$ is the stiffness of an associated spring that links the trajectory of the discrete particle to the one of its fluid particle partner. Note that when $k_{\text{spring}} = 0$ this fluid particle partner is identical to the companion fluid particle described above. On the other hand, when $k_{\text{spring}}$ is very large, we are in the situation where the fluid particle partner is quickly pulled back towards the discrete particle, which corresponds to the case considered in the discrete time setting (re-initialization of the fluid particle partner at each time step). For intermediate value of $k_{\text{spring}}$, the evolution of $\delta r_p^f(t)$ governs how $t_0$ is updated, since we have $t_0 = \tau_{\delta r_p^f}^{\eta}$ where $\tau_{\delta r_p^f}^{\eta}$ are the instants when the discrete particle and its fluid particle partner enter a zone of strong interaction, i.e., when $|\delta r_p^f(\tau_{\delta r_p^f}^{\eta})| \leq \eta_K$. This two-step stochastic model is sketched in Fig. 8.6.

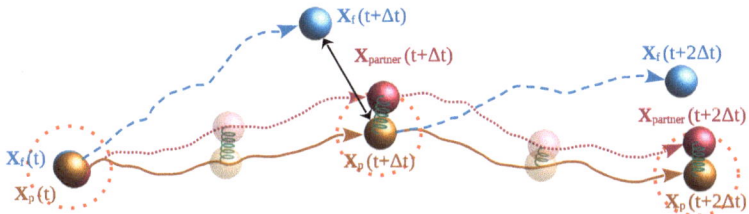

**Fig. 8.6** Sketch of the two-step model in continuous time. A fluid-particle partner is associated to each discrete particle trajectory, cf. Eq. (8.64), and is used to generate new companion fluid particles for the two-step model. The correct spatial correlations are still between the velocity of the fluid seen and the velocity of the companion fluid particle, as shown by the black arrows. This companion fluid partner is unchanged as long as the distance between the discrete particle and its partner remains larger than the size of the zone of strong interactions. However, whenever the particle partner enters a zone of strong interaction, a new companion fluid-particle is generated, as indicated by the dashed blue lines

With the complete formulation of the new two-step stochastic model, we can express the timescale of the velocity of the fluid seen that results from it. First, the evolution equation for the auto-correlation of the velocity of the fluid seen, $\mathcal{R}_s(t) = \langle U_s(\tau + t)\, U_s(\tau)\rangle$, is derived from Eq. (8.63) and is

$$\frac{d\mathcal{R}_s(t)}{dt} = -\frac{\mathcal{R}_s(t)}{T_L} - \frac{\mathcal{R}_s(t)}{T_E}\frac{2}{3}A t^{-1/3} \, , \tag{8.65}$$

where we have taken $t_0 = 0$. By direct integration, we get

$$\log\left(\frac{\mathcal{R}_s}{\mathcal{R}_s(0)}\right) = -\frac{t}{T_L} - \frac{A\, t^{2/3}}{T_E} \, . \tag{8.66}$$

Using $\mathcal{R}_s(0) = u_f^2$, we obtain therefore the same expression as the one used directly with the Markov chains with $\Delta t$ as the elapsed time, cf. Eq. (8.60). However, to work out the timescale of the velocity of the fluid seen $T_L^*$, we need to go back to its proper definition as the integral of the auto-correlation function for the velocity of the fluid seen, $\rho_{\mathcal{R}_s}(t) = \mathcal{R}_s(t)/u_f^2$. This gives

$$T_L^* = \int_0^{+\infty} \rho_{\mathcal{R}_s}(t')\, dt' = \int_0^{+\infty} \exp\left(-\frac{t'}{T_L} - \frac{A\, t'^{2/3}}{T_E}\right) dt' \, , \tag{8.67}$$

which, by making the change of variable $s = t'/T_L$, can be written as

$$T_L^* = T_L \int_0^{+\infty} \exp\left(-s - \frac{C_T\, A}{T_L^{1/3}} s^{2/3}\right) ds \, . \tag{8.68}$$

To use the same notation as in Sect. 6.1.1, this means that the equivalent of the Csanady factor, $T_L^*/T_L$ is

$$\mathcal{f}(\langle U_r \rangle, T_L, T_E, \langle \epsilon_f \rangle) = \int_0^{+\infty} \exp\left(-s - \frac{C_T A}{T_L^{1/3}} s^{2/3}\right) ds . \tag{8.69}$$

From these expressions of $A$ and of $T_L$, we get $A/T_L^{1/3} = (C_0/2)^{1/3} (\langle U_r \rangle/u_f)^{2/3}$ and the above integral is actually only a function of the ratio $\langle U_r \rangle/u_f$

$$\mathcal{f}(\langle U_r \rangle/u_f) = \int_0^{+\infty} \exp\left(-s - C_T(C_0/2)^{1/3} \left(\frac{\langle U_r \rangle}{u_f}\right)^{2/3} s^{2/3}\right) ds , \tag{8.70}$$

which is in line with the functional forms used in the various Csanady-like formulas presented previously. Some additional properties of the timescale of the velocity of the fluid seen resulting from this two-step model, cf. Eq. (8.68), can be given. First, since the integrand function is smaller than $\exp(-s)$, whose integral over $[0, +\infty]$ is equal to 1, then we get the following bounding interval $0 \leq \mathcal{f}(\langle U_r \rangle/u) \leq 1$. When the mean velocity drift $\langle U_r \rangle$ is negligible (i.e., $\langle U_r \rangle = 0$ in Eq. (8.68)), we retrieve the fluid-like behavior and $T_L^*(\langle U_r \rangle = 0) = T_L$, or $\mathcal{f}(\langle U_r \rangle = 0) = 1$. Second, $\mathcal{f}$ is a decreasing function of $\langle U_r \rangle/u_f$ and its behavior in the limit of large velocity drifts can be assessed. In that limit, we can neglect the first term in the exponential function in Eq. (8.68), corresponding to the Lagrangian step, which gives

$$T_L^*(\langle U_r \rangle/u_f \gg 1) \simeq T_L \int_0^{+\infty} \exp\left(-\frac{C_T A}{T_L^{1/3}} s^{2/3}\right) ds . \tag{8.71}$$

Writing $A = \tilde{C}\left(\langle U_r \rangle^{2/3}/\langle \epsilon_f \rangle^{1/3}\right)$, where $\tilde{C}$ is a constant which is now to be determined, we obtain

$$T_L^*(\langle U_r \rangle/u_f \gg 1) \simeq \frac{u_f T_E}{\langle U_r \rangle} \frac{1}{\tilde{C}^{3/2}} \left(\frac{2}{C_E}\right)^{1/2} \int_0^{+\infty} \exp\left(-v^{2/3}\right) dv . \tag{8.72}$$

Since $\int_0^{+\infty} \exp\left(-v^{2/3}\right) dv = 3\sqrt{\pi}/4$, we can therefore choose the constant $\tilde{C}$ so that

$$\frac{1}{\tilde{C}^{3/2}} \left(\frac{2}{C_E}\right)^{1/2} \frac{3\sqrt{\pi}}{4} = 1 , \tag{8.73}$$

showing that we retrieve the expected scaling of $T_L^*$ for very large velocity slips,

$$T_L^*(\langle U_r \rangle/u_f \gg 1) \simeq \frac{u_f T_E}{\langle U_r \rangle} = \frac{L_E}{\langle U_r \rangle} . \tag{8.74}$$

This corresponds to the limit situation of frozen turbulence, already stated with the formalism developed from the Extended Kolmogorov Hypothesis and given in Eq. (8.39b), since $\langle U_r \rangle T_L^*(\langle U_r \rangle / u_f \gg 1) = L_E$.

If the limit $T_L^*(\langle U_r \rangle = 0)$ is correct, the expression of $T_L^*$ from the two-step model needs to be refined for very small values of $\langle U_r \rangle$ (note that, as such, Eq. (8.70) predicts that the derivative of $F$ with respect to $\langle U_r \rangle$ is infinite at $\langle U_r \rangle = 0$). Indeed, in the derivations leading to Eq. (8.68), we not only retained a single zone of strong interaction but we also assumed that the Eulerian step was applied as soon as $t > 0$. This amounts to neglecting the time spent in the zone of strong interaction. Yet, by assuming that the size of the zone of strong interaction is of the order of the Kolmogorov scale $\eta_K$, there is always a minimum time, which we can estimate as $t_{min} \simeq \eta_K / \langle U_r \rangle$, to consider before switching on the additional terms coming from the time-rescaling model (when $t \leq t_{min}$, only the Lagrangian step can be retained). In the previous derivations, we implicitly assumed very high Reynolds-number turbulent flows, where $\eta_K$ becomes vanishingly small, and non-zero values of $\langle U_r \rangle$, so that $t_{min} \simeq 0$. It is, however, much sounder to regard $\eta_K$ as a small but non-zero value, which means that $t_{min}$ increases as $\langle U_r \rangle \to 0$. When $t_{min}$ becomes, for instance, of the order of the Lagrangian timescale $T_L$, then the Eulerian part of the two-step model would only be applied when velocities are already nearly uncorrelated and would only slightly modify the value of $T_L^*$ in comparison to $T_L$. Said otherwise, the two-step model is applied and is effective for the time range which is typically between $t_{min}$ and $T_L$ (or another similar timescale of large-scale motions) and the above derivations are valid provided that we have $U_r \geq \eta_K / T_L$, while we can expect that $T_L^* \simeq T_L$ when $\langle U_r \rangle \leq \eta_K / T_L$. It results that $\mathcal{F}$, or $T_L^* / T_L$, should have a zero-slope behavior in the immediate vicinity of $\langle U_r \rangle \simeq 0$.

Another remark is in order. Even though it is difficult to work out an analytical formula for the resulting timescale $T_L^*$, it must be noted that the above expressions were obtained in the limit where $k_{spring}$ is small so that $t_0$ was taken as a constant. When this is not the case, it follows from the description of the model that Eq. (8.63) involves successive terms such as $t - \tau_\eta(\delta r_p^f)$ where $\tau_\eta(\delta r_p^f)$ are random numbers (similar to threshold or first-passage times), which makes the above derivations intractable. Only numerical estimations of $T_L^*$ seem accessible, which is why the model was also described in the discrete-time setting.

### 8.3.2.4 Extension to the General 3D Non-homogeneous Case

Should they prove worth pursuing, these developments need to be generalized to three-dimensional situations. Actually, the same remarks made at the end of Sect. 8.3.1 for the propositions from the Extended Kolmogorov hypothesis apply here. Given that the Eulerian time scales are different depending on whether we consider the direction aligned with the mean relative velocity or a transverse one, cf. Eq. (8.40), care should be taken to distinguish between directions when using the two-step model, as in Eq. (8.63). The safest way is probably to make a change in the frame of reference so as to have directions aligned or transverse with respect to the mean relative velocity. Note that this implies a local change in the frame of reference that is, however, not dependent on each particle (it depends only on mean properties) and is the same for all the particles contained in a small volume around a point of interest. Thus, there is no need to account for non-inertial effects.

## 8.4     Accounting for the Signature of Coherent Structures

When modeling turbulence, two main standpoints are typically followed. The first one, referred to as the statistical approach, is represented by the Kolmogorov theories (K41 and K62, see Sect. 5.3) as well as similar analyses centered around finding proper scaling for quantities of interest. With the development of numerical tools (DNS), regarded as numerical experiments and used in complement of actual experiments, another viewpoint has emerged in which instantaneous flow patterns are identified and analyses of turbulence properties carried out in terms of the dynamics of these structures. This corresponds to the geometrical approach, which centers around identifying a set of structures whose characteristics could explain key turbulence features (e.g., the so-called hairpin vortices in near-wall boundary layers on which there is now a vast literature). Surprisingly, a middle-of-the-road formulation has rarely been considered. In this third approach, one starts with a given picture of turbulence including models for a selected number of fully-characterized coherent structures and develops a statistical model where these structures are explicitly treated. In that sense, the objective is neither to average out their presence (as in the statistical approach) nor to predict their formation (as in the geometrical approach) but to account for their dynamical role in a statistical description. In a nutshell, we wish to mimic given coherent structures using stochastic models.

### 8.4.1   A Twofold Picture of Turbulence

To illustrate these ideas, we develop a twofold description in which turbulent flows are regarded as composed of two regions with different statistical behavior:

1. a random background flow, occupying most of the flow domain, well-described by the K41 theory (including, perhaps, some features of the K62 refined theory), and where the PDF models presented in the previous sections provide a satisfactory description of one-particle and one-point statistics;
2. a collection of coherent structures (exhibiting stable, ordered and long-life patterns) taken to be intense vortex filaments, whose statistical description needs to be detailed and which can account for intermittency and deviations of high-order moments from Kolmogorov predictions.

Since we are drawing a roughly-cut sketch to exemplify the new approach, we limit ourselves to simple scaling arguments to support the selection of vortex filaments and express their characteristic features.

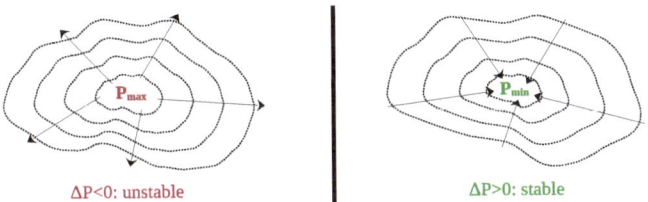

$\Delta P < 0$: unstable $\qquad\qquad\qquad\qquad\qquad$ $\Delta P > 0$: stable

**Fig. 8.7** Representation of an unstable flow structure with positive pressure (left side) at the center and where the pressure-gradient force is oriented outward and a stable one with negative pressure (right side) at the center and where the pressure-gradient force is oriented inward

Retaining vortex tubes is related to simple equilibrium reasoning based on the pressure-gradient force and, thus, on the minimum of potentials (recalling that pressure is basically a potential), as illustrated in Fig. 8.7.

Indeed, the positive pressure side (i.e., $P_f \geq \langle P_f \rangle$) is not stable and follows approximately a Gaussian distribution. On the other hand, the negative pressure side (i.e., $P_f \leq \langle P_f \rangle$) corresponds to more stable flow patterns and this part of the pressure distribution deviates from Gaussianity [4]. In short, stable structures are associated with high vorticity, often under the form of vortex tubes. In the present description, we do not consider the classical vortex structures, such as vortex sheets, whose study has received a great deal of attention in past decades [5] but some intense events inherent to the turbulence cascade although they do not explain the mean turbulent kinetic energy dissipation rates (see detailed analyses in [6,7] about this difference). The most intense vorticity events are obtained by considering velocity differences of the order of the large-scale fluctuating velocity $u_f$ over the smallest size which is of the order of the Kolmogorov length scale $\eta_K$. This yields the typical vorticity of these intense vortex tubes as $\Omega_{f,vt} \sim u_f/\eta_K$. This estimate is also obtained by considering that the kinetic energy of large scales $u_f^2$ is converted entirely into small-scale rotational energy:

$$r^2 \Omega_f^2 \simeq u_f^2 \,, \text{ which for } r \simeq \eta_K \implies \Omega_{f,vt} \sim \frac{u_f}{\eta_K} \,. \tag{8.75}$$

From the equation satisfied by the enstrophy $\Omega_f^2$ (with $\Omega_f^2 = \boldsymbol{\Omega}_f \cdot \boldsymbol{\Omega}_f$), which is

$$\frac{D\Omega_f^2}{Dt} = 2\,\Omega_{f,i}\,S_{f,ij}\,\Omega_{f,j} + \nu_f\,\Omega_{f,i}\,\Delta\Omega_{f,i}$$

$$= 2\,\Omega_{f,i}\,S_{f,ij}\,\Omega_{f,j} - \nu_f\left(\frac{\partial\Omega_{f,i}}{\partial x_j}\right)\left(\frac{\partial\Omega_{f,i}}{\partial x_j}\right) + \nu_f\Delta(\Omega_f^2)\,, \tag{8.76}$$

where $S_{f,ij} = 1/2\left(\partial U_{f,i}/\partial x_j + \partial U_{f,j}/\partial x_i\right)$ is the flow strain rate, we see that the stretching term $\Omega_{f,i}\,S_{f,ij}\,\Omega_{f,j}$ must be positive to balance sink terms due to molecular viscosity and maintain the structure for a longer-than-average life time.

If we regard these vortex tubes as relatively stable structures maintained out of equilibrium due to the energy flux from the background flow, we are referring to 'dissipative structures' that are by-products of the turbulence cascade [7] and we can apply the theorem of minimum entropy production rate [8,9] for such structures (with the 'entropy' here embodied by the enstrophy $\Omega_f^2$). This means that, for $\Omega_f^2$ remaining approximately constant, the entropy production or stretching term, $\Omega_{f,i} S_{f,ij} \Omega_{f,j}$, is minimum inside the vortex filaments. It follows that the vorticity $\mathbf{\Omega}_f$ is essentially aligned, or correlated, with the intermediate eigenvalue of the strain rate tensor $S_{f,ij}$ and that, in these structures, this eigenvalue is positive. This positive value of the intermediate eigenvalue is in line with results from DNS studies [6,7].

When velocities are rescaled to the velocity $u_f$ of the large scales, the typical length of dissipative structures is the Taylor length scale $\lambda_f$. Pursuing the idea of transforming kinetic energy coherently into rotational energy, we can estimate the length of a vortex tube by imagining that an initial sphere of radius $\lambda_f$ is squeezed into a tube until the cross-section radius of that tube is of the order of $\eta_K$. From the incompressibility of the flow, which entails that volume is conserved, we have that the typical length of a vortex tube is $l_{f,vt} \sim \lambda_f^3/\eta_K^2$ and is therefore of the order of the fluid large scales $l_{f,vt} \sim L_f$, as found in fundamental analyses [7]. The emerging simplified picture is sketched in Fig. 8.8 with a region of high vorticity (but small dissipation) inside the tubes and a region of high dissipation (but reduced vorticity) around these tubes. The life-time, or typical time scale, of these structures is harder to estimate and, to the authors' knowledge, has not been considered in the fundamental DNS studies dedicated to the role of vortex tubes, reflecting that probing turbulence from a Lagrangian viewpoint has only emerged in recent years. For the present concern, we refer to it as $\tau_{f,vt}$, whose value needs to be derived from first principles or extracted from DNS investigations (a first, and rather rough, guess could be to assume that $\tau_{f,vt} \sim L_f/u_f$, which is the eddy-turn-over time of large scales).

To assess the macroscopic role played by the vortex tubes, the key point is to estimate the number of these tubes in a box of volume $L_f^3$ or, conversely, the probability to encounter such a structure. If we consider a locally homogeneous flow

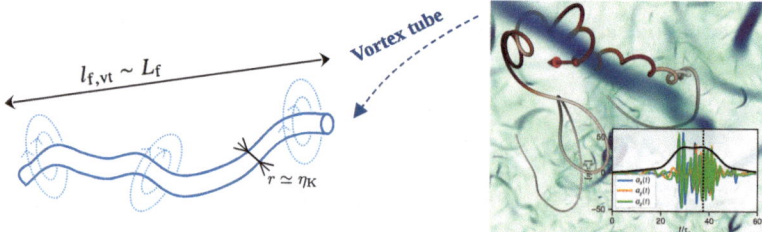

**Fig. 8.8** Representation of a vortex filament with its characteristic spatial scales (left) and an illustration from DNS simulations showing the interaction between tracer particles and vortex filaments (trajectories are colored according to the particle acceleration, shown in the inset on the right). Reprinted with permission from [10]. © 2019 Springer Nature. All rights reserved

domain, this probability is the same as the volume fraction, $\alpha_{f,vt}$, occupied by the vortex tubes. In the spirit of Kolmogorov-like theories, we imagine then a process of successive breakdowns whereby a large-scale particle with size $L_f$ and velocity $u_f$ yields a number $n_{f,vt}$ of small-scale particles having a rotation velocity $\Omega_{f,vt}$ and, at least, one typical length of the order of $\eta_K$ (as for the cross-section radius of the intense vortex filaments). From the conservation of angular momentum $r^2\Omega_f$, we get

$$L_f^2 \left(\frac{u_f}{L_f}\right) = u_f L_f \simeq n_{f,vt} \left(\eta_K^2 \Omega_{f,vt}\right) = n_{f,vt} \left(\eta_K u_f\right) , \tag{8.77}$$

which gives

$$n_{f,vt} \sim \left(\frac{L_f}{\eta_K}\right) \quad \text{and} \quad \alpha_{f,vt} = n_{f,vt} \left(\frac{\eta_K^2 L_f}{L_f^3}\right) \sim \frac{\eta_K}{L_f} . \tag{8.78}$$

We can then evaluate the contribution of the vortex tubes to the budget of the first moments of the dissipation rate (or the enstrophy). Since the dissipation rate associated to a vortex tube can be written as $\epsilon_{f,vt} = \nu_f(u_f^2/\eta_K^2) = \langle\epsilon_f\rangle(\lambda_f^2/\eta_K^2)$, we obtain that the contribution of the dissipation rate due to vortex tubes to the budget of the mean dissipation rate is negligible since

$$\langle\epsilon_{f,vt}\rangle = \langle\epsilon_f\rangle \frac{\lambda_f^2}{\eta_K^2} \frac{\eta_K}{L_f} \implies \langle\epsilon_{f,vt}\rangle \sim \langle\epsilon_f\rangle \left(\frac{L_f}{\eta_K}\right)^{-1/3} . \tag{8.79}$$

However, for the second-order moment of the dissipation rate, we get that

$$\langle\epsilon_{f,vt}^2\rangle = \langle\epsilon_f\rangle^2 \frac{\lambda_f^4}{\eta_K^4} \frac{\eta_K}{L_f} \implies \langle\epsilon_{f,vt}^2\rangle \sim \langle\epsilon_f\rangle^2 \left(\frac{L_f}{\eta_K}\right)^{1/3} , \tag{8.80}$$

which shows that the vortex tubes provide the main contribution to the variance of the dissipation rate (similar results hold for the enstrophy). Note that we retrieve a scaling relation in line with Kolmogorov K62 hypothesis with an exponent $\mu = 1/3$. To modify the crude estimation of the number of vortex tubes and its volumetric fraction to account for a Kolmogorov exponent $\mu$ different from the value above (usually $\mu$ is estimated at a slightly lower value than $1/3$, often $\mu \simeq 0.2 - 0.25$), it can be assumed that $\alpha_{vt}$ scales as $(\eta_K/L_f)^{4/3-\mu}$, from which the same estimations yield that we have now

$$\langle\epsilon_f^2\rangle \sim \langle\epsilon_f\rangle^2 (L_f/\eta_K)^\mu . \tag{8.81}$$

Additional results of interest concern Lagrangian statistics, e.g. the acceleration of a fluid particle $\mathbf{A}_f = d\mathbf{U}_f/dt$. As already indicated, the K41 scaling for the variance of a fluid particle acceleration is $a_K^2 \sim \left(\langle\epsilon_f\rangle^3/\nu_f\right)^{1/2}$. When a fluid particle is caught by a vortex tube, the magnitude of its acceleration can be estimated as $a_{f,vt} \sim u_f^2/\eta_K$, from which we get that $a_{f,vt}^2 \sim a_K^2 \left(\lambda_f L_f\right)/\eta_K^2$. The second-order moment of a fluid particle acceleration deviates therefore from the K41 prediction $\langle \mathbf{A}_f^2 \rangle \sim a_K^2$, since we obtain for the vortex tube contribution that

$$\langle a_{f,vt}^2 \rangle \sim a_K^2 \frac{\lambda_f}{\eta_K} \sim a_K^2 \, Re_{\lambda_f}^{1/2} \,, \tag{8.82}$$

a relation found in [11] but not explained by classical scaling [12]. Note that if we retain the more general estimate $\alpha_{f,vt} \simeq (\eta_K/L_f)^{4/3-\mu}$, the predicted scaling becomes $\langle a_{f,vt}^2 \rangle \sim a_K^2 \, Re_{\lambda_f}^{3\mu/2}$, with a smaller exponent, in line with more recent probes [13]). Drawing on these results, it is interesting to note that the effects of coherent structures are different on Eulerian and Lagrangian statistics. At a fixed location, these structures are swept by, making little marks on Eulerian statistics, whereas fluid particles can be trapped in them for longer times (of the order of $\tau_{f,vt}$) leaving therefore potentially more significant traces on Lagrangian statistics (see an interesting account in [13]).

### 8.4.2  Putting the Twofold Description into Practice

In spite of its crudely-cut features (for instance, we only considered one coherent structure and only one type of vortex tubes instead of a distribution of sizes, intensity, length, etc.), this twofold model is useful to reveal that, not only classical signatures of intermittency are retrieved, but also that complex scaling of fluid particle statistics can be captured. This is relevant for stochastic modeling since descriptions such as the one presented above are easily implemented in the PDF framework. In practice, it means that we consider two stochastic models for the velocity of the fluid seen by discrete particles, one for the background 'structure-less flow' based typically on a Langevin model as introduced in preceding sections (we call this model, 'model BG') and one model for the flow in the near vicinity of a vortex tube typically by generating a Burgers-like flow once the vortex tube orientation and intensity are generated (we call this model, 'model VT'). To that effect, we introduce a parent process, $S(t)$, attached to each particle, which governs the random switches from one flow region to the other: say $S(t) = 1$ when the particle is considered in the background flow and where the model BG is applied for $\mathbf{U}_s$, and $S(t) = 2$ when the particle is captured by one vortex tube and where the model VT is applied for $\mathbf{U}_s$. This parent process can be modeled by a generalized Poisson process jumping at random times between $S(t) = 1$ and $S(t) = 2$ at the end of duration times (i.e., the times spent in each flow type) which, according to the properties of Poisson processes, are random variables following exponential laws determined by two timescales: $\tau_{f,BG}$ the mean duration within the background flow and $\tau_{f,vt}$ the mean duration within a vortex tube. Note that $\tau_{f,vt}$ was already introduced above, while $\tau_{f,BG}$ is derived from the volumetric fraction of the vortex

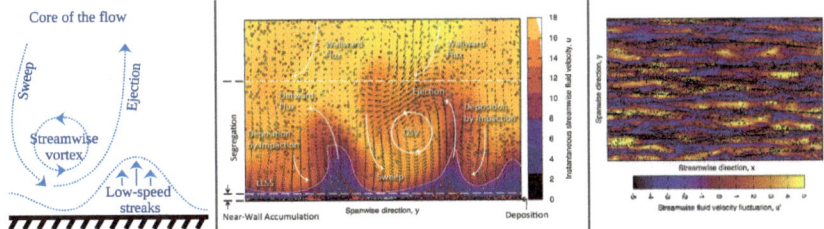

**Fig. 8.9** Representation of near-wall turbulent structures (left) together with side view (middle) and top view (right) of a DNS simulation showing the vorticity regions as well as the particle dynamics in the near-wall region. Reprinted with permission from [16]. © 2017 Springer-Verlag Wien. All rights reserved

tubes $\alpha_{f,vt}$ as for inter-collision times in the kinetic theory (this is, thus, a foretaste of developments to come in Sect. 12). These conditions define the $2 \times 2$ transition matrix $\mathbf{M}(\Delta t)$ which governs the Markov chain $S(k\Delta t)$ at discrete times $t = k\Delta t$ ($k \in \mathbb{N}$) with $M_{ij}(\Delta t)$ the probability to have $S(t + \Delta t) = j$ at time $t + \Delta t$ conditioned on $S(t) = i$ at time $t$:

$$
\mathbf{M}(\Delta t) = \begin{pmatrix} e^{-\Delta t/\tau_{f,BG}} & 1 - e^{-\Delta t/\tau_{f,BG}} \\ 1 - e^{-\Delta t/\tau_{f,vt}} & e^{-\Delta t/\tau_{f,vt}} \end{pmatrix}. \tag{8.83}
$$

To the best of the authors' knowledge, the development of such structure-based stochastic models is still at an early stage (though there are similarities with ideas put forward in [14]), apart from one model of particle transport in wall-boundary layers for deposition problems [15]. In this model, a random combination of three typical structures of the near-wall region, namely sweeps, ejections and diffusion, was used to capture the particle flux towards the wall where a sticking condition was enforced for particles touching the surface. The model represents a simplified picture based on insights provided by DNS results, cf. Fig. 8.9.

The construction of this model is in line with the twofold description introduced above and is applied using a zonal decomposition: the background 'structure-less' flow is identified with the bulk of the flow while the three selected structures are explicitly simulated only in a thin zone in the immediate vicinity of a wall surface (say between $0 \leq z^+ \leq 100$, with $z^+ = z u_{f,*}/\nu_f$ the wall-distance $z$ normalized by the so-called wall units with $u_{f,*}$ the fluid friction velocity, see accounts in [17,18]). The random interplay of structures is clearly visible in the trajectories of discrete particles and satisfactory statistics for the particle deposition rate are retrieved over the whole range of particle inertia, as shown in Fig. 8.10.

This good agreement for the deposition rate can lead us to believe that the newly-proposed model is validated for particle deposition studies. If we remember however the discussion developed in Sect. 7.5, we know that this result is not enough to reach such a conclusion since it must be complemented by the demonstration that the same model respects known fluid profiles in the tracer-particle limit. In particular,

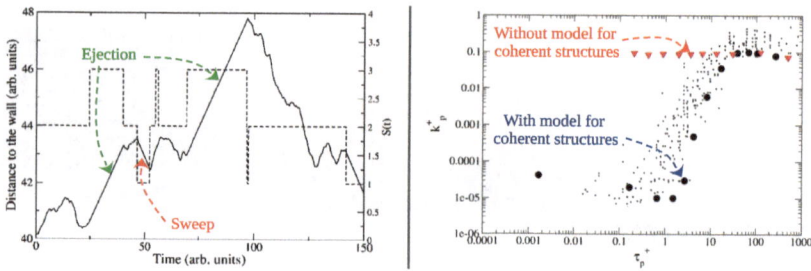

**Fig. 8.10** Results obtained with a stochastic model for near-wall coherent structures showing the effect of the random switches between coherent structures on particle trajectories (left) and on the deposition rate (right). Reprinted with permission from [15]. © 2008 AIP Publishing. All rights reserved

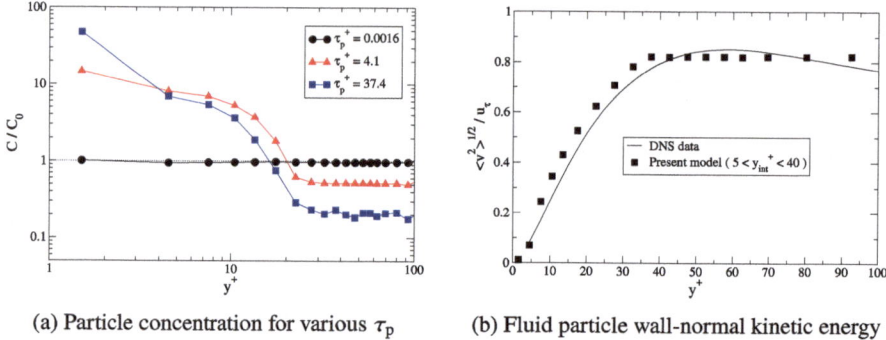

(a) Particle concentration for various $\tau_p$     (b) Fluid particle wall-normal kinetic energy

**Fig. 8.11** Results obtained with the model for near-wall coherent structures: profiles of particle concentration for different particle inertia timescale $\tau_p$ (**a**); profile of the wall-normal kinetic energy for fluid particles compared to DNS results (**b**). Reprinted with permission from [15]. © 2008 AIP Publishing. All rights reserved

checking that the fluid particle concentration remains equal to one is essential to accept the stochastic model as corresponding to a valid PDF description of a fluid system (see the discussion in [19, section 4.7]). This was indeed carried out in the analysis of the new model based on the random interplay of structures [15] as shown in Fig. 8.11a where it is evident that the concentration of fluid particles remains uniform in the domain when particle inertia, measured by its relaxation timescale $\tau_p$, goes to zero. Once this is verified, a further assessment is provided by the profile of the wall-normal kinetic energy displayed in Fig. 8.11b which shows that the model reproduces very well the typical DNS profile. As a more qualitative but still interesting remark, the profiles obtained for discrete particles with non-vanishing inertia plotted in Fig. 8.11a indicate that the model predicts near-wall particle accumulation which is one of the signatures of the particle concentration effect in turbulent boundary layers. Note that these concentration profiles were obtained by replacing the deposition wall-boundary condition by an elastic rebound one to avoid the effect of the wall sink term created by the irreversible sticking condition and to correspond to situations often considered when studying particle dynamics and concentration effects in wall boundary layers.

### 8.4.3 On Different Ways to Address Complexity

In more general terms, it is worth underlying that these structure-based models represent a new way to address complexity: instead of developing a single model which is likely to become complex itself, the leading idea is to capture complexity as the result of random changes between sub-models, each of which remaining simple, which are used to describe the components of an heterogeneous system. In that sense, this introduces us to similar concerns and approaches in so-called complex fluids, which are now addressed.

## References

1. S. Pope, *Turbulent Flows* (Cambridge University Press, Cambridge, 2000). https://doi.org/10.1017/CBO9780511840531
2. J.P. Minier, Phys. Fluids **33**(2), 023312 (2021). https://doi.org/10.1063/5.0039249
3. J.P. Minier, S. Chibbaro, S.B. Pope, Phys. Fluids **26**(11), 113303 (2014). https://doi.org/10.1063/1.4901315
4. A. Vincent, M. Meneguzzi, J. Fluid Mech. **225**, 1 (1991). https://doi.org/10.1017/S0022112091001957
5. P.G. Saffman, *Vortex Dynamics* (Cambridge University Press, Cambridge, 1995). https://doi.org/10.1017/CBO9780511624063
6. J. Jiménez, A.A. Wray, P.G. Saffman, R.S. Rogallo, J. Fluid Mech. **255**, 65 (1993). https://doi.org/10.1017/S0022112093002393
7. J. Jimenez, A.A. Wray, J. Fluid Mech. **373**, 255 (1998). https://doi.org/10.1017/S0022112098002341
8. S. Kjelstrup, D. Bedeaux, *Non-Equilibrium Thermodynamics of Heterogeneous Systems* (World Scientific, Singapore, 2008). https://doi.org/10.1142/11729
9. S. Kjelstrup, D. Bedeaux, E. Johannessen, J. Gross, *Non-equilibrium Thermodynamics for Engineers* (World Scientific, Singapore, 2010). https://doi.org/10.1142/10286
10. L. Bentkamp, C.C. Lalescu, M. Wilczek, Nature Commun. **10**(1), 3550 (2019). https://doi.org/10.1038/s41467-019-11060-9
11. P.K. Yeung, S.B. Pope, J. Fluid Mech. **207**, 531 (1989). https://doi.org/10.1017/S0022112089002697
12. P. Yeung, Ann. Rev. Fluid Mech. **34**(1), 115 (2002). https://doi.org/10.1146/annurev.fluid.34.082101.170725
13. A.S. Lanotte, L. Biferale, G. Boffetta, F. Toschi, J. Turbulence **14**(7), 34 (2013). https://doi.org/10.1080/14685248.2013.839882
14. A.R. Kerstein, J. Fluid Mech. **392**, 277 (1999). https://doi.org/10.1017/S0022112099005376
15. M. Guingo, J.P. Minier, Phys. Fluids **20**(5) (2008). https://doi.org/10.1063/1.2908934
16. C. Marchioli, Acta Mech. **228**(3), 741 (2017). https://doi.org/10.1007/s00707-017-1803-x
17. C. Henry, J.P. Minier, G. Lefèvre, Adv. Colloid Interface Sci. **185**, 34 (2012). https://doi.org/10.1016/j.cis.2012.10.001
18. J.P. Minier, Progress Energy Combustion Sci. **50**, 1 (2015). https://doi.org/10.1016/j.pecs.2015.02.003
19. S.B. Pope, Progress Energy Combustion Sci. **11**(2), 119 (1985). https://doi.org/10.1016/0360-1285(85)90002-4

# Dispersed Two-Phase Flows and Complex Fluids

**9**

**Abstract**

Though they are also referred to as 'soft matter', we designate by 'complex fluids' fluids whose microstructure contains specific components (the solute) having sizes and relaxation timescales much larger than those of the atoms/molecules of the 'simple fluid' (the solvent) in which they are embedded. The differences between the dynamical behavior of simple and complex fluids is worth describing. In simple fluids, the microstructure remains regular and locally well-described by classical equilibrium statistical mechanics regardless of whether the material is flowing or is at rest. At the hydrodynamical level of description, this is reflected by the linear force-flux relations, such as the Newton, Fick and Fourier laws. In complex fluids, e.g. polymeric fluids, colloidal suspensions, liquid crystals, etc., the slow and non-equilibrium response of the solute usually implies more involved rheological laws. To avoid formulating these rheological properties at the macroscopic level of description, one approach consists in adding structural variables to the state vector and proposing a coarse-grained model to capture the solute response to external forces. Given the correspondences with the PDF approach, it is interesting to investigate the connections between complex fluids and turbulent dispersed two-phase flows.

**Chapter Content** After describing similarities and differences with complex fluids in Sect. 9.1, we analyze how fluctuations and random terms enter the modeling picture in Sect. 9.2. We then discuss how to account for Brownian noise in Sect. 9.3.

© The Author(s) 2025
J.-P. Minier et al., *Understanding Turbulent Systems*,
Lecture Notes in Physics 1039, https://doi.org/10.1007/978-3-031-84466-9_9

## 9.1    Similarities and Differences with Complex Fluids

When we consider typical models for complex fluids, such the Rouse model for polymers [1], Brownian dynamics for colloids [2] or the Maier-Saupe theory for liquid crystals [3], it appears that complex fluids and turbulent dispersed two-phase flows have marked similarities. First, we are considering similar physical entities, with those handled in complex fluids appearing as smaller-size versions of the ones considered in dispersed two-phase flows. This correspondence is displayed in Fig. 9.1, which exhibits the obvious couples: polymers and flexible fibers; colloids and discrete particles; rod-like molecules and rod-like particles. Second, there are also similarities between the main physical processes at play: fluctuations and dispersive effects on the one hand, particle-particle interactions and tendency to self-assemble on the other hand. For complex fluids, fluctuations are related to thermal noise while 'turbulence noise' is present in one-particle PDF descriptions of dispersed two-phase flows (note that repulsive particle-particle interactions have also a dispersive effect). The tendency to self-assemble is manifested by phase transitions in soft matter systems, e.g., solid/liquid/gas transitions in fluids, isotropic/nematic liquid crystals, paramagnetic/ferromagnetic states (if we regard them as a complex fluid), while attractive forces govern the formation of aggregates in disperse two-phase flows.

There are, nevertheless, differences in the overall objective that is pursued. Indeed, even if a specific model is applied to describe the solute dynamics, the purpose in soft matter is to close the equations governing the evolution of a complex fluid treated as a single entity. This is typically the case of polymeric fluids where the simulation of polymer motion is used to obtain the additional stress tensor, through the Kramers expression, which is added to the (Newtonian) stress tensor of the solvent. In other words, the aim is to model the complete system, solvent+solute, treated as one equivalent, but now complex, fluid. This is not what is being done in dispersed two-phase flow modeling where the continuous (the solvent) and the dispersed (the solute) phases are treated separately as two physical systems in

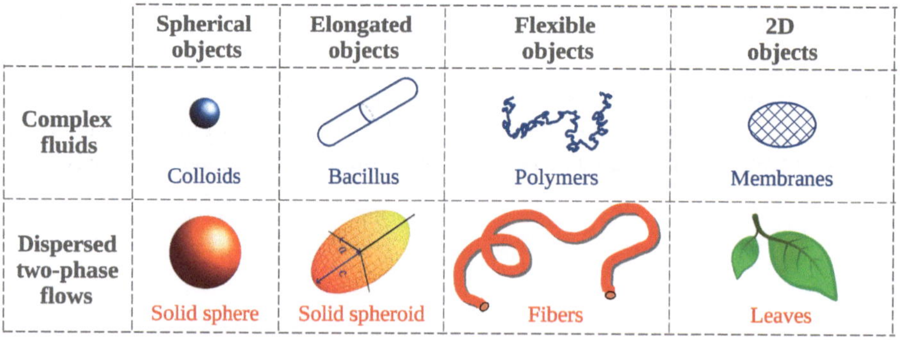

|  | Spherical objects | Elongated objects | Flexible objects | 2D objects |
|---|---|---|---|---|
| Complex fluids | Colloids | Bacillus | Polymers | Membranes |
| Dispersed two-phase flows | Solid sphere | Solid spheroid | Fibers | Leaves |

**Fig. 9.1**  Correspondence between complex fluids and disperse two-phase flows

interaction through the exchange of mass, momentum and energy. Said otherwise, complex fluids are handled as an equivalent 'homogeneous two-component system' in which particle inertia is not considered significant. In effect, most, if not all, formulations of complex fluids neglect the solute inertia and rely on the Brownian limit. This raises, however, interesting issues in the formulation of such fast-variable techniques as well as in the expression of random terms and corresponding Fokker-Planck type of closures.

## 9.2   Neither Complete Order Nor Disorder: Thermal Versus Turbulence Noise

To bring out the differences related to how random terms are introduced in complex fluids and in turbulent dispersed two-phase flows, it is interesting to consider polymers and flexible fibers. We start with the chosen mechanical description before discussing the stochastic model applied in each case.

For polymers, it is sufficient to retain the classical Rouse model in which a polymer is described by a chain of $N_b$ beads connected by $N_b - 1$ Hookean springs (see bottom left panel in Fig. 9.2) [1]. This is a phenomenological model where the beads represent large segments of monomers so that a continuum description of the solvent is justified while the springs represent 'entropic forces' used to account for the elimination of many degrees of freedom from the actual polymer.

For flexible fibers, even though a continuous representation using curvilinear coordinates is possible (as in the slender body theory or in the Cosserat equations [4]), a typical approach consists in representing a flexible fiber as a chain of $N_{rod}$ connected rigid rods (see bottom right panel in Fig. 9.2). This allows to capture fiber bending and twisting motions while preserving their connectivity through constraints applied at the hinges between two adjacent rigid rods. In a way, this description amounts to a discretized version of the continuous one, each rod being a stiffened segment of the fiber. Note that a description in terms of beads connected

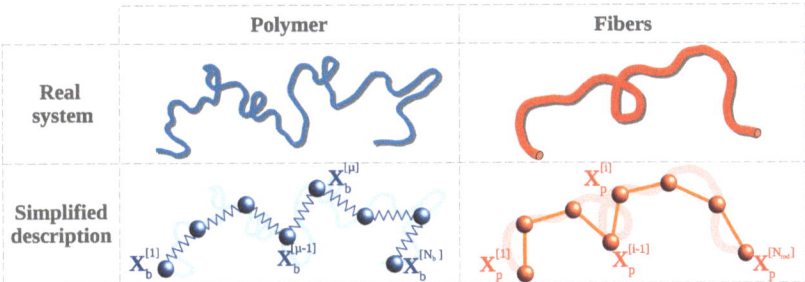

**Fig. 9.2** Comparison between models for flexible objects: polymers are represented by a chain of $N_b$ beads connected by Hookean springs (in complex fluids) while fibers can be simplified to $N_{rod}$ connected rigid rods (in dispersed two-phase flows)

by springs (e.g. based on a finitely-extendable-nonlinear-elastic—FENE—model) is also possible, with the rigid-segment description obtained as the limit of infinite spring stiffness [5,6]. To capture the evolution of such a rod chain, we can choose to follow the center of mass of each segment, in which case each elementary mechanical object is a non-spherical rigid particle. Similarly to the polymer bead-spring description, another choice consists in following the equivalent of beads which are here the hinges around which two adjacent rods can rotate. For the sake of having two similar descriptions and without any loss of generality for the present analysis, we retain the latter mechanical representation (see Fig. 9.2).

In polymer theory, bead inertia is usually neglected and the Rouse model is formulated in the frame of Brownian dynamics. This means that the particle state vector $\mathbf{Z}_b$ attached to each polymer gathers the bead positions, i.e., $\mathbf{Z}_b = (\mathbf{X}_b^{[\mu]})_{\mu=1,\dots,N_b}$. In the absence of external forces and leaving out the interactions between polymers (dilute polymer systems), the evolution equations for $\mathbf{Z}_b$ are (for $\mu = 1, \dots, N_b$)

$$
d\mathbf{X}_b^{[\mu]} = \mathbf{U}_f\left(t, \mathbf{X}_b^{[\mu]}(t)\right) dt + \frac{\tau_p}{m_p} \mathbf{F}_{\text{spring}}^{R,[\mu]} dt + \sqrt{\frac{2k_B\,\Theta_f\,\tau_p}{m_p}} d\mathbf{W}^{[\mu]}, \qquad (9.1)
$$

with $m_p$ the bead mass and where the friction coefficient is written as $m_p/\tau_p$. In this equation, $\mathbf{F}_{\text{spring}}^{R,[\mu]}$ represents the forces acting on the bead labeled $[\mu]$ due to the connecting springs, as expressed in the Rouse model. We introduce first a formulation in which springs can have different stiffness (i.e., $\mathbf{k}_{\text{spring}} = (k_{\text{spring}}^{[\mu]})$ for $\mu = 1, \dots, N_b - 1$), which for a generic model that depends on the stiffness $\mathbf{k}_{\text{spring}}$ gives the following expression for the force $\mathbf{F}_{\text{spring}}^{[\mu]}$ acting on the bead $[\mu]$

$$
\mathbf{F}_{\text{spring}}^{[\mu]} = \begin{cases} k_{\text{spring}}^{[1]}\left(\mathbf{X}_b^{[2]} - \mathbf{X}_b^{[1]}\right) & \text{if } \mu = 1, \\ k_{\text{spring}}^{[\mu]}\left(\mathbf{X}_b^{[\mu+1]} - \mathbf{X}_b^{[\mu]}\right) - k_{\text{spring}}^{[\mu-1]}\left(\mathbf{X}_b^{[\mu]} - \mathbf{X}_b^{[\mu-1]}\right) & \text{if } 1 < \mu < N_b, \\ -k_{\text{spring}}^{[N_b-1]}\left(\mathbf{X}_b^{[N_b]} - \mathbf{X}_b^{[N_b-1]}\right) & \text{if } \mu = N_b. \end{cases}
$$

$$(9.2)$$

In the Rouse model, all springs have the same stiffness $k_{\text{spring}}^R$ so that the vector $\mathbf{k}_{\text{spring}}^R$ is a vector of identical values (i.e., $k_{\text{spring}}^{R,[\mu]} = k_{\text{spring}}^R$, $\forall\mu$), which means that in Eq. (9.1) we have $\mathbf{F}_{\text{spring}}^{R,[\mu]} = \mathbf{F}_{\text{spring}}^{[\mu]}(k_{\text{spring}}^R)$.

For flexible fibers, rod inertia is not necessarily neglected leading to a particle state vector with includes the bead positions and velocities, i.e., $\mathbf{Z}_p = (\mathbf{X}_p^{[i]}, \mathbf{U}_p^{[i]})_{i=1,N_{\mathrm{rod}}}$. A simple dynamical model has then the form

$$d\mathbf{X}_p^{[i]} = \mathbf{U}_p^{[i]} , \tag{9.3a}$$

$$d\mathbf{U}_p^{[i]} = -\frac{\mathbf{U}_p^{[i]} - \mathbf{U}_f\left(t, \mathbf{X}_p^{[i]}(t)\right)}{\tau_p} \, dt + \frac{1}{m_p}\mathbf{F}_{\mathrm{spring}}^{[i]}(\lambda) \, dt + \sqrt{\frac{2k_B \, \Theta_f}{m_p \, \tau_p}} \, d\mathbf{W}^{[i]} ,$$
$$\tag{9.3b}$$

where the vector $\lambda/m_p$ can be thought of as a set of spring stiffness. When the flexible fiber is made up by connected rigid rods, $\lambda$ represents the Lagrange multipliers used to enforce the non-extensibility of each rod, obtained from the constraint that $|\mathbf{X}_p^{[i+1]} - \mathbf{X}_p^{[i]}|$ remains constant (for $i = 1, \ldots, N_{\mathrm{rod}} - 1$). Note that Brownian effects are retained in Eqs. (9.3) but are expressed by a white-noise term appearing in the velocity equation, cf. Eq. (9.3b), rather than directly in the position equation, cf. (9.3), as in Brownian dynamics (see further discussion in Sect. 9.3).

Regardless of how thermal noise is accounted for, it is seen from Eq. (9.1) and Eq. (9.3b) that its introduction is straightforward and consists in an isotropic diffusion term in front of a Wiener process. A crucial point is that, at the bead or particle level of description, thermal noise is independent between distinct locations. This means that the formulation given, for instance, in Eq. (9.1) remains exact regardless of the fact that we track one or a set of $N$ particles.

Far different is the situation with respect to turbulent flows in the case of non-fully resolved velocity fields. Note that this situation does not arise for polymers transported by laminar flows (most of the time, even simple laminar shear flows), where the velocities at the bead locations, $(\mathbf{U}_f(t, \mathbf{X}_b^{[\mu]}(t)))_{\mu=1,N_b}$ in Eq. (9.1), are known. The same remains valid for the velocity equation of each segment or bead of the flexible fiber, cf. Eq. (9.3b), but only provided that we are dealing with fully-resolved turbulent flows by which we have access to the values of the instantaneous velocities at the bead locations, $(\mathbf{U}_f(t, \mathbf{X}_p^{[i]}(t)))_{i=1,N_{\mathrm{rod}}}$. In the case of high Reynolds-number turbulent flows or when we have only access to a reduced statistical description of the velocity field, we are then faced with the difficulty of having to reconstruct the set of the $N_{\mathrm{rod}}$ fluid velocities seen, $(\mathbf{U}_s^{[i]}(t) = \mathbf{U}_f(t, \mathbf{X}_p^{[i]}(t)))_{i=1,N_{\mathrm{rod}}}$. In that case, the Langevin models studied in preceding sections cannot be applied since it corresponds to a one-particle PDF model and, therefore, to a one-point type of closure. As already emphasized, there is no spatial information in present one-particle model for the velocity of the fluid seen. It is thus clear that a set of Langevin models for $(\mathbf{U}_s^{[i]}(t))_{i=1,N_{\mathrm{rod}}}$, with a drift vector $\mathbf{A}_s^{[i]}$ and diffusion matrix $\mathbf{B}_s^{[i]}$ (which need not be precised for the present discussion), such as

$$d\mathbf{U}_s^{[i]}(t) = \mathbf{A}_s^{[i]} \, dt + \mathbf{B}_s^{[i]} \, d\mathbf{W}^{[i]} , \tag{9.4}$$

but with independent Wiener processes, $(\mathbf{W}^{[i]}(t))_{i=1,N_{\mathrm{rod}}}$, between each rod segment or bead [i] would grossly misrepresent the spatial correlations which are one of the hallmarks of turbulent flows. In other words, we are dealing with 'turbulence noise' which is neither order nor complete disorder. For general non-homogeneous turbulent flows where velocities can deviate from Gaussianity, generating a set of $N$ fluid velocities seen, $(\mathbf{U}_s^{[i]}(t))_{i=1,N}$, that respect known correlations in time (this is already achieved) but also in space (this is an open issue) is a challenging task.

Comparing the situations of polymers in laminar flows and flexible fibers in non-fully resolved turbulent flows is therefore interesting to illustrate that modeling 'turbulence noise' is far more complicated than adding a $k_B\,\Theta_f$ diffusion term to a deterministic mechanical model.

### 9.2.1  A Numerical Example for Trumbells

To illustrate this discussion, we present in Fig. 9.3 numerical results obtained in [7] with a coarse-grained model of a polymer chain in which only three beads are retained and which, for this reason, is called a trumbbell. This corresponds apparently to a much-simplified description of a polymer chain but tracking the locations of the three beads in a given fluid flow allows to capture many of a polymer intricate motions, such as bending and twisting, as well as the stretched or coiled configurations.

In the study whose results are displayed in Fig. 9.3, the three beads making up one trumbbell are connected with rigid rods which means that when the beads are aligned stretching of the trumbbell is not simulated but could be easily included in the description by replacing the rigid rods by springs. On the other hand,

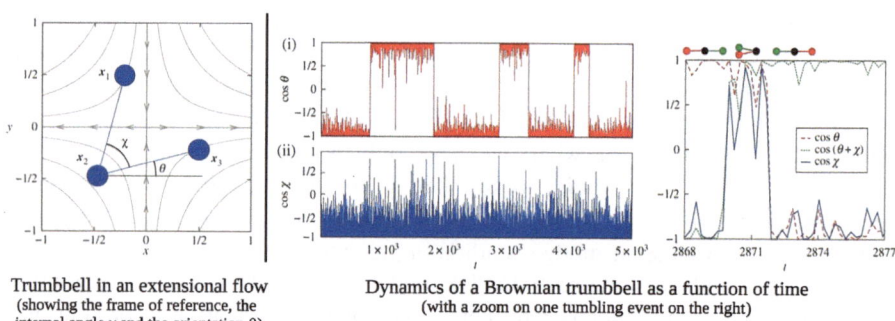

Trumbbell in an extensional flow
(showing the frame of reference, the
internal angle χ and the orientation θ)

Dynamics of a Brownian trumbbell as a function of time
(with a zoom on one tumbling event on the right)

**Fig. 9.3** Simulation of a trumbbell in an extensional fully-resolved flow and subject to Brownian effects [7]. By following the evolution of the angles $\theta$ and $\chi$ shown on the left side, the traces left by a tumbling event, when the trumbbell goes from one stretched configuration to the inverted one by going through an intermediate coiled configuration, are captured and can be reconstituted, as shown in the right-hand side. Reprinted with permission from [7]. © 2016 The Royal Society. All rights reserved

more complicated motions such as internal loops and possible entanglement of the polymer with itself in a coiled configuration cannot be captured. Nevertheless, this coarse-grained description is sufficiently rich, for instance compared to a dumbbell where only two connected beads are considered, and interesting features are reproduced. Only particle positions are retained in the particle state vector, which means that $\mathbf{Z}_b = (\mathbf{X}_b^{[\mu]})_{\mu=1,2,3}$ with evolution equations as in Eqs. (9.1) using infinite stiffness (see the comment following Eqs. (9.3) on the use of the spring stiffness as Lagrange multipliers to represent rigid rods). For our purpose, the relevant point is that introducing Brownian motion in a such a description of a polymer chain is straightforward and consists simply in adding three separate Wiener processes as done on the right-hand side of Eqs. (9.1). When such a trumbbell is placed in a given (laminar) fluid flows which is here an extensional flow, it tends to be extended or pulled towards a stretched configuration but there is a competition with pure disorder effects induced by the separate random contributions to the three bead positions. By tracking the time-evolution of the defining angles shown in the left-hand side of Fig. 9.3, it is seen that the model can reproduce tumbling events, when one trumbbell goes from one stretched configuration to the inverted stretched alignment by passing through a coiled configuration, as shown in the right-hand side of Fig. 9.3 (note that the positions of the red and green beads represented above the right-hand side plot are inverted after going through an intermediate coiled configuration). Stimulated by such possibilities, we may be tempted to carry out similar models for fibers in a turbulent fluid flow. However, in the case of a non-fully-resolved turbulent flow, we would be at a loss to do so since we would have to model the three contributions of the random or flow-unresolved parts in the three bead position evolution equations. Contrary to Brownian motion, these are correlated events and require a three-point PDF description which is out of reach of present model formulations. This example confirms therefore that the bottleneck is not about the mechanical description of a flexible fiber but in the generation of a physically-meaningful 'turbulence noise'.

## 9.2.2   Focusing on Brownian Motion

In the present work we focus essentially on small discrete particles treated as point particles or small round objects rather than elongated ones. In that sense, the above issue is less sensitive or, at least, does not appear as our first concern. Yet, if we go back to the couple made up by colloids and discrete particles, cf. Fig. 9.1, we can wonder whether the description of colloidal suspensions, in particular Brownian motion and the diffusive limit (Brownian dynamics), is contained in present disperse two-phase flow models. In the next section, we consider the different ways to represent Brownian motion in discrete particle dynamical models and how one goes from one formulation based on particle position and velocity to the classical expression of Brownian motion with the Einstein diffusion coefficient introduced in the particle position equation. These considerations are helpful to introduce the notion of fast-variable elimination techniques and the over-damped limit of Langevin equations in a simple setting before being discussed in more details in Chap. 10.

## 9.3 Different Ways to Account for Brownian Motion

The introduction of Brownian effects in particle stochastic models is fairly well-established and goes back to the contributions of [8] and [9]. To account for Brownian motion, it is best to consider first Brownian particles in a fluid at rest. Following Langevin's original point of view [9], the classical formulation consists in writing the particle velocity equation as the sum of a linear return-to-equilibrium term due to fluid friction and a random one written in terms of the increments of a Wiener process. In a one-dimensional notation, this gives

$$dX_p = U_p \, dt \, , \tag{9.5a}$$

$$dU_p = -\frac{U_p}{\tau_p} \, dt + K \, dW \, , \tag{9.5b}$$

where $K$ is a diffusion coefficient to be determined. From statistical mechanics and the equipartition of energy, we know that $1/2 \, m_p \langle U_p^2 \rangle$ reaches a constant value in the long-time limit which is equal to $1/2 \, k_B \, \Theta_f$. Then, Ito calculus yields that

$$\frac{d \langle U_p^2 \rangle}{dt} = 0 \quad \Longrightarrow \quad K^2 = \frac{2 \langle U_p^2 \rangle}{\tau_p} = \frac{2 k_B \, \Theta_f}{m_p \, \tau_p} \, , \tag{9.6}$$

which is the form already given in Eq. (9.3b). Note that this a straightforward expression of the fluctuation-dissipation theorem, as already noted in Sect. 4.3.

When $\tau_p$ and $K$ are constants, $U_p$ is a simple OU process and the exact behavior of the position of a Brownian particle can be worked out before taking the limit $\tau_p \to 0$ (for example by substituting $r$ with $X_p$, $T_L$ with $\tau_p$, and $\sqrt{C_0 \langle \epsilon_f \rangle}$ with $K$ in Eq. (8.54)). There is, however, an often-used short-cut method which consists in neglecting the particle acceleration in the limit of small inertia (manifested here by small values of the relaxation timescale $\tau_p$). For example, if we consider a general evolution equation with the fluid velocity seen in the drag force (when the fluid is not at rest) and a force $F_p$ due to external actions or to other particles,

$$dU_p = \frac{U_s - U_p}{\tau_p} \, dt + F_p \, dt + K \, dW \, , \tag{9.7}$$

we get from $dU_p \simeq 0$ that the particle position equation, Eq. (9.5a), becomes

$$dX_p = U_s \, dt + (\tau_p F_p) \, dt + (\tau_p K) \, dW \, . \tag{9.8}$$

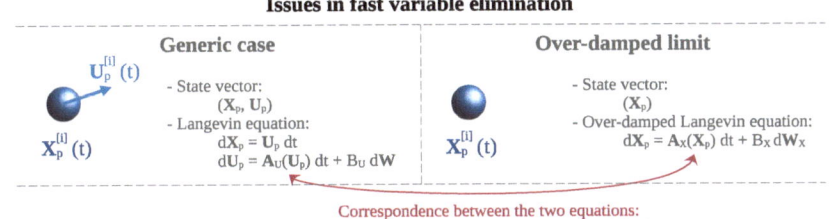

**Fig. 9.4** Representation of the key issue for the over-damped Langevin

This corresponds to the diffusive behavior of Brownian particles, from which we retrieve the classical Einstein relation for the position diffusion coefficient since

$$\mathcal{D} = \frac{1}{2}\left(\tau_{\mathrm{p}}K\right)^2 = \frac{k_{\mathrm{B}}\,\Theta_{\mathrm{f}}}{3\pi\,d_{\mathrm{p}}\rho_{\mathrm{f}}\nu_{\mathrm{f}}}\,, \tag{9.9}$$

and the classical formulation for the Brownian particle position equation

$$\mathrm{d}X_{\mathrm{p}} = U_{\mathrm{s}}\,\mathrm{d}t + \tau_{\mathrm{p}}F_{\mathrm{p}}\,\mathrm{d}t + \sqrt{2\mathcal{D}}\,\mathrm{d}W\,. \tag{9.10}$$

This is the basis of the slaving principle (introduced in Sect. 5.2), obtained by this heuristic formulation of fast-variable elimination techniques (see also Fig. 9.4). When the degrees of freedom are separated into slow and fast variables (here, the particle position and velocity, respectively), and when the drift and diffusion coefficients are functions of the slow variables, the same reasoning is assumed to remain valid leading to a diffusive limit for the slow variable. In the present case, this still gives Eq. (9.8) with $\tau_{\mathrm{p}} = \tau_{\mathrm{p}}(t, X_{\mathrm{p}})$ and $K = K(t, X_{\mathrm{p}})$.

Though intuitive, this short-cut method of the slaving principle raises two concerns. The first one is the loose manner with which the limit $\tau_{\mathrm{p}} \to 0$ is taken and how terms such as $(\tau_{\mathrm{p}}K)$ and $(\tau_{\mathrm{p}}F_{\mathrm{p}})$ are regarded in that limit. To get around this difficulty, we introduce a small parameter $\chi$ and consider the following model

$$\mathrm{d}X_{\mathrm{p}}^{(\chi)} = U_{\mathrm{p}}^{(\chi)}\,\mathrm{d}t\,, \tag{9.11a}$$

$$\mathrm{d}U_{\mathrm{p}}^{(\chi)} = \frac{U_{\mathrm{s}} - U_{\mathrm{p}}^{(\chi)}}{\chi\,\tau_{\mathrm{p}}}\,\mathrm{d}t + \frac{1}{\chi}F_{\mathrm{p}}\,\mathrm{d}t + \frac{K}{\chi}\,\mathrm{d}W\,, \tag{9.11b}$$

where the timescale varies as $\chi\,\tau_{\mathrm{p}}$ while the force and the diffusion coefficient scale as $F_{\mathrm{p}}/\chi$ and $K/\chi$, respectively. This allows to formalize the previous manipulations and give a more precise meaning to the limit of vanishing particle inertia since we have $\tau_{\mathrm{p}} \to 0$ while $K \to +\infty$ but with $\tau_{\mathrm{p}} K$ remaining constant. Other parametrizations have been proposed but have been shown to be equivalent or even of a less general application for time-dependent force, i.e. when $F_{\mathrm{p}} = F_{\mathrm{p}}(t, X_{\mathrm{p}})$

(see a comprehensive discussion of this point in [10, section 9.3]). The short-cut method of fast-variable elimination takes now a more satisfactory expression since we have

$$U_{\mathrm{p}}^{(\chi)}\mathrm{d}t \simeq U_{\mathrm{s}}\,\mathrm{d}t + \left(\tau_{\mathrm{p}} F_{\mathrm{p}}\right)\,\mathrm{d}t + \left(K\,\tau_{\mathrm{p}}\right)\mathrm{d}W\,, \tag{9.12}$$

leading, in a more rigorous manner, to the diffusive limit of $\mathrm{d}X_{\mathrm{p}}^{(\chi)}$ when $\chi \to 0$, which is Eq. (9.8). Note that the particle mean kinetic energy $\langle (U_{\mathrm{p}}^{(\chi)})^2 \rangle$ scales now as $1/2K^2\,\tau_{\mathrm{p}}/\chi$, cf. Eq. (9.6), so that it becomes infinite in the limit $\chi \to 0$ but we retrieve that the diffusive coefficient, $\chi\,\tau_{\mathrm{p}}\,\langle (U_{\mathrm{p}}^{(\chi)})^2 \rangle)$ tends to a finite limit, which is the Einstein diffusion coefficient.

The second concern is about the validity of the assumption that the diffusive limit can be derived by neglecting the particle acceleration, i.e. by enforcing $\mathrm{d}U_{\mathrm{p}}^{(\chi)} = 0$. In particular, does this short-cut method remain valid when the coefficients are functions of the particle position and, therefore, of the slow variable itself?

### 9.3.1   A First Glimpse into Model Reduction

Through these relatively simple manipulations of Brownian motion, we are touching a much broader issue. Indeed, going from Eqs. (9.11) to the limit model equation, Eq. (9.8) or (9.10), means that the dimensional of the state vector is reduced: instead of handling a state vector made up by particle position and velocity $\mathbf{Z}_{\mathrm{p}} = (\mathbf{X}_{\mathrm{p}}, \mathbf{U}_{\mathrm{p}})$, we are only tracking $\mathbf{Z}_{\mathrm{p}} = (\mathbf{X}_{\mathrm{p}})$ in the diffusive limit. In that sense, the technical questions related to the proper way to apply fast-variable elimination are connected to the all-important underlying theme of model reduction which appears therefore at the forefront of our investigations and which is addressed in more details in Chap. 10.

## References

1. P.E. Rouse Jr., J. Chem. Phys. **21**(7), 1272 (1953). https://doi.org/10.1063/1.1699180
2. J.C. Chen, A.S. Kim, Adv. Colloid Interface Sci. **112**(1–3), 159 (2004). https://doi.org/10.1016/j.cis.2004.10.001
3. M.J. Stephen, J.P. Straley, Rev. Mod. Phys. **46**(4), 617 (1974). https://doi.org/10.1103/RevModPhys.46.617
4. A. Lindner, M. Shelley, in *Fluid-Structure Interactions in Low-Reynolds-Number Flows* (Royal Society of Chemistry, 2015), pp. 168–192. http://pubs.rsc.org/en/content/chapter/bk9781849738132-00168/978-1-78262-849-1
5. M. Somasi, B. Khomami, N.J. Woo, J.S. Hur, E.S. Shaqfeh, J. Non-Newtonian Fluid Mech. **108**(1–3), 227 (2002). https://doi.org/10.1016/S0377-0257(02)00132-5
6. C. Henry, G. Krstulovic, J. Bec, Phys. Rev. E **98**(2), 023107 (2018). https://doi.org/10.1103/PhysRevE.98.023107
7. E.L.C.V.M. Plan, D. Vincenzi, Proc. R. Soc. A: Math. Phys. Eng. Sci. **472**(2194), 20160226 (2016). https://doi.org/10.1098/rspa.2016.0226

8. A. Einstein, et al., Ann. Phys. **17**(549–560), 208 (1905)
9. P. Langevin, CR Acad. Sci. (Paris) **146**, 530 (1908)
10. J.P. Minier, Phys. Rep. **665**, 1 (2016). https://doi.org/10.1016/j.physrep.2016.10.007

# Fast-Variable Elimination and Local or Non-local Constitutive Relations

# 10

**Abstract**

Where do we stand now? Since Chap. 5, we have been following up the choice to adopt a Markovian approach, which led us to a particle state vector made up by the discrete particle positions, their velocities and the velocity of the fluid seen $\mathbf{Z}_p = (\mathbf{X}_p, \mathbf{U}_p, \mathbf{U}_s)$. In Chaps. 6 and 8, we brought forward the physical reasons that justify to represent the velocity of the fluid seen by a Langevin model and proposed new research directions. We can wrap up our progress by saying that $\mathbf{Z}_p$ is modeled as a stochastic diffusion process. This implies that each stochastic particle is characterized by the knowledge of a 9-component vector while the PDF $p(t; \mathbf{y}_p, \mathbf{V}_p, \mathbf{V}_s)$ is the solution of a Fokker-Planck equation in a 9-dimensional space. We are thus evolving in a high-dimensional sample space, whose dimension is likely to increase if we consider additional phenomena requiring to add extra variables. The different ways to account for Brownian motion described in the preceding chapter suggest something enticing: can we eliminate some variables from the above description and derive a reduced description? What are the physical arguments allowing such projections into a reduced space to be made and, when justified, what is the form of SDEs characterizing this reduced stochastic model?

**Chapter Content** After further precisions given on this search in Sect. 10.1, recent results on the over-damped limit of Langevin models are recalled in Sect. 10.2. These results form the basis for the important analysis presented in Sect. 10.3 which investigates the conditions allowing local closure laws to be derived from non-local ones. These developments are first obtained for fluid particle scalars and velocity and are extended to discuss the concept of turbulent viscosities and the consequences on turbulent dispersed two-phase flow modeling in Sect. 10.4.

© The Author(s) 2025
J.-P. Minier et al., *Understanding Turbulent Systems*,
Lecture Notes in Physics 1039, https://doi.org/10.1007/978-3-031-84466-9_10

## 10.1    The Search for Reduced Descriptions

These questions appear also, with perhaps even stronger impetus, for the set of equations satisfied by the mean fields corresponding to the moments of the PDF. In that respect, it is instructive to consider the first two particle-velocity moments obtained from the PDF methodology recalled in Sect. 4.1.1, which are

$$\frac{\partial \alpha_p \rho_p}{\partial t} + \frac{\partial \left( \alpha_p \rho_p \langle U_{p,k} \rangle \right)}{\partial x_k} = 0 \, , \tag{10.1a}$$

$$\frac{\partial \langle U_{p,i} \rangle}{\partial t} + \langle U_{p,k} \rangle \frac{\partial \langle U_{p,i} \rangle}{\partial x_k} = -\frac{1}{\alpha_p \rho_p} \frac{\partial \left( \alpha_p \rho_p \langle u_{p,i} \, u_{p,k} \rangle \right)}{\partial x_k} + \langle A_{p,i} \rangle \, . \tag{10.1b}$$

The first equation, Eq. (10.1a), is the mass continuity equation, while Eq. (10.1b) is the mean particle momentum equation involving $\langle u_{p,i} \, u_{p,k} \rangle$, which is referred to as the particle kinetic tensor. In turn, this particle kinetic tensor is the solution of a transport equation, triggering a succession of additional transport equations for various velocity second-order moments (to be revisited later in Sect. 10.4.2). In this approach, it is tempting to close the above system of equations by proposing a local expression of the particle kinetic tensor based on the equivalent of a particle turbulent viscosity, or similar local closure laws. When addressed as such, it is difficult not to lose track of the implicit assumptions that could support such reductions. It is nevertheless essential to be aware of these assumptions to assess the true physical content of the resulting closed set of mean-field equations.

In the following, we address these issues by considering first the tracer-particle limit and revisiting results for fluid particles in turbulence. The correspondence between the formulations obtained in the over-damped Langevin limit and local or non-local closures is indeed an important aspect at the forefront of research on these topics. The recent results obtained in the fluid limit turn out to be general properties and are helpful to discuss consequences for turbulent dispersed two-phase flows.

## 10.2    The Over-Damped Langevin Limit in Turbulent Flows

In [1, section 9.3], it is shown that the most general way to derive the limit of over-damped Langevin equations is, for a particle state vector which includes particle position and velocity $\mathbf{Z} = (\mathbf{X}, \mathbf{U})$, to consider the following system of equations

$$dX_i = U_i \, dt \, , \tag{10.2a}$$

$$dU_i = G_{ij}(t, \mathbf{X}(t)) \left[ U_j - \Phi_j(t, \mathbf{X}(t)) \right] dt + \sigma_{ij}(t, \mathbf{X}(t)) \, dW_j \, , \tag{10.2b}$$

and to study the limit $\chi \to 0$ of a parametrized version where a small parameter $\chi$ is used to represent the desired scaling of the drift and diffusion coefficients

$$dX_i^{(\chi)} = U_i^{(\chi)} \, dt \, , \tag{10.3a}$$

$$dU_i^{(\chi)} = \frac{1}{\chi} G_{ij}(t, \mathbf{X}^{(\chi)}(t)) \left[ U_j^{(\chi)} - \Phi_j(t, \mathbf{X}^{(\chi)}(t)) \right] dt + \frac{1}{\chi} \sigma_{ij}(t, \mathbf{X}^{(\chi)}(t)) \, dW_j \, . \tag{10.3b}$$

As already indicated with Brownian particles in Sect. 9.3, the introduction of the parameter $\chi$ formalizes the limit we wish to take: using a scalar version for the sake of simplicity, this means that we consider $G^{-1} \to 0$ and $\sigma \to +\infty$ but so that $G^{-1}\sigma$ remains constant. In the above system of equations, this is done by multiplying directly the drift and diffusion coefficients appearing in Eq. (10.2) by $1/\chi$ while the same notations are kept for these coefficients in Eq. (10.3b). In a physics-oriented analysis, it is best to identify first a small parameter and describe then how the drift and diffusion coefficients scale with it, as shown below for the case of fluid particles in turbulent flows. However, we are first concerned with the correct mathematical limit of the parametrized system, Eqs. (10.3), in which $\mathbf{G}$ and $\sigma$ are now regarded as regular functions. In these equations, the specific dependencies of the drift and diffusion coefficients were kept but are left out from now on for the sake of simplicity. Moreover, the matrix $\mathbf{G}$ is assumed to be invertible and $\Phi(t, \mathbf{x})$ is a known field that is also considered as regular.

It is demonstrated in [1, section 9.3] that, in the limit $\chi \to 0$, the over-damped Langevin limit model is a stochastic diffusion process for particle positions which is

$$dX_i = \Phi_i \, dt - \frac{\partial G_{ij}^{-1}}{\partial x_k} A_{kj} dt + \left( G^{-1}\sigma \right)_{ij} dW_j \, , \tag{10.4}$$

where the matrix $\mathbf{A}$ stands for $A_{ij} = \lim_{\chi \to 0} \left( \chi \langle U_i^{(\chi)} U_j^{(\chi)} \rangle \right)$ and is the solution of the matrix equation

$$\mathbf{GA} + \mathbf{AG}^{\perp} = -\sigma\sigma^{\perp} \, . \tag{10.5}$$

To apply these results to the case of fluid particles in turbulent flows, for which we use the notation $\mathbf{Z}_f = (\mathbf{X}_f, \mathbf{U}_f)$, the relevant choice for the field $\Phi(t, \mathbf{x})$ is to take

$$\phi_i = \langle U_{f,i} \rangle + G_{ij}^{-1} \left( \frac{1}{\rho_f} \frac{\partial \langle P_f \rangle}{\partial x_j} \right) \, , \tag{10.6}$$

which means that $\langle \mathbf{U}_f \rangle$ and $\langle P_f \rangle$ are treated as regular fields whereas the finite nature of the matrix $\mathbf{A}$ translates that the kinetic energy grows unbounded as $1/\chi$ (this is in line with the remark made concerning Brownian motion in Sect. 9.3). More precisely, this corresponds to a situation where the first-order moment of the velocity field remains regular but where its second-order moments grow as $1/\chi$, implying that $\mathbf{A}$ is also given by $A_{ij} = \lim_{\chi \to 0} \left( \chi \langle u_i^{(\chi)} u_j^{(\chi)} \rangle \right)$ that is as functions of the velocity fluctuating components which appear as the rapidly-varying variables. The introduction of the mean-pressure gradient in $\phi$ is explained by the fact that it should scale as $1/\chi$ to compensate the divergence of the Reynolds stresses in the mean Navier-Stokes equations if we want the mean velocity field to remain regular. To justify this scaling, we consider the GLM, cf. Eqs. (6.11) or (6.21), which we rewrite with a slightly more general form as

$$dX_{\mathrm{f},i} = U_{\mathrm{f},i}\, dt \,, \tag{10.7a}$$

$$dU_{\mathrm{f},i} = -\frac{1}{\rho_{\mathrm{f}}} \frac{\partial \langle P_{\mathrm{f}} \rangle}{\partial x_i}\, dt + G_{ij} \left( U_{\mathrm{f},j} - \langle U_{\mathrm{f},j} \rangle \right) dt + \sigma_{ij}\, dW_j \,, \tag{10.7b}$$

where the diffusion coefficient is noted $\sigma_{ij}$ as in Eq. (10.2). With the expression given in Eq. (10.6) for the field $\Phi(t, \mathbf{x})$ entering Eqs. (10.3), the parametrized version of the GLM takes the following form

$$dX_{\mathrm{f},i}^{(\chi)} = U_{\mathrm{f},i}^{(\chi)}\, dt \,, \tag{10.8a}$$

$$dU_{\mathrm{f},i}^{(\chi)} = -\frac{1}{\chi} \frac{1}{\rho_{\mathrm{f}}} \frac{\partial \langle P_{\mathrm{f}} \rangle}{\partial x_i}\, dt + \frac{1}{\chi} G_{ij} \left[ U_{\mathrm{f},j}^{(\chi)} - \langle U_{\mathrm{f},j} \rangle \right] dt + \frac{1}{\chi} \sigma_{ij}\, dW_j \,. \tag{10.8b}$$

At the moment, we are still assuming that the fields $\langle P_{\mathrm{f}} \rangle$, $\langle U_{\mathrm{f}} \rangle$ and $\sigma$ (which depends typically on $\langle \epsilon_{\mathrm{f}} \rangle$) remain regular. For these conditions to hold and also in order to have that $\langle \mathbf{U}_{\mathrm{f}}^{(\chi)} \rangle = \langle \mathbf{U}_{\mathrm{f}} \rangle$, the equation satisfied by the mean velocity field $\langle \mathbf{U}_{\mathrm{f}}^{(\chi)} \rangle$ must be meaningful. This equation writes

$$\frac{\partial \langle U_{\mathrm{f},i}^{(\chi)} \rangle}{\partial t} + \langle U_{\mathrm{f},j}^{(\chi)} \rangle \frac{\partial \langle U_{\mathrm{f},i}^{(\chi)} \rangle}{\partial x_j} = -\frac{1}{\chi} \left\{ \frac{1}{\rho_{\mathrm{f}}} \frac{\partial \langle P_{\mathrm{f}} \rangle}{\partial x_i} + \frac{\partial \left[ \chi \langle u_{\mathrm{f},i}^{(\chi)} u_{\mathrm{f},j}^{(\chi)} \rangle \right]}{\partial x_j} \right\} \,, \tag{10.9}$$

which shows that, for the left-hand side of Eq. (10.9) to make sense when $\chi \to 0$, the term inside the bracket on the right-hand side of Eq. (10.9) must vanish. Two important consequences can be drawn. First, since $A_{ij} = \lim_{\chi \to 0} \left( \chi \langle u_i^{(\chi)} u_j^{(\chi)} \rangle \right)$ is a local finite expression given by the solution of the matrix equation, Eq. (10.5), the mean pressure-gradient term must indeed scale as $1/\chi$, as on the right-hand side of Eq. (10.8b), which justifies its presence in the definition of $\Phi(t, \mathbf{x})$ in Eq. (10.6). Second, the now properly-scaled mean pressure gradient must balance the divergence of the tensor $A_{ij}$, which means that we must have

$$\frac{1}{\rho_{\mathrm{f}}} \frac{\partial \langle P_{\mathrm{f}} \rangle}{\partial x_i} + \frac{\partial A_{ij}}{\partial x_j} = 0 \,. \tag{10.10}$$

This important point is discussed at length in [1, section 9.3] and is revisited below when the behavior of particle velocities in such parametrized models is investigated in Sect. 10.3.2. Given that $A_{ij}$ stands for an ersatz of the Reynolds-stress tensor, note that this condition on the mean pressure gradient is not altogether surprising and amounts to a Bernoulli-like formulation since Eq. (10.10) states basically that $\langle P_f \rangle + \text{Tr}(\mathbf{A})$ is constant in the limit $\chi \rightarrow 0$.

With these conditions, the over-damped Langevin model for fluid particles is

$$
dX_{f,i} = \langle U_{f,i} \rangle \, dt + G_{ij}^{-1} \left( \frac{1}{\rho_f} \frac{\partial \langle P_f \rangle}{\partial x_j} \right) dt - \frac{\partial G_{ij}^{-1}}{\partial x_k} A_{kj} dt + \left( G^{-1} \sigma \right)_{ij} dW_j ,
$$
(10.11)

which is easily rearranged as follows

$$
dX_{f,i} = \langle U_{f,i} \rangle \, dt + G_{ij}^{-1} \left\{ \frac{1}{\rho_f} \frac{\partial \langle P_f \rangle}{\partial x_j} + \frac{\partial A_{jk}}{\partial x_k} \right\} dt
$$
$$
- \frac{\partial}{\partial x_k} \left[ G_{ij}^{-1} A_{jk} \right] dt + \left( G^{-1} \sigma \right)_{ij} dW_j .
$$
(10.12)

Since the term inside the brackets on the right-hand side of Eq. (10.12) vanishes due to the enforced condition on the mean pressure gradient, cf. Eq. (10.10), we obtain the expression for the diffusive model of particle positions as

$$
dX_{f,i} = \langle U_{f,i} \rangle \, dt - \frac{\partial}{\partial x_k} \left[ G_{ij}^{-1} A_{jk} \right] dt + \left( G^{-1} \sigma \right)_{ij} dW_j .
$$
(10.13)

The diffusion matrix, $\mathbf{D}$ in the corresponding Fokker-Planck equation is then

$$
\mathbf{D} = \frac{1}{2} \left( \mathbf{G}^{-1} \sigma \right) \left( \mathbf{G}^{-1} \sigma \right)^{\perp} .
$$
(10.14)

After some tedious but straightforward manipulations (detailed in [1, section 9.3]), we obtain the final form of the diffusion equation for fluid particle positions in the over-damped limit as

$$
dX_{f,i} = \langle U_{f,i} \rangle \, dt + \frac{\partial D_{ik}}{\partial x_k} dt + B_{ik} \, dW_k ,
$$
(10.15)

with $\mathbf{B}$ a matrix such that $\mathbf{BB}^{\perp} = 2\mathbf{D}$. Note that $\mathbf{B} = \mathbf{G}^{-1}\sigma$ is, of course, a solution but other square root solutions of $\mathbf{D}$ can also be retained since they are equivalent in a probabilistic weak formulation. In the case where $\mathbf{G}$ and $\sigma$ are isotropic tensors, i.e. $G_{ij} = -1/T_L \delta_{ij}$ and $\sigma_{ij} = \sigma \delta_{ij}$, Eq. (10.15) takes the simple form

$$dX_{f,i} = \langle U_{f,i} \rangle \, dt + \frac{\partial}{\partial x_i} \left[ \frac{1}{2} (T_L \, \sigma)^2 \right] dt + (T_L \, \sigma) \, dW_i . \tag{10.16}$$

It is important to be aware that the additional drift terms, $\partial D_{ik}/\partial x_k$ in Eq. (10.15), appear only when the fast variable to be eliminated depends on the slow variable where the white-noise terms emerge. In other words, they appear when $U_f$ is the first-order derivative of $X_f$, as in the general Langevin model considered in Eqs. (10.2) or in the GLM in Eqs. (10.7). To illustrate this point, let us consider an extended model for fluid particles, where the particle state vector becomes $Z_f = (X_f, U_f, A_f)$ with $A_f$ the fluid particle acceleration, written as

$$dX_{f,i} = U_{f,i} \, dt , \tag{10.17a}$$

$$dU_{f,i} = -\frac{1}{\rho_f} \frac{\partial \langle P_f \rangle}{\partial x_i} \, dt + G_{ij} \left( U_{f,j} - \langle U_{f,j} \rangle \right) dt + \gamma_{f,i} \, dt , \tag{10.17b}$$

$$d\gamma_{f,i} = -\frac{\gamma_{f,i}}{\tau^A} \, dt + K^A \, dW_i . \tag{10.17c}$$

In Eq. (10.17b), $\gamma_f$ is now a colored noise of zero mean representing the rapidly-varying part of the particle acceleration, which means that

$$A_{f,i} = -\frac{1}{\rho_f} \frac{\partial \langle P_f \rangle}{\partial x_i} + G_{ij} \left( U_{f,j} - \langle U_{f,j} \rangle \right) + \gamma_{f,i} . \tag{10.18}$$

In the spirit of the Kolmogorov theory, cf. Sect. 5.3, we have retained a simple Langevin, or OU, process with an isotropic expression for $\gamma_f$ in Eq. (10.17c). Further application of Kolmogorov theory tells us that we can take this process as being locally in equilibrium, so that the diffusion coefficient in Eq. (10.17c) is given by $(K^A)^2 \simeq 2\langle \gamma_{f,i}^2 \rangle / \tau^A$ and the timescale $\tau^A$ is of the order of the local Kolmogorov timescale, $\tau^A \simeq \tau_K$. Given that $\langle \gamma_{f,i}^2 \rangle \simeq \langle \epsilon_f \rangle / \tau_K$, we are therefore in line with the criterion stated at the beginning of this section if we want to eliminate $\gamma_f$ as a fast variable in the limit of very small $\tau^A$, since we have $\tau^A \rightarrow 0$ and $K^A \rightarrow +\infty$ but such that $\tau^A K^A \rightarrow \sqrt{C_0 \langle \epsilon_f \rangle}$ where we have used the Kolmogorov constant $C_0$. As a side remark, this explains that the GLM, as given in the general expression Eqs. (10.7) or in the more applicable ones in Eqs. (6.11) or (6.21), is actually the outcome of the application of such elimination techniques when the timescale of the rapid part of a fluid particle acceleration is taken to zero. For our present concern, the point worth underlining, however, is that, although $\tau^A$ and $K^A$ are space-dependent functions, no extra term appears in Eq. (10.17b) and we simply substitute the colored noise process $\gamma_f \, dt$ for a white-noise one, such as $\sqrt{C_0 \langle \epsilon_f \rangle} \, dW$.

In a more physics-oriented perspective, these results for the overdamped Langevin limit call for some comments which, for the sake of simplicity, are based on the case of a scalar diffusivity, cf. Eq. (10.16), to avoid more cumbersome tensor notations.

(1) It is seen that the short-cut derivation used in Sect. 9.3 for Brownian particles is valid but only in situations where we know that $\tau_p$ is not a space-dependent function. We can now answer the second concern raised in that section and point out that this naive derivation, which consists in neglecting the acceleration term, is flawed in the sense that it leads to spurious drift terms in the resulting equation for particle positions. This is revealed with fluid particles, as pointed out now.

(2) The rigorous derivation of the diffusive limit for fluid particles shows that the elimination of $\mathbf{U}_f$ leads to a diffusive and a drift term in the position equation

$$dX_{f,i} = \langle U_{f,i} \rangle \, dt + \frac{\partial \Gamma_{ft}}{\partial x_i} \, dt + \sqrt{2\Gamma_{ft}} \, dW_i \; . \tag{10.19}$$

where $\Gamma_{ft} = (T_L \sigma)^2 / 2$. This is to be compared to the naive formulation

$$dX_{f,i}^{(naive)} = \langle U_{f,i} \rangle \, dt + \sqrt{2\Gamma_{ft}} \, dW_i \; , \tag{10.20}$$

where the drift term $\partial \Gamma_{ft}/\partial x_i$ is therefore missing. In sample space, the correct Fokker-Planck equation for $p(t; \mathbf{y})$ corresponding to Eq. (10.19) is

$$\frac{\partial p}{\partial t} = -\frac{\partial \left[ \langle U_{f,i} \rangle \, p \right]}{\partial y_i} - \frac{\partial}{\partial y_i} \left[ \left( \frac{\partial \Gamma_{ft}}{\partial x_i} \right) p \right] + \frac{\partial^2 \left[ \Gamma_{ft} \, p \right]}{\partial y_i \partial y_i} \; , \tag{10.21}$$

where the diffusion matrix involves only diagonal (even isotropic) terms when dealing with a scalar diffusivity (the general case with a full second-order diffusivity tensor is addressed below). Since each particle corresponds to the same amount of mass, the mean concentration $\langle c_f \rangle (t, \mathbf{x})$ is proportional to the particle-position PDF and satisfies in physical space the same equation as the above Fokker-Planck one, so that we have

$$\frac{\partial \langle c_f \rangle}{\partial t} = -\frac{\partial \left[ \langle U_{f,i} \rangle \, \langle c_f \rangle \right]}{\partial x_i} - \frac{\partial}{\partial x_i} \left[ \left( \frac{\partial \Gamma_{ft}}{\partial x_i} \right) \langle c_f \rangle \right] + \frac{\partial^2 \left[ \Gamma_{ft} \, \langle c_f \rangle \right]}{\partial x_i \partial x_i} \; . \tag{10.22}$$

Both equations show that an uniform particle concentration field is maintained, as it should be for incompressible flows (given that $\partial \langle U_{f,i} \rangle / \partial x_i = 0$). In contrast, the naive diffusive limit, expressed by Eq. (10.20), produces a spurious drift term due to a non-compensated variable diffusion coefficient $\Gamma_{ft}(\mathbf{x})$ and induces artificial accumulations or depletions of mass in the domain which is at variance with the basic property of incompressible flows.

(3) It can be observed that the diffusive limit of over-damped Langevin models has been derived by following only the trajectory point of view and using Ito stochastic calculus. In other words, we have not relied on the projection-

operator formalism, which is often believed to be needed (see the discussion in [2, chapter 8]) but which appears therefore as not strictly mandatory. Yet, these results are in line with the way fluctuations are proposed to be added in the frame of the GENERIC (which stands for 'general equation for the nonequilibrium reversible-irreversible coupling') approach (see [3, section 1.2.5]) and with subsequent derivations based on the rather formal projection-operator techniques (see [3, section 6.3.3]). The present approach works the other way around and obtains the Fokker-Planck equation for the slow variables after having derived their SDEs. In physical terms, present ideas are perhaps best cast in the terminology of force-flux relations introduced in Sect. 6.2.1. As an illustration, let us consider again the simple toy model used in Sect. 5.2 with a slow variable $X_{\text{slow}}$ whose time-rate-of-change is a fast-variable $X_{\text{fast}}$, i.e. $dX_{\text{slow}}/dt = X_{\text{fast}}$. The exact but unclosed equation for the PDF of $X_{\text{slow}}$, $p(t; y)$, is

$$\frac{\partial p(t; y)}{\partial t} = -\frac{\partial}{\partial y}\left[\langle X_{\text{fast}} | X_{\text{slow}}=y\rangle \, p(t; y)\right] . \tag{10.23}$$

When $X_{\text{fast}}$ becomes a fast variable in the sense studied above, that is with $\langle X_{\text{fast}}^2 \rangle \, T_{X_{\text{fast}}}$ tending towards a finite coefficient $\Gamma_{X_{\text{fast}}}(x)$ which can depend on the local value of the slow variable $X_{\text{slow}}$, we have

$$\langle X_{\text{fast}} | X_{\text{slow}}=y\rangle \, p(t; y) = -\Gamma_{X_{\text{fast}}}(y)\frac{\partial p(t; y)}{\partial y} , \tag{10.24}$$

which, when introduced in Eq. (10.23), gives the same Fokker-Planck equation

$$\frac{\partial p(t; y)}{\partial t} = \frac{\partial}{\partial y}\left[\Gamma_{X_{\text{fast}}}(y)\frac{\partial p(t; y)}{\partial y}\right] . \tag{10.25}$$

From Eq. (10.24), the 'flux', $\langle X_{\text{fast}} | X=y\rangle$, can also be written as

$$\langle X_{\text{fast}} | X_{\text{slow}}=y\rangle = -\Gamma_{X_{\text{fast}}}(y)\frac{\partial \ln[p(t; y)]}{\partial y} , \tag{10.26}$$

and appears to be proportional to the gradient of an 'entropic force', since the statistical entropy is $S \simeq \ln(p)$. At this stage, it is tempting to treat random effects as being due to an equivalent 'entropic force' which, per unit mass, is

$$F_{\text{rand}} = \frac{\Gamma_{X_{\text{fast}}}}{T_{X_{\text{fast}}}}\frac{\partial \ln(p)}{\partial x_i} . \tag{10.27}$$

First, the expression in Eq. (10.27) is to be regarded in a formal sense since we are in the limit when $T_{X_{\text{fast}}} \to 0$, so that this entropic force is very large and actually meaningful when considering its integrated effect $F_{\text{rand}} T_{X_{\text{fast}}}$ (similarly to the Brownian motion case, cf. Eq. (9.8) in Sect. 9.3). Second, this correspondence should be taken with some care as the two formulations do not have the same mathematical support. Indeed, the first formulation, in Eqs. (10.19) or (10.21), corresponds to a well-established SDE in which the drift and diffusion coefficients are functions of the stochastic process $X_{\text{slow}}$ or, perhaps, of statistics derived from the PDF in the extended sense of McKean SDEs (cf. Sect. 4.3.3). However, the second formulation, given by $F_{\text{rand}}$ in Eq. (10.27), is written in terms of the PDF itself and does not fall into the category of mathematically well-defined drift or diffusion terms in SDEs.

## 10.3 Non-local and Local Constitutive Relations

There is an interesting connection between the introduction of white-noise effects in the SDEs of the variables retained in the particle state vector and the existence of local or non-local closures in the corresponding transport equations for the moments of these variables. This point is already addressed in some details in [1, section 10.3] and is of interest for general particle stochastic systems. In consequence, the application of the fast-variable elimination techniques presented in Sect. 10.2 and the results that have been obtained are helpful to determine in which transport equations a local closure can appear and, if so, what form it takes.

To work out how this connection is manifested, we follow a two-step approach in which we discuss first in Sect. 10.3.1 what happens to scalars associated to fluid particles still assuming that the mean velocity field remains regular, before addressing in Sect. 10.3.2 the more difficult question of how this velocity field evolves when we consider that it contains a rapidly-varying component (in a sense to be defined) that is eliminated. In a way, the first situation corresponds to a weak approach in which we assess the effect of the elimination of the fluid particle velocity treated as a fast variable on some test functions which are transported scalars, while the second situation corresponds to a strong approach of the statistics of the velocity field itself.

### 10.3.1 Fast-Variable Elimination of Transported Scalars

We consider here the case of a conserved scalar $\Phi_f$ with two sets of fluid-particle-attached variables. Note that we could also consider reactive scalars but, since the reactive source terms play no role in the developments to follow, we can limit ourselves to conserved scalars. When the particle velocity is retained in the particle state vector, $\mathbf{Z}_f = (\mathbf{X}_f, \mathbf{U}_f, \Phi_f)$, and is represented by a Langevin model, we have

$$dX_{f,i} = U_{f,i}\, dt\, , \tag{10.28a}$$

$$dU_{f,i} = -\frac{1}{\rho_f}\frac{\partial\langle P_f\rangle}{\partial x_i}\, dt + G_{ij}\left(U_{f,j} - \langle U_{f,j}\rangle\right)\, dt + \sqrt{C_0\langle\epsilon_f\rangle}\, dW_i\, , \tag{10.28b}$$

$$d\Phi_f = 0\, . \tag{10.28c}$$

We are handling a Lagrangian velocity-scalar PDF, $p(t; \mathbf{y}_f, \mathbf{V}_f, \Psi_f)$, where $\Psi_f$ is the sample space variable corresponding to $\Phi_f$, which is the solution of a Fokker-Planck equation. By expressing the corresponding Eulerian velocity-scalar PDF, $p(t, \mathbf{x}; \mathbf{V}_f, \Psi_f)$, which follows the same equation (cf. Sect. 4.1), it is easy to derive the equation satisfied by the mean scalar field $\langle\Phi_f\rangle(t, \mathbf{x})$

$$\frac{\partial\langle\Phi_f\rangle}{\partial t} + \langle U_{f,i}\rangle\frac{\partial\langle\Phi_f\rangle}{\partial x_i} = -\frac{\partial}{\partial x_i}\left[\langle u_{f,i}\phi_f\rangle\right]\, , \tag{10.29}$$

where $\phi_f = \Phi_f - \langle\Phi_f\rangle$ is the scalar fluctuation. In Eq. (10.29), $\langle u_{f,i}\phi_f\rangle$ stands for the scalar flux which satisfies its own transport equation

$$\frac{\partial\langle u_{f,i}\phi_f\rangle}{\partial t} + \frac{\partial\langle u_{f,k}u_{f,i}\phi_f\rangle}{\partial x_k} =$$
$$- \langle u_{f,k}\phi_f\rangle\frac{\partial\langle U_{f,i}\rangle}{\partial x_k} - \langle u_{f,i}u_{f,k}\rangle\frac{\partial\langle\phi_f\rangle}{\partial x_k} + G_{ij}\langle u_{f,j}\phi_f\rangle\, , \tag{10.30}$$

which is also derived from the same Fokker-Planck equation. The key point for the present discussion is that the scalar flux $\langle u_{f,i}\phi_f\rangle$ is a non-local quantity in physical space since it is the solution of a transport equation.

On the other hand, when the particle velocity is regarded as a fast-variable and eliminated with the techniques outlined above, the particle state vector is reduced to $\mathbf{Z}_f = (\mathbf{X}_f, \Phi_f)$ and the evolution equations have the form

$$dX_{f,i} = \langle U_{f,i}\rangle dt + \frac{\partial\Gamma_{ft,ik}}{\partial x_k}dt + B_{ft,ik}\, dW_k\, , \tag{10.31a}$$

$$d\Phi_f = 0\, . \tag{10.31b}$$

In Eq. (10.31a), the diffusive matrix $\mathbf{B}_{ft}$ corresponds to $(2\mathbf{\Gamma}_{ft})^{1/2}$, which means that $\mathbf{B}_{ft}\mathbf{B}_{ft}^\perp = 2\mathbf{\Gamma}_{ft}$, while the diffusivity tensor is $\mathbf{\Gamma}_{ft} = 1/2\mathbf{G}^{-1}\boldsymbol{\sigma}\boldsymbol{\sigma}^\perp(\mathbf{G}^\perp)^{-1}$ with $\boldsymbol{\sigma} = \sqrt{C_0\langle\epsilon_f\rangle}\,\boldsymbol{\delta}$ (this is the same expression as the general diffusivity tensor given above, cf. Eq. (10.14), but we use here the notation $\mathbf{\Gamma}_{ft}$ for this specific choice of $\boldsymbol{\sigma}$). The Fokker-Planck equation for $p(t; \mathbf{y}_f, \Psi_f)$ is then

$$\frac{\partial p}{\partial t} = -\frac{\partial\left[\langle U_{f,i}\rangle p\right]}{\partial y_i} + \frac{\partial}{\partial y_i}\left[\Gamma_{ft,ij}\frac{\partial p}{\partial y_j}\right]\, . \tag{10.32}$$

Note that Eqs. (10.19) and (10.21) correspond to the case where $\mathbf{G}$ and $\boldsymbol{\sigma}$ are isotropic tensors. From Eq. (10.32), we obtain the transport equation satisfied by the mean scalar field $\langle \Phi_f \rangle (t, \mathbf{x})$ which is

$$\frac{\partial \langle \Phi_f \rangle}{\partial t} + \langle U_{f,i} \rangle \frac{\partial \langle \Phi_f \rangle}{\partial x_i} = \frac{\partial}{\partial x_i} \left[ \Gamma_{ft,ij} \left( \frac{\partial \langle \phi_f \rangle}{\partial x_j} \right) \right] . \tag{10.33}$$

The important outcome is that the scalar flux $\langle u_{f,i} \phi_f \rangle$ is no longer the solution of a transport equation, as in Eq. (10.30), but is expressed locally in physical space with an eddy-diffusivity type of relation which is

$$\langle u_{f,i} \phi_f \rangle = -\Gamma_{ft,ij} \left( \frac{\partial \langle \Phi_f \rangle}{\partial x_j} \right) . \tag{10.34}$$

Therefore, we have moved from a non-local closure of the scalar flux when particle velocity is explicitly modeled to a local closure when it is eliminated (as displayed in Fig. 10.1). This corresponds to an interesting model reduction, reflected by the number of moment equations we have to consider: instead of four coupled transport equations, Eq. (10.29) plus Eqs. (10.30), when particle velocity is treated as a slow-varying variable (at least, on par with the scalar variables), we have only one moment equation in closed form, namely Eq. (10.33), when particle velocity is eliminated. One of the interests of these developments based on fast-variable elimination techniques is to provide an approach through which the continuous evolution from a non-local to a local closure is obtained as the timescale of the fluid particle velocity becomes close to zero.

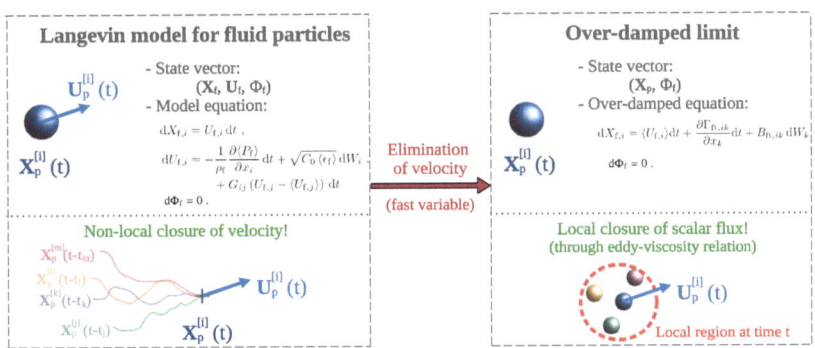

**Fig. 10.1**  Sketch of the shift from a non-local to a local closure constitutive relation obtained as the over-damped limit of a Langevin model for fluid particles

## 10.3.2 Fast-Variable Elimination of the Velocity

Up to now, we have assumed that the mean velocity remains finite in the fast-variable limit, which, in a way, amounted to regard it as an externally-provided statistic. We can now wonder what happens when we treat the mean velocity as the first-order moment of particle velocities even in the fast-variable limit. More precisely, we ask whether the mean Navier-Stokes equation remains tractable and, if so, which form it takes. Having demonstrated in Sect. 10.3.1 that eliminating particle velocity leads to a local closure for the scalar flux based on an eddy-diffusivity type of relation, we intuitively expect that the mean velocity remains finite and that its governing equation involves local closure laws based on a turbulent viscosity. To show that this is indeed the case proves however much more difficult than for scalars since it requires careful bookkeeping of the unbounded terms that cancel each other out in the fast-variable limit.

To help us in this task, we follow up on the ideas introduced in [1, section 9.3.3] which rely on the notion of a two-timescale description of turbulent motions and exchanges. In classical turbulence formulations [4], only one timescale $T^{\text{dis}}$, scaling as $k_f/\langle \epsilon_f \rangle$, is retained to model the rate of turbulent kinetic energy production and dissipation, as well as of the energy transfer between the components of the Reynolds-stress tensor (the redistribution terms which are conservative in the sense that they neither produce nor dissipate turbulent kinetic energy), as illustrated in Fig. 10.2a. In the new description, we introduce a second timescale, $T^{\text{red}}$, which governs the redistribution of turbulent kinetic energy between the components of the Reynolds-stress tensor and is taken as much smaller than $T^{\text{dis}}$ which is

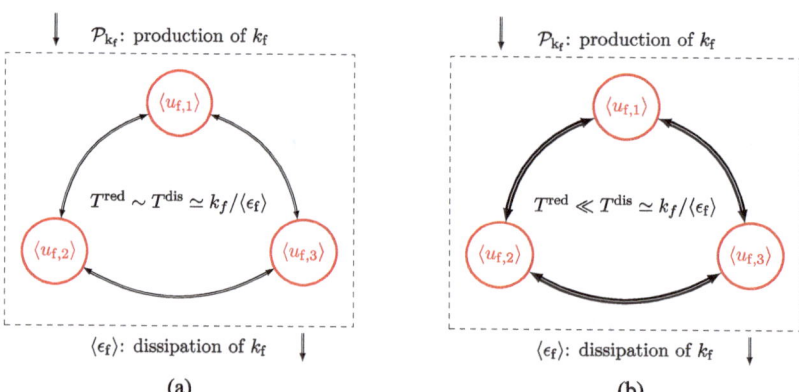

(a)                                              (b)

**Fig. 10.2** Sketch of the differences between classical models and the new one: in classical models (on the left), one timescale is used to describe the energy production/dissipation and its redistribution between the components of the Reynolds stress tensor (indicated by the same thickness of the arrows representing these energy transfers); in the new model (on the right), two timescales are used to describe separately the energy production/dissipation and its redistribution between the components of the Reynolds stress tensor (indicated by arrows with different thickness). (**a**) Classical models with one timescale. (**b**) New model with two timescales

still governing the turbulent kinetic energy production and dissipation rates. This second model is sketched in Fig. 10.2b. Broadly speaking, this new model is meant to introduce a decomposition of particle velocity into two components, with one component varying rapidly (to be taken as the fast variable) compared to the second one (to be taken as the slow variable), so that effects due to the fast variables relax towards local equilibrium values while the slow variables evolve through transport equations.

To formalize these ideas, we consider for $\mathbf{Z}_f = (\mathbf{X}_f, \mathbf{U}_f)$ the particle stochastic system written as

$$dX_{f,i} = U_{f,i}\, dt \, , \tag{10.35a}$$

$$
\begin{aligned}
dU_{f,i} = &-\frac{1}{\rho_f}\frac{\partial \langle P_f \rangle}{\partial x_i}\, dt - \frac{1}{T^{dis}}\left[U_{f,i} - \langle U_{f,i}\rangle\right] dt \\
&- \frac{1}{\rho_f}\frac{\partial \langle \widetilde{P}_f \rangle}{\partial x_i}\, dt - \frac{1}{T^{red}}\left[U_{f,i} - \langle U_{f,i}\rangle\right] dt + \sqrt{C_0 \langle \widetilde{\epsilon}_f \rangle}\, dW_i \, .
\end{aligned}
\tag{10.35b}
$$

Drawing on the above description, the decomposition on the right-hand side of Eq. (10.35b) reproduces the distinction between the two timescales, $T^{dis}$ and $T^{red}$, while a simplified version based on an isotropic formulation of the return-to-equilibrium terms is used in Eq. (10.35b) to bring out the physical ideas. For reasons that are clarified below, the pressure-gradient terms is split into two contributions and a yet-to-be-defined dissipation rate $\langle \widetilde{\epsilon}_f \rangle$ is written in the diffusion coefficient. We need to go one step further and express the timescales and how we can obtain different relative values. Since these two timescales are typically written as functions of $k_f/\langle \epsilon_f \rangle$, two different dissipation rates of the turbulent kinetic energy, labeled $\langle \epsilon_f^{slow}(\chi) \rangle$ and $\langle \epsilon_f^{fast}(\chi) \rangle$, are introduced for that purpose, corresponding to two different scaling with respect to a small parameter $\chi$, as detailed below. Remember that the turbulent kinetic energy dissipation rate $\langle \epsilon_f \rangle$ is external to the description based on $\mathbf{Z}_f = (\mathbf{X}_f, \mathbf{U}_f)$ and appears therefore as the proper quantity to tune to obtain a different scaling of the timescales. This leads to the following parametrized system

$$dX_{f,i}^{(\chi)} = U_{f,i}^{(\chi)}\, dt \, , \tag{10.36a}$$

$$
\begin{aligned}
dU_{f,i}^{(\chi)} = &-\frac{1}{\rho_f}\frac{\partial \langle P_f^{(\chi)} \rangle}{\partial x_i}\, dt - \frac{1}{2}\frac{\langle \epsilon_f^{slow}(\chi) \rangle}{k_f^{(\chi)}}\left[U_{f,i}^{(\chi)} - \langle U_{f,i}^{(\chi)}\rangle\right] dt \\
&- \frac{1}{\rho_f}\frac{\partial \langle \widetilde{P}_f^{(\chi)} \rangle}{\partial x_i}\, dt + \widetilde{G}_{ij}^{(\chi)}\left[U_{f,j}^{(\chi)} - \langle U_{f,j}^{(\chi)}\rangle\right] dt + \sqrt{C_0 \langle \epsilon_f^{fast}(\chi) \rangle}\, dW_i \, .
\end{aligned}
\tag{10.36b}
$$

The timescale $T^{dis}$ governing the rate at which turbulent kinetic energy is dissipated is now written as $k_f^{(\chi)}/\langle \epsilon_f^{slow}(\chi) \rangle$ and is used on the first line on the right-hand side of Eq. (10.36b). Similarly, the timescale $T^{\text{red}}$ with which turbulent kinetic energy is exchanged between the velocity components is used on the second line on the right-hand side of Eq. (10.36b), which means that $(\widetilde{\mathbf{G}}^{(\chi)})^{-1}$, which extends this timescale to the non-isotropic formulation, varies as $k_{\text{f}}^{(\chi)}/\langle\epsilon_{\text{f}}^{\text{fast}}(\chi)\rangle$. Before precising the specific relations of $\langle\epsilon_{\text{f}}^{\text{slow}}(\chi)\rangle$ and $\langle\epsilon_{\text{f}}^{\text{fast}}(\chi)\rangle$ with respect to the small parameter $\chi$, further clarifications are in order.

First, when $\chi = 1$, we take the two timescales as being equal (meaning that $\langle\epsilon_{\text{f}}\rangle = \langle\epsilon_{\text{f}}^{\text{slow}}(\chi)\rangle = \langle\epsilon_{\text{f}}^{\text{fast}}(\chi)\rangle$) and gather the two pressure gradients terms into one ($\langle P_{\text{f}}\rangle = \langle P_{\text{f}}^{(\chi)}\rangle + \langle\widetilde{P}_{\text{f}}^{(\chi)}\rangle$) to retrieve the classical form of a GLM. The mean Navier-Stokes, or Reynolds, equation is then

$$\frac{\partial\langle U_{\text{f},i}\rangle}{\partial t} + \langle U_{\text{f},j}\rangle\frac{\partial\langle U_{\text{f},i}\rangle}{\partial x_j} = -\frac{1}{\rho_{\text{f}}}\frac{\partial\langle P_{\text{f}}\rangle}{\partial x_i} - \frac{\partial\langle u_{\text{f},i}u_{\text{f},j}\rangle}{\partial x_j} , \tag{10.37}$$

where the Reynolds stress tensor, $\langle u_{\text{f},i}u_{\text{f},j}\rangle$, is the solution of its transport equation

$$\frac{\partial\langle u_{\text{f},i}u_{\text{f},j}\rangle}{\partial t} + \langle U_{\text{f},k}\rangle\frac{\partial\langle u_{\text{f},i}u_{\text{f},j}\rangle}{\partial x_k} + \frac{\partial\langle u_{\text{f},i}u_{\text{f},j}u_{\text{f},k}\rangle}{\partial x_k} =$$
$$- \langle u_{\text{f},i}u_{\text{f},k}\rangle\frac{\partial\langle U_{\text{f},j}\rangle}{\partial x_k} - \langle u_{\text{f},j}u_{\text{f},k}\rangle\frac{\partial\langle U_{\text{f},i}\rangle}{\partial x_k}$$
$$+ G_{ik}\langle u_{\text{f},k}u_{\text{f},j}\rangle + G_{jk}\langle u_{\text{f},k}u_{\text{f},i}\rangle + C_0\langle\epsilon_{\text{f}}\rangle\delta_{ij} , \tag{10.38}$$

where $\mathbf{G} = -1/2\langle\epsilon_{\text{f}}\rangle/k_{\text{f}}\,\boldsymbol{\delta} + \widetilde{\mathbf{G}}$. The rationale behind splitting the matrix $\mathbf{G}$ into two parts, as done in Eq. (10.36b), is to pave the way for a decomposition between a 'slow' and a 'fast' velocity components. To that effect, we attribute different roles to two parts of the matrix $\mathbf{G}$ according to the decomposition

$$G_{ij} = \underbrace{-\frac{1}{2}\frac{\langle\epsilon_{\text{f}}\rangle}{k_{\text{f}}}\delta_{ij}}_{\text{slow}} + \underbrace{\widetilde{G}_{ij}}_{\text{fast}} . \tag{10.39}$$

The 'slow' part governs the rate of kinetic energy dissipation and is the term entering the transport equation for $k_{\text{f}}$, whereas the 'fast' part governs the rate of energy redistribution between velocity components and is therefore subject to the condition

$$2\,\text{Tr}(\widetilde{\mathbf{G}}\mathbf{R}_{\text{f}}) + 3\,C_0\langle\epsilon_{\text{f}}\rangle = 0 . \tag{10.40}$$

This association of the rapidly-varying component to the redistributive part of the matrix $\mathbf{G}$ is key to decouple (or loosely couple) the different turbulent motions and to allow the evolution equation for the turbulent kinetic energy $k_{\text{f}}$ to be unaffected when we consider redistribution as an infinitely fast process (as compared to the evolution of $k_{\text{f}}$). Note that the decomposition used in Eq. (10.39) is not the only one possible and is made here by reference to the SLM, as will be seen below. More general decompositions of the type

$$G_{ij} = G_{ij}^{\text{slow}} + G_{ij}^{\text{fast}} . \qquad (10.41)$$

are possible, provided that Eq. (10.40) be respected for the fast part $\mathbf{G}^{\text{fast}}$ while the slow part $\mathbf{G}^{\text{slow}}$ must be such that we have $\text{Tr}(\mathbf{G}^{\text{slow}}\mathbf{R}_f) = -2\langle\epsilon_f\rangle$. For the sake of simplicity, we pursue with the simple form in Eq. (10.39).

Although they bear some similarities, note that the decomposition in Eq. (10.39) is different from the one introduced in Eq. (6.19). In essence, the decomposition in Eq. (6.19) separates between isotropic contributions scaling as $\langle\epsilon_f\rangle/k_f$ (Rotta-like terms) from other contributions which scale typically with the mean fluid velocity gradients. The important point is that the different contributions in Eq. (6.19), which enter the drift and diffusion coefficients of the GLM, are meant to capture both redistributive and dissipative terms, as indicated by the last line on the right-hand side of Eq. (10.38). This is not the philosophy behind the present decomposition introduced in Eq. (10.39) which aims precisely to distinguish between one part inducing dissipation and another one acting only as a redistributive term. This explains that, for the SLM which includes only a single Rotta-like contribution, the matrix $\mathbf{G}$ is not decomposed according to Eq. (6.19) but can be split according to Eq. (10.39) into two parts playing different roles, with $\widetilde{G}_{ij} = -3C_0/4\langle\epsilon_f\rangle/k_f\,\delta_{ij}$.

To put these ideas into practice, we introduce specific scaling relations in terms of a small parameter $\chi$ and make use of the difference between the two timescales.

If we write $k_f^{(1)}$ and $\langle\epsilon_f^{(1)}\rangle$ to refer to the level of turbulent kinetic energy and dissipation rates, respectively, coming from the 'regular' solution, for instance the one obtained with $\chi = 1$ and equipped with its corresponding modeled equation for $\langle\epsilon_f^{(1)}\rangle$, we then choose $\langle\epsilon_f^{\text{fast}}(\chi)\rangle$ to scale as $1/\chi^2$, that is $\langle\epsilon_f^{\text{fast}}(\chi)\rangle \simeq 1/\chi^2\,\langle\epsilon_f^{(1)}\rangle$. Given that $\widetilde{\mathbf{G}}^{(\chi)}$ scales as $\langle\epsilon_f^{\text{fast}}(\chi)\rangle/k_f^{(\chi)}$ and that we already know from the results of the fast-variable techniques recalled earlier that we can expect that $k_f^{(\chi)} \simeq 1/\chi\,k_f^{(1)}$, we have then $\widetilde{\mathbf{G}}^{(\chi)} = \langle\epsilon_f^{\text{fast}}(\chi)\rangle/k_f^{(\chi)}\widetilde{\mathbf{G}} \sim 1/\chi\,\widetilde{\mathbf{G}}$ where $\widetilde{\mathbf{G}} \sim \widetilde{\mathbf{G}}^{(1)}$. As shown earlier, we can also consider that the mean pressure gradient due to $\langle\widetilde{P}_f^{(\chi)}\rangle$ scales as $1/\chi$, as will be confirmed below, and can therefore be written as $\langle\widetilde{P}_f^{(\chi)}\rangle \sim 1/\chi\,\langle\widetilde{P}_f\rangle$.

For the timescale at which turbulent kinetic energy is dissipated, we choose $\langle\epsilon_f^{\text{slow}}(\chi)\rangle$ to scale as $k_f^{(\chi)}$, i.e. we take $\langle\epsilon_f^{\text{slow}}(\chi)\rangle \sim 1/\chi\,\langle\epsilon_f^{(1)}\rangle$. This is in line with the notion that $\langle\epsilon_f^{\text{slow}}(\chi)\rangle$ represents the 'true' dissipation rate, as can be seen on the right-hand side of Eq. (10.38) which indicates that $\langle\epsilon_f^{\text{slow}}(\chi)\rangle$ varies indeed as the production terms, $\langle u_{f,i}u_{f,k}\rangle\partial\langle U_{f,j}\rangle/\partial x_k$, and thus as $1/\chi$ since the mean velocity field and its gradients remain of order 1. This implies that the timescale of turbulent kinetic energy dissipation, $T^{\text{dis}} = 2k_f^{(\chi)}/\langle\epsilon_f^{\text{slow}}(\chi)\rangle$ on the first line on the right-hand side of Eq. (10.36b), is of order 1 with respect to the small parameter $\chi$, i.e., $k_f^{(\chi)}/\langle\epsilon_f^{\text{slow}}(\chi)\rangle \sim k_f^{(1)}/\langle\epsilon_f^{(1)}\rangle$. Accordingly, we can take that the pressure gradient due to $\langle P_f^{(\chi)}\rangle$ scales as 1, i.e., $\langle P_f^{(\chi)}\rangle = \langle P_f\rangle$ with $\langle P_f\rangle \sim \langle P_f^{(1)}\rangle$. It also follows that by taking the trace of the transport equations satisfied by $\langle u_{f,i}^{(\chi)} u_{f,j}^{(\chi)} \rangle$ (which is similar to Eq. (10.38)), we eliminate the effects of the energy redistribution process and, since all remaining terms scale as $1/\chi$, we can multiply the resulting equation by $\chi$ to retrieve the classical transport equation for $k_f$ (which is here $k_f^{(1)}$)

$$\frac{\partial k_f}{\partial t} + \langle U_{f,k} \rangle \frac{\partial k_f}{\partial x_k} + \frac{\partial T(k_f)}{\partial x_k} = \mathcal{P}_f - \langle \epsilon_f \rangle , \tag{10.42}$$

where $T(k_f) = \langle u_{f,k} \sum_i (u_{f,i}^2)/2 \rangle$ is the convective flux due to the fluctuating part of the slowly varying part of the velocity.

Note that the superscripts used for $\langle \epsilon_f^{slow}(\chi) \rangle$ and $\langle \epsilon_f^{fast}(\chi) \rangle$ refer in fact to their relative variation or to the order of variation as a function of $1/\chi$ rather than to their intrinsic value, since we have for both that $\langle \epsilon_f^{slow}(\chi) \rangle, \langle \epsilon_f^{fast}(\chi) \rangle \ll \langle \epsilon_f^{(1)} \rangle$ but $\langle \epsilon_f^{fast}(\chi) \rangle \ll \langle \epsilon_f^{slow}(\chi) \rangle$. In other words, the variation of $\langle \epsilon_f^{slow}(\chi) \rangle$ is slow when compared to the one of $\langle \epsilon_f^{fast}(\chi) \rangle$.

We can summarize the different choices for $\langle \epsilon_f^{slow}(\chi) \rangle$ and $\langle \epsilon_f^{fast}(\chi) \rangle$

$$\frac{\langle \epsilon_f^{fast}(\chi) \rangle}{\langle \epsilon_f^{(1)} \rangle} \sim \frac{1}{\chi^2} , \quad \frac{\langle \epsilon_f^{slow}(\chi) \rangle}{\langle \epsilon_f^{(1)} \rangle} \sim \frac{1}{\chi} , \tag{10.43}$$

and the resulting scaling relations in terms of the small parameter $\chi$ as follows

$$\frac{k_f^{(\chi)}}{k_f^{(1)}} \sim \frac{1}{\chi} , \quad \frac{\widetilde{\mathbf{G}}^{(\chi)}}{\widetilde{\mathbf{G}}^{(1)}} \sim \frac{1}{\chi} , \quad \frac{\langle \widetilde{P}_f^{(\chi)} \rangle}{\langle \widetilde{P}_f^{(1)} \rangle} \sim \frac{1}{\chi} \frac{\langle P_f^{(\chi)} \rangle}{\langle P_f^{(1)} \rangle} \sim 1 . \tag{10.44}$$

Relying on these scaling relations, we rewrite Eqs. (10.36) as

$$dX_{f,i}^{(\chi)} = U_{f,i}^{(\chi)} dt , \tag{10.45a}$$

$$dU_{f,i}^{(\chi)} = -\frac{1}{\rho_f} \frac{\partial \langle P_f \rangle}{\partial x_i} dt - \frac{1}{2} \frac{\langle \epsilon_f^{(1)} \rangle}{k_f^{(1)}} \left[ U_{f,i}^{(\chi)} - \langle U_{f,i}^{(\chi)} \rangle \right] dt$$

$$- \frac{1}{\chi} \frac{1}{\rho_f} \frac{\partial \langle \widetilde{P}_f \rangle}{\partial x_i} dt + \frac{1}{\chi} \widetilde{G}_{ij} \left[ U_{f,j}^{(\chi)} - \langle U_{f,j}^{(\chi)} \rangle \right] dt + \frac{1}{\chi} \sqrt{C_0 \langle \epsilon_f^{(1)} \rangle} \, dW_i , \tag{10.45b}$$

and apply the fast-variable elimination techniques described above when $\chi \to 0$.

The return-to-equilibrium term on the first line on the right-hand side of Eq. (10.45b) has been written as $\langle \epsilon_f^{(1)} \rangle / k_f^{(1)}$ and we have left out a possible constant appearing in front of this expression. This is not a concern since the terms on the first line on the right-hand side of Eq. (10.45b) play no role in the elimination process which is entirely governed by the terms scaling as $1/\chi$, that is by the terms on the second line on the right-hand side of Eq. (10.45b).

The equation satisfied by the mean velocity field $\langle \mathbf{U}_f^{(\chi)} \rangle$ is

$$\frac{\partial \langle U_{f,i}^{(\chi)} \rangle}{\partial t} + \langle U_{f,j}^{(\chi)} \rangle \frac{\partial \langle U_{f,i}^{(\chi)} \rangle}{\partial x_j} = -\frac{1}{\rho_f} \frac{\partial \langle P_f \rangle}{\partial x_i} - \frac{1}{\chi} \left\{ \frac{1}{\rho_f} \frac{\partial \langle \widetilde{P_f} \rangle}{\partial x_i} + \frac{\partial \left[ \chi \langle u_{f,i}^{(\chi)} u_{f,j}^{(\chi)} \rangle \right]}{\partial x_j} \right\} ,$$

$$(10.46)$$

where the equivalent of the Reynolds stress tensor, $A_{ij} = \chi \langle u_{f,i}^{(\chi)} u_{f,j}^{(\chi)} \rangle$ is a local term which, in the limit $\chi \to 0$, is the solution of the matrix equation

$$\widetilde{\mathbf{G}} \mathbf{A} + \mathbf{A} \widetilde{\mathbf{G}}^\perp = -C_0 \langle \epsilon_f^{(1)} \rangle \, \boldsymbol{\delta} .$$

$$(10.47)$$

It is instructive to work out the resulting form of Eq. (10.46) corresponding to two different closures of $\widetilde{\mathbf{G}}$.

First, if we retain the SLM, the decomposition used in Eq. (10.36b) indicates that

$$\widetilde{G}_{ij}^{(\chi)} = -\frac{3 C_0}{4} \frac{\langle \epsilon_f^{fast}(\chi) \rangle}{k_f^{(\chi)}} \delta_{ij} \implies \widetilde{G}_{ij} \simeq -\frac{3 C_0}{4} \frac{\langle \epsilon_f^{(1)} \rangle}{k_f^{(1)}} \delta_{ij} \implies A_{ij} = \frac{2}{3} k_f^{(1)} \delta_{ij} .$$

$$(10.48)$$

As done in Sect. 10.2 with Eqs. (10.9)–(10.10), we can take $\widetilde{P}_f$ in Eq. (10.46) so that $\widetilde{P}_f + 2/3 k_f^{(1)}$ is constant. The equation satisfied by $\langle \mathbf{U}_f \rangle$ obtained as the limit of the one satisfied by $\langle U_f^{(\chi)} \rangle$ when $\chi \to 0$ is thus

$$\frac{\partial \langle U_{f,i} \rangle}{\partial t} + \langle U_{f,j} \rangle \frac{\partial \langle U_{f,i} \rangle}{\partial x_j} = -\frac{1}{\rho_f} \frac{\partial \langle P_f \rangle}{\partial x_i} .$$

$$(10.49)$$

This is the equivalent of the Euler equations obtained as the first-order approximation in the derivation of the equations for hydrodynamics.

Second, if we consider a more elaborate closure of the matrix $\mathbf{G}$ in GLMs, such as the IP (Isotropization of Production) model [4, chapter 12], we have

$$\widetilde{G}_{ij}^{(\chi)} = \left( -\frac{3 C_0}{4} - \frac{1}{2} C_2 \frac{\mathcal{P}^{(\chi)}}{\langle \epsilon_f^{fast}(\chi) \rangle} \right) \frac{\langle \epsilon_f^{fast}(\chi) \rangle}{k_f^{(\chi)}} \delta_{ij} + C_2 \frac{\partial \langle U_{f,i} \rangle}{\partial x_j}$$

$$= \frac{\langle \epsilon_f^{fast}(\chi) \rangle}{k_f^{(\chi)}} \left[ \left( -\frac{3 C_0}{4} - \frac{1}{2} C_2 \frac{\mathcal{P}^{(\chi)}}{\langle \epsilon_f^{fast}(\chi) \rangle} \right) \delta_{ij} + C_2 \frac{k_f^{(\chi)}}{\langle \epsilon_f^{fast}(\chi) \rangle} \frac{\partial \langle U_{f,i} \rangle}{\partial x_j} \right] ,$$

$$(10.50)$$

where $\mathcal{P}^{(\chi)}$ is the turbulent kinetic energy production term (varying as $1/\chi$). With the different scaling relations given above, this yields

$$\widetilde{G}_{ij} \simeq -\frac{3\,C_0}{4}\frac{\langle\epsilon_f^{(1)}\rangle}{k_f^{(1)}}\left[\delta_{ij} + \chi\frac{4C_2}{3C_0}\left(\frac{\mathcal{P}^{(1)}}{2\langle\epsilon_f^{(1)}\rangle}\delta_{ij} - \frac{k_f^{(1)}}{\langle\epsilon_f^{(1)}\rangle}\frac{\partial\langle U_{f,i}\rangle}{\partial x_j}\right)\right]. \qquad (10.51)$$

We can then make a first-order approximation of the matrix $\widetilde{G}^{-1}$ in terms of the small parameter $\chi$ to obtain the first-order approximation of $\mathbf{A}$ from Eq. (10.47). Given that $A_{ij}$ is a symmetrical tensor and leaving out the exact calculation of the constants appearing in front of the terms of the series, we obtain

$$A_{ij} \simeq \frac{2}{3}k_f^{(1)}\delta_{ij} - \chi\left[C_{A,1}\frac{k_f^{(1)}}{\langle\epsilon_f^{(1)}\rangle}\mathcal{P}^{(1)}\delta_{ij} - C_{A,2}\frac{(k_f^{(1)})^2}{\langle\epsilon_f^{(1)}\rangle}S_{ij}\right], \qquad (10.52)$$

where $C_{A,1}$ and $C_{A,2}$ are two positive constants and $S_{ij}$ the strain rate tensor, i.e., $S_{ij} = 1/2\left(\partial\langle U_{f,i}\rangle/\partial x_j + \partial\langle U_{f,j}\rangle/\partial x_i\right)$.

So far, we have used $A_{ij}$ as an ersatz of the Reynolds stress. This is correct for the leading term of $A_{ij}$ which is constant with respect to the parameter $\chi$, corresponding to the infinite part of the kinetic energy to be compensated by the fast-varying pressure gradient, as shown with the derivation of the Euler equations in Eq. (10.49). Indeed, this component of the turbulent kinetic energy is dominated by the fast variations of the 'rapid part' of the velocity. When considering the next terms, such as the ones scaling as $\chi$, it is however important to account not only for this rapid part of the velocity but also for the 'slow part', following the ideas behind the decomposition in Eqs. (10.35) and (10.36). This leads to introduce a decomposition of $A_{ij}$ into two terms, $A_{ij}^{\text{fast}}$ and $A_{ij}^{\text{slow}}$, so that we can expect that

$$A_{ij}^{\text{fast}} + A_{ij}^{\text{slow}} \simeq \frac{2}{3}\widetilde{k}_f\,\delta_{ij} \qquad (10.53)$$

with $\widetilde{k}_f$ a value of the kinetic energy of the order of $k_f^{(1)}$, without being necessarily identical. It is then important to notice that the solution of the matrix equation, Eq. (10.47), corresponds in fact to the fast part of the ersatz of the Reynolds stress tensor, that is $A_{ij}^{\text{fast}}$, since this is an equilibrium equation (in accordance with the slaving principles set forth with the Markovian approach in Sect. 5.2) while the slow part can be associated with the term $\chi\,\langle u_{f,i}^{(\chi)}u_{f,j}^{(\chi)}\rangle$. Combining Eqs. (10.53) and (10.52) gives then

$$A_{ij}^{\text{slow}} \simeq \frac{2}{3}\widetilde{\delta k_f}\,\delta_{ij} + \chi\left[C_{A,1}\frac{k_f^{(1)}}{\langle\epsilon_f^{(1)}\rangle}\mathcal{P}^{(1)}\delta_{ij} - C_{A,2}\frac{(k_f^{(1)})^2}{\langle\epsilon_f^{(1)}\rangle}S_{ij}\right]. \qquad (10.54)$$

where $\widetilde{\delta k}_f = \widetilde{k}_f - k_f^{(1)}$ is a trace of the rapidly-varying part to be eliminated, leaving the significant term as the one with the brackets on the right-hand side of Eq. (10.54) (which justifies the label slow to refer to this term). By making the same evaluation of $\widetilde{P}_f$ as in the zero-order above (the Euler equation) to compensate the term $\widetilde{\delta k}_f$ in Eq. (10.54), we obtain from Eq. (10.46) that the equation satisfied by $\langle U_f \rangle$ as the limit of the one for $\langle U_f^{(\chi)} \rangle$ when $\chi \to 0$ is now

$$\frac{\partial \langle U_{f,i} \rangle}{\partial t} + \langle U_{f,j} \rangle \frac{\partial \langle U_{f,i} \rangle}{\partial x_j} = -\frac{1}{\rho_f} \frac{\partial \langle P_f \rangle}{\partial x_i} + \frac{\partial}{\partial x_j} \left[ \nu_{ft} \left( \frac{\partial \langle U_{f,i} \rangle}{\partial x_j} + \frac{\partial \langle U_{f,j} \rangle}{\partial x_i} \right) \right],$$
(10.55)

where the term involving $C_{A,1}$ has been added to the pressure $\langle P_f \rangle$ and where $\nu_{ft}$ is referred to as the turbulent viscosity. It is seen that $\nu_{ft}$ is expressed by $\nu_{ft} = C_{ft} k_f^2 / \langle \epsilon_f \rangle$ with $C_{ft}$ a positive constant. From the direct application of Eq. (10.54), we would have $C_{ft} = C_{A,2}$ but more careful derivations could yield different values of the constants entering the above equations. This result constitutes a consistent outcome since, with the present scaling of $k_f$ (as $1/\chi$) and of $\langle \epsilon_f \rangle$ (as $1/\chi^2$), it is indeed the ratio $k_f^2 / \langle \epsilon_f \rangle$ that remains of order 1. With the result stated for the remaining equation for the turbulent kinetic energy, cf. Eq. (10.42), it can be noted that we retrieve the essence of the classical $k - \epsilon$ turbulence model [4].

Therefore, the first-order approximation in terms of the small parameter $\chi$ leads to a form which is similar to the hydrodynamics equations where transport coefficients, such as the molecular viscosity, appear in the first-order approximation of the Chapman-Enskog iterative approach. In that sense, the length $k_f^{3/2} / \langle \epsilon_f \rangle$ could be taken as the equivalent of the 'mean free path' but this is a mere analogy since the picture of free flights and collisions does not apply. This length represents much more the characteristic length over which velocities remain significantly correlated.

**To Summarize** To the best of the authors' knowledge, the developments presented throughout Sect. 10.3 represent first steps into what is still an uncharted territory. While the derivation of closures based on a turbulent diffusivity for scalar fluxes in Sect. 10.3.1 seems already to rest on solid foundations, the application of similar ideas on the velocity field itself is more delicate and the results obtained in Sect. 10.3.2 are therefore to be regarded as preliminary attempts. They need to be revisited but, as such, they serve as examples of possible approaches and it is hoped that they will stimulate developments along similar lines.

For our discussion, the important conclusion is that we started with a non-local expression of the Reynolds stress tensor, solution of a transport equation, cf. Eq. (10.38), influenced by upstream values, but have ended up with a local closure consistent with models based on the turbulent-viscosity concept in the fast-variable limit.

## 10.4    Consequences on Macroscopic Closure Expressions

### 10.4.1  The Limitations of the Turbulent Viscosity Concept

Technical in nature, these results are meant to describe the steps towards the diffusive regime obtained as the over-damped limit of the Langevin equation for particle velocities. They demonstrate that consistent results can be derived for velocity statistics by manipulating the white-noise limit while still handling SDEs. Yet, they should not be regarded as providing arguments for the concept of a turbulent viscosity or diffusivity. Given the length at which we have been obliged to go, it can even be said that they point to their lack of universal validity. There is indeed an essential difference with classical kinetic theory. When going from molecular motions to the Navier-Stokes equations, the ratio of the molecular relaxation timescale to a characteristic fluid time $\mathcal{L}/\delta U$ (where $\delta U$ is a typical velocity difference over $\mathcal{L}$ the size of the domain considered) is of the order of $Kn \times Ma$, where $Kn$ and $Ma$ are the Knudsen and Mach numbers, respectively, which are very small in continuous incompressible flows. The Newton and Fick laws are therefore excellent approximations. However, this is not so in turbulence since the typical timescale ratio, $k_f/\langle\epsilon_f\rangle \partial\langle U_f\rangle/\partial x$, is of the order of 1–5 rather than being small [4].

Due to their similarity with molecular viscosity, closure relations relying on turbulent viscosity have been considered very early and have a long history in turbulence modeling. Their attractiveness comes also from the model reduction they allow and, perhaps for this reason, they prove to be a die-hard notion. It may be fair to say that the limitations outlined above should not necessarily rule them out for some practical applications where the errors induced can be carefully assessed with respect to other turbulent flow simulation issues. This explains that they are still considered in turbulence modeling [4, chapter 10]. Interestingly enough, up-to-date formulations of turbulent viscosity are derived also through model reduction, by starting from the transport equations for the Reynolds-stress tensor components and introducing what is referred to as a 'weak-equilibrium assumption' which states that transport terms are proportional to the local value of the difference between turbulent kinetic energy and dissipation (see the detailed discussion in [4, section 11.9]). This weak-equilibrium assumption is however applied without physical justification and without elaborating a clear-cut criterion with which to assess its range of validity. In that sense, it is believed that the formulation developed in Sects. 10.2 and 10.3 offers a more physically-justified perspective based on fast-variable elimination techniques triggered by monitoring the velocity timescales. By evaluating these physical timescales with respect to an observation time lapse, say $\Delta t$ (which corresponds to the time step in numerical simulations), we can rely on a criterion with which to govern the now-continuous shift from one model formulation to a reduced one. Note that we take the fast-variable limit only for the redistributive part of the turbulent kinetic energy exchanges and do not explicitly enforce a weak-equilibrium assumption for the slowly-varying components of turbulent motions. As mentioned in the preceding paragraph, this approach is also useful to point out that the concept of a turbulent viscosity has actually limited physical support.

In particle-laden turbulent flows, these limitations are compounded by the timescales attached to the discrete particles (typically $\tau_p$) which, apart from some very specific situations, are comparable to the fluid ones or, at least, not necessarily much larger or smaller. In short, there is even less reason to expect that physical quantities arising from transport, such as the particle kinetic tensor $\langle u_{p,i}\, u_{p,j}\rangle$ or the fluid-particle correlation tensor $\langle u_{s,i}\, u_{s,j}\rangle$, can be expressed by relations involving turbulent viscosities. We are now in a position to take a broader standpoint and clarify that the real issue lies elsewhere.

## 10.4.2 Local and Non-local Closures in Dispersed Two-Phase Flow

More than the specific formulas for the turbulent viscosity, what is at stake is whether we can use local closure relations for the statistics of interest or we have to handle them as non-local physical quantities obtained as the solution of their own transport equations. At the beginning of Sect. 10.3, we mentioned the connection between the introduction of white-noise terms in the SDEs of the particle state vector and the appearance of local or non-local terms in the transport equations satisfied by the fields corresponding to the moments of these variables. In fact, this connection is an echo of the choice of a Markovian approach, as explained in Sect. 5.2, and, loosely speaking, we are looking at two sides of the same coin. However, since there may be some confusion about these correspondences and given that they are of direct concern for two-phase flow modeling, it is important to clarify this interplay.

To that end, we follow up on the description given in [1, section 10.3] and use the same isotropic formulation of the PDF model for $\mathbf{Z}_p = (\mathbf{X}_p, \mathbf{U}_p, \mathbf{U}_s)$,

$$dX_{p,i} = U_{p,i}\, dt \,, \tag{10.56a}$$

$$dU_{p,i} = \frac{U_{s,i} - U_{p,i}}{\tau_p}\, dt \,, \tag{10.56b}$$

$$dU_{s,i} = -\frac{U_{s,i} - \Phi_i}{T_L^*}\, dt + \underbrace{\sigma_{s,ij}\, dW_j}_{\text{fast term}} \,, \tag{10.56c}$$

where we have retained the reduced particle momentum equation in Eq. (10.56b) with a constant $\tau_p$ and used the same relaxation timescale $T_L^*$ in each direction in Eq. (10.56c), for the sake of simplicity. This allows us to bring out the key physical quantities, namely these two timescales $\tau_p$ and $T_L^*$. In Eq. (10.56c), $\Phi$ is a field quantity (containing typically the mean fluid velocity at the particle location and additional terms depending on the choice of the stochastic model for $\mathbf{U}_s$), while $\sigma_s$ is the diffusion matrix selected in the model of $\mathbf{U}_s$. The precise expressions of these drift and diffusion coefficients play no role in the present analysis and we can proceed with this simplified but nevertheless general formulation. What is relevant is the corresponding structure of the set of transport equations for the second-order velocity moments equations which we write using a compact notation as follows

$$\frac{D_p \langle u_{p,i} u_{p,j} \rangle}{Dt} = \mathcal{D}_{u_{p,i} u_{p,j}} + \mathcal{P}_{u_{p,i} u_{p,j}}$$

$$+ \frac{1}{\tau_p} [ \underbrace{\langle u_{p,i} u_{s,j} \rangle + \langle u_{p,j} u_{s,i} \rangle}_{\text{non-local term}} -2 \langle u_{p,i} u_{p,j} \rangle ], \qquad (10.57a)$$

$$\frac{D_p \langle u_{s,i} u_{p,j} \rangle}{Dt} = \mathcal{D}_{u_{p,i} u_{s,j}} + \mathcal{P}_{u_{p,i} u_{s,j}}$$

$$+ \frac{1}{\tau_p} \underbrace{\langle u_{s,i} u_{s,j} \rangle}_{\text{non-local term}} - \left( \frac{1}{\tau_p} + \frac{1}{T_L^*} \right) \langle u_{s,i} u_{p,j} \rangle , \qquad (10.57b)$$

$$\frac{D_p \langle u_{s,i} u_{s,j} \rangle}{Dt} = \mathcal{D}_{u_{s,i} u_{s,j}} + \mathcal{P}_{u_{s,i} u_{s,j}} - \frac{2}{T_L^*} \langle u_{s,i} u_{s,j} \rangle + \underbrace{(\sigma_s \sigma_s^\perp)_{ij}}_{\text{local term}} . \qquad (10.57c)$$

In these equations, $D_p/Dt = \partial/\partial t + \langle U_{p,k} \rangle \partial/\partial x_k$ while $\mathcal{D}_{u_{k,i} u_{l,j}}$ and $\mathcal{P}_{u_{k,i} u_{l,j}}$ (with $k, l \in \{p, s\}$) stand for the diffusion and production terms entering the second-order equations and whose details we can leave out for the present discussion (see the complete formulations in [5, section 8.5.3] or in [6] for example). The important point is the structure of the second-order equations. It is seen that the complete system in Eqs. (10.57) exhibits a typical hierarchical structure in terms of the local and non-local terms appearing on the right-hand side of Eqs. (10.57). More specifically, the particle kinetic tensor $\langle u_{p,i} u_{p,j} \rangle$ is a non-local physical quantity which depends on $\langle u_{p,i} u_{s,j} \rangle$ which, in turn, is non-locally dependent on $\langle u_{s,i} u_{s,j} \rangle$ whose closure is again non-local since it is given by the solution of the transport equation in Eq. (10.57c). Only the last quantity on the right-hand side of Eq. (10.57c) has a local expression, since this is the first level at which the white-noise term of Eq. (10.56c) appears in these second-order moments.

It may be worth emphasizing that the issue discussed in this section is not related to whether we choose to describe particle-laden flows following the PDF approach with the set of SDEs for the variables entering the particle state vector, as in Eqs. (10.56), or prefer to consider the set of transport equations for the first- and second-order moments of the same variables, as in Eqs. (10.57). The key point is where, or at what level, the shift from a rapid but ordinary process to a white-noise one is made in the SDEs or, conversely, where, or in which transport equation, a non-local term is replaced by a local closure law in the transport equations. To exemplify this mechanism, let us consider what happens when white-noise terms enter the particle system at another level. For instance, when the velocity of the fluid seen $U_s$ is regarded as a rapidly-varying variable with respect to particle inertia (i.e. $T_L^* \ll \tau_p$), we can use the fast-variable elimination techniques presented in Sect. 10.2 and the system in Eqs. (10.56) is reduced to

$$dX_{p,i} = U_{p,i} \, dt \, , \tag{10.58a}$$

$$dU_{p,i} = -\frac{U_{p,i}}{\tau_p} \, dt + \frac{1}{\tau_p} \Phi_i \, dt + \frac{1}{\tau_p} \underbrace{\left( T_L^* \sigma_{s,ij} \right) dW_j}_{\text{fast term}} \, . \tag{10.58b}$$

In line with the remark made in Sect. 10.2 and illustrated with the extended system in Eqs. (10.17), note that there are no additional drift terms in the equation for $\mathbf{U}_p$ due to the elimination of the fast variable $\mathbf{U}_s$. For this reduced model, the transport equations for the particle kinetic tensor become

$$\frac{D_p \langle u_{p,i} \, u_{p,j} \rangle}{Dt} = \mathcal{D}_{u_{p,i} u_{p,j}} + \mathcal{P}_{u_{p,i} u_{p,j}} - \frac{2}{\tau_p} \langle u_{p,i} \, u_{p,j} \rangle + \underbrace{\left( \sigma_p \, \sigma_p^\perp \right)_{ij}}_{\text{local term}}, \tag{10.59}$$

where $\sigma_p$ is the diffusion matrix in the SDE for $\mathbf{U}_p$ given by $\sigma_p = 1/\tau_p(T_L^* \sigma_s)$. Since the fast-variable elimination is local in time and space, it is seen that the white-noise term initially present in the equation of $\mathbf{U}_s$ in Eq. (10.56c) has been 'shifted upward' and appears now in the equation of $\mathbf{U}_p$, cf. Eq. (10.58b). Correspondingly, the transport equation for the particle kinetic tensor is now closed with the right-hand side of Eq. (10.59) involving only local terms. From a more physical viewpoint, the above criterion used to derive this reduced description, namely that $T_L^* \ll \tau_p$, is helpful to indicate that this corresponds to a situation involving only bullet-like particles, with respect to which turbulence can be seen as noise. This is however a very special case of little physical and practical interest.

### 10.4.3 Ensuring Consistency in the Tracer-Particle Limit

In Sect. 6.5, we raised the issue of ensuring consistency between two descriptions: the one chosen for the fluid phase, on the one hand, and the one selected for the dispersed phase, on the other hand. Indeed, it was underlined that present PDF models for the dispersed phase revert to fluid ones in the tracer-particle limit. Therefore, when the fluid mean fields entering the retained PDF model for the particle phase are obtained by choosing a turbulence model, it is essential to guarantee that this turbulence model be identical to the limit one obtained from the particle-phase PDF model when particle inertia vanishes. Said differently, we have in effect a double prediction for the same fluid statistics and we must assess that these duplicate fields corresponds to the same single physical quantity (this argument was set forth in the numerical context in Sect. 7.3). Any discrepancy in that regard would manifest a flaw in the overall formulation for the fluid and particle system since we would have two different predictions for one physical variable.

With the developments presented in Sect. 10.3, we have now enough material to answer the question brought up in Sect. 6.5 concerning the theoretical consistency of hybrid formulations where different models are used for the fluid and for the particle phases. In Sect. 6.2.2, the connection between Lagrangian stochastic models based on a GLM and the transport equations for the second-order fluid velocity moments was expressed, and a similar connection was analyzed just above in Sect. 10.4.2 through the simplified formulation of the SDEs in Eqs. (10.56) and the corresponding transport equations in Eqs. (10.57) for the second-order moments of the velocities contained in the particle state vector $\mathbf{Z}_\mathrm{p}$. Given this ubiquity in the fluid- as well as in the particle-phase descriptions, it is useful to consider a reduced but general expression of the particle state vector, $\mathbf{Z} = (\mathbf{X}, \mathbf{U})$, with evolution equations as

$$dX_i = U_i \, dt \tag{10.60a}$$

$$dU_i = F_i \, dt \tag{10.60b}$$

where $\mathbf{F}$ on the right-hand side of Eq. (10.60b) stands for the time rate of change of the particle velocity while the incremental notation $F_i \, dt$ is used to loosely denote the fact that $\mathbf{F}$ can contain white-noise terms. The specific expression of the force-per-unit-mass terms $\mathbf{F}$ is not of essence here and we can leave this point aside and consider that we handle them using classical manipulations by resorting to Stratonovich calculus, cf. Sect. 4.3.1, to avoid Ito calculus which requires special care if white-noise terms are involved in the expression of $\mathbf{F}$. What is however an essential point is that, as soon as the particle velocity is included in the state vector, Eq. (10.60a) indicates that transport is treated without approximation. As shown in previous sections, this is enough to imply that we obtain the first two moments of the velocity field as the solutions of the following system of transport equations

$$\frac{\partial \langle U_i \rangle}{\partial t} + \langle U_k \rangle \frac{\partial \langle U_i \rangle}{\partial x_k} + \frac{\partial \langle u_i \, u_k \rangle}{\partial x_k} = \langle F_i \rangle , \tag{10.61a}$$

$$\frac{\partial \langle u_i \, u_j \rangle}{\partial t} + \langle U_k \rangle \frac{\partial \langle u_i \, u_j \rangle}{\partial x_k} + \frac{\partial \langle u_i \, u_j \, u_k \rangle}{\partial x_k}$$
$$+ \langle u_i \, u_k \rangle \frac{\partial \langle U_j \rangle}{\partial x_k} + \langle u_j \, u_k \rangle \frac{\partial \langle U_i \rangle}{\partial x_k} = \langle u_i \circ F_j \rangle + \langle u_j \circ F_i \rangle , \tag{10.61b}$$

where all the convective terms have been written on the left-hand side of Eqs. (10.61) and where the symbol in $\langle u_i \circ F_j \rangle$ designates Stratonovich calculus.

Therefore, all the formulations of classical (or stochastic) mechanics, such as the one in Eqs. (10.60), entail that the equivalent of the Reynolds stress tensor $\langle u_i \, u_k \rangle$ (whatever the choice of the velocity components, be there fluid or particle ones) is always a non-local term given as the solution of a transport equation. We can now observe that the results derived in Sect. 10.3.2 have far-reaching consequences. Indeed, if we go through the intricate developments in that section, it is worth emphasizing that quite stringent assumptions had to be made to derive that $\langle u_i\, u_k \rangle$ becomes a local term expressed through a force-flux relation based on a turbulent viscosity. In particular, we had to assume that fluid velocities, or at least a part of them, become infinitely fast variables driven by vanishing timescales but having at the same time very large fluctuations. However, when we retain $\mathbf{Z}_p = (\mathbf{X}_p, \mathbf{U}_p, \mathbf{U}_s)$ as the relevant state vector for the description of the particle phase, then all the variables in $\mathbf{Z}_p$ are taken as slow ones, in accordance with the Markovian principles in Sect. 5.2. In the limit of vanishing inertia, $\mathbf{U}_p$ and $\mathbf{U}_s$ tend towards the velocity of a fluid particle which, for this reason, must be treated also as a slow variable.

There is no avoiding the conclusion that hybrid formulations, in which the fluid phase is described with a turbulence model based on a turbulent viscosity concept (be it the $k - \epsilon$ in the RANS approach or the Smagorinsky closure in LES) and the particle phase described with the PDF models presented in preceding chapters, are inconsistent. This is not a moot point since it indicates that some sort of spurious forcing would be acting in the particle evolution equations. A corresponding simple image is that, by artificially forcing a spring to oscillate around what is not its natural resting position, we would induce unphysical fluctuations.

## 10.4.4  Modeling in the Making

In this chapter, we have seen that the search for reduced descriptions can be carried out with the particle SDEs and, moreover, that this turns out to be helpful to reveal the physical nature of the assumptions we introduce as well as the limitations they imply. As already underlined on a few occasions in previous chapters, there is a correspondence or even an equivalence in a weak sense (if we limit ourselves to descriptions in terms of only the first two moments) between the PDF and the moment points of view. From a modeling perspective, it is however clear that the PDF approach, manifested by its formulation in terms of the set of SDEs for the particle state vector, cf. Eqs. (10.56), is much more amenable to accommodate new physical phenomena than the moment approach, cf. Eqs. (10.57) (to which the equations for the first-order moments must be added). In turbulent dispersed two-phase flow modeling, the later one consists indeed in a large number of highly coupled PDEs (25 if we add to the second-order moment equations in Eqs. (10.57) the mass continuity and mean particle momentum equations in Eqs. (10.1)) and this per class of particle diameters when we are dealing with the usual situation of poly-dispersed flows. In the moment approach, it is therefore tempting to enforce a model reduction, typically by resorting to a relation based on a turbulent viscosity for some (or even all) of the second-order velocity moments appearing in Eqs. (10.57). The developments presented in this chapter bring out that this corresponds to making implicitly an assumption about the rapidly-varying behavior of some of the variables, such as the fluid or particle velocities. Unless we limit ourselves to some particular situations, this is at variance with key characteristics of turbulent particle-laden flows, as illustrated with the example in Eqs. (10.58) and (10.59). In fact, such steps stray away from the Markovian approach which was demonstrated in Chap. 5 to provide a reliable framework and expose us to the risk of ill-based formulations.

The inescapable conclusion is that there are no possible reduced descriptions if we wish to retain the same physical content and are interested in developing all-purpose turbulence models with the benefit of being ensured that we evolve in a mathematically-sound and well-posed framework. For these reasons, we regard the present PDF approach, with its current formulation in terms of general Langevin-type of stochastic models for the complete particle state vector $\mathbf{Z}_p = (\mathbf{X}_p, \mathbf{U}_p, \mathbf{U}_s)$, as the basis for extensions when accounting for new phenomena, such as particle back effects on the fluid in Chap. 11 and particle collisions in Chap. 12.

## References

1. J.P. Minier, Phys. Rep. **665**, 1 (2016). https://doi.org/10.1016/j.physrep.2016.10.007
2. C. Gardiner, *Stochastic Methods*, 4th edn. (Springer, Berlin, 2009). https://link.springer.com/book/9783540707127
3. H.C. Öttinger, *Beyond Equilibrium Thermodynamics* (Wiley, 2005). https://doi.org/10.1002/0471727903
4. S. Pope, *Turbulent Flows* (Cambridge University Press, 2000). https://doi.org/10.1017/CBO9780511840531
5. J.P. Minier, E. Peirano, Phys. Rep. **352**(1–3), 1 (2001). https://doi.org/10.1016/S0370-1573(01)00011-4
6. E. Peirano, J.P. Minier, Phys. Rev. E **65**(4), 046301 (2002). https://doi.org/10.1103/PhysRevE.65.046301

# Modeling Particle Back Effects on Turbulent Fluid Flows

11

2

8

**Abstract**

Up to now, we have focused on the description of particle dynamics while the fluid flow characteristics were calculated separately. In that sense, the fluid flow was treated as unaffected by particles and, since only the actions from the fluid to the particles are addressed in this approach, this is referred to as one-way coupling. Conversely, when particle effects on the fluid need to be considered in the fluid-phase description, we are dealing with two-way coupling. Two-way coupling involves two specific effects. The first one is the volumetric fraction occupied by the particles, $\alpha_p$ and the corresponding volume fraction of the fluid $\alpha_f$ (with $\alpha_f = 1 - \alpha_p$) which may have to be included in the formulation of the governing equations. The second specific effect is the exchange of mean momentum and kinetic energy from particles to the fluid flow, which can become as important as exchanges between different parts of the fluid flow itself. In this chapter, we concentrate on this second effect and analyze the source terms accounting for these exchanges and their consequences on the stochastic particle dynamical model.

**Chapter Content** The different modes of coupling between fluids and particles are first sorted out in Sect. 11.1. Then, the source terms accounting for the exchange of momentum and kinetic energy between the two phases in the fluid mean field equations are formulated in Sect. 11.2. These source terms call for a modification of the Langevin model for the velocity of the fluid seen which is analyzed in Sect. 11.3. A study of the limit of these source terms when particle inertia vanishes is developed in Sect. 11.4. This study provides useful feedback on the formulation of the particle dynamical model whose complete form is proposed in Sect. 11.5. Finally, a challenging numerical test case involving two-way coupling is presented in Sect. 11.6.

© The Author(s) 2025                                                                                          233
J.-P. Minier et al., *Understanding Turbulent Systems*,
Lecture Notes in Physics 1039, https://doi.org/10.1007/978-3-031-84466-9_11

## 11.1  One-Way or Two-Way Coupling Between Fluids and Particles

Strictly speaking, we are always concerned by two-way coupling effects. Their relevance in present statistical descriptions of turbulent particle-laden flows is however to be evaluated also from a statistical perspective. For example, if we consider dilute flows, in which the particle volume fraction is low enough, two-way coupling effects can be neglected altogether. This is the context in which we have evolved so far. From a physical point of view, a simple way to assess two-way coupling is provided by the local mass fraction ratio, $\alpha_p \rho_p / \alpha_f \rho_f$, which represents at a given location the probable particle mass fraction compared to the probable fluid mass fraction. If this ratio is not negligible but, for instance, of order one, then we expect that momentum and kinetic energy exchanges between particles and the fluid need to be accounted for in the fluid phase description. For particles much heavier than the fluid, $\rho_p \gg \rho_f$, this ratio can indeed be of order one while the volume fraction $\alpha_p$ is still quite low. In that case, the source terms representing momentum and kinetic energy exchanges between the phases are to be retained while the volumetric effect can be disregarded so that $\alpha_f \simeq 1$. These flows can be described as moderately dense particle-laden flows, whereas at higher particle-number densities, that is at higher particle volume fractions, particle collisions or particle-particle interactions come into play (they are discussed in Chap. 12), as illustrated in Fig. 11.1. All these distinctions leading to neglect this or that two-

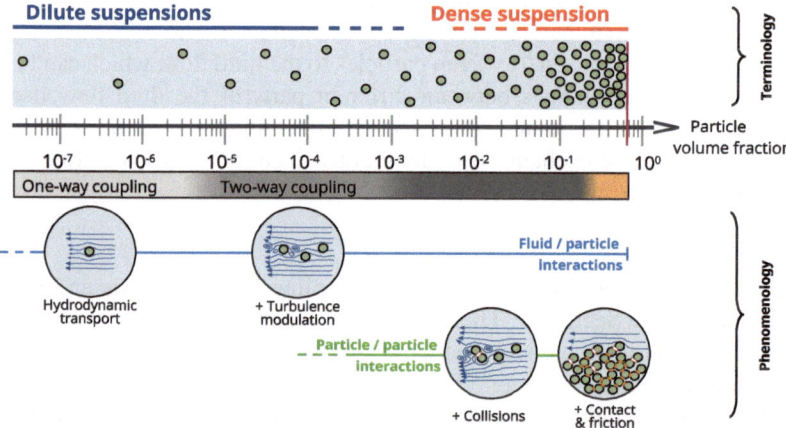

**Fig. 11.1** Sketch of the various two-phase flow regimes encountered for particle-laden flows and of the key physical phenomena using the particle volume fraction as an indicator. As the particle volume fraction increases, we go from dilute flows to moderately-dense ones and up to very dense or compact suspensions as with granular matter. In correspondence, the influences between the fluid and particle phases are indicated as one-way or two-way coupling while the fundamental interactions involved in these situations are pointed out. The inter-particle interactions, which can be regarded as a self-interaction for the particle phase, are often referred to as four-way coupling, although describing the various situations in terms of the key physical interactions at play provides a sounder overview of the issues at stake

way coupling effects have given rise to a range of particle-laden flow classifications which are useful for practical concerns since they help to simplify computations (or, sometimes, simply make them tractable). They are not, however, our primary concern here.

Our central objective is the investigation of the physical content of modeling proposals and how physical effects can be included in the present PDF approach. Thus our main question is: how do two-way coupling effects modify the modeling picture? By two-way coupling effects, we mean here both the volumetric effect and the changes in the fluid flow mean fields as the result of momentum and kinetic energy exchanges with particles. Actually, it is obvious that the volumetric effect, which is a direct outcome of the particle PDF description and is therefore already contained in the present formulation, does not raise any difficulties. On the other hand, the situation is different for the supplementary terms due to mean momentum and kinetic energy exchanges that are added in the fluid mean-field equations. Do we need to modify present formulations of the velocity of the fluid seen and, if so, how? Related to this issue is the study of the tracer-particle limit of such stochastic models and the analysis of the consistency (or lack of it) between the fluid description and the limit of the particle one when particle inertia vanishes. For this reason, we revisit first the source terms in the fluid mean-field equations before addressing how stochastic models for $\mathbf{U}_s$ are to be extended and what is the most general form of the complete particle dynamical model.

## 11.2    Source Terms in the Fluid Moment Equations

Basically, particle actions on fluid flows reflects Newton's third law: if a fluid exerts a force $\mathbf{F}_{f \to p}$ on a discrete particle, this particle exerts an opposite force $\mathbf{F}_{p \to f} = -\mathbf{F}_{f \to p}$ on the same fluid. The evaluation of particle back effects on a fluid flow requires however to sort out the various forces acting on particles in order to select the ones that are specifically due to the action of the fluid onto these particles. More precisely, only the forces due to the perturbations induced by discrete particles must be retained for this evaluation, while external forces as well as 'forces' or acceleration terms of fluid elements must be left out. This is where the careful analysis of the forces exerted on discrete particles in terms of 'undisturbed' and 'disturbed' fluid velocity fields, as carried out in Sect. 3.3, becomes relevant. To exemplify this distinction, we write the particle momentum equation as

$$m_p \frac{d\mathbf{U}_p}{dt} = m_f \frac{D\mathbf{U}_f}{Dt} + \mathbf{F}_{ext} + \mathbf{F}_{drag} + \mathbf{F}_{added-mass} + \mathbf{F}_{Basset} + \mathbf{F}_{lift} , \qquad (11.1)$$

where $\mathbf{F}_{ext}$ stands for external forces (such as buoyancy or, for instance, electrostatic forces should we treat charged particles in an externally-provided electrical field). Then, the forces acting back on the fluid are here the drag, added-mass, Basset and lift ones. If other effects due to the disturbance fluid velocity fields induced by particles are present, then the total force $\mathbf{F}_{p\rightarrow f}$ acting back on the fluid can be written as

$$\mathbf{F}_{p\rightarrow f} = -\left( m_p \frac{d\mathbf{U}_p}{dt} - m_f \frac{D\mathbf{U}_f}{Dt} - \mathbf{F}_{\text{ext}} \right) . \tag{11.2}$$

In the present work, we retain the drag force and neglect other effects, so that we can write that $\mathbf{F}_{p\rightarrow f} = -m_p \mathbf{A}_p^D$ where the drag acceleration term $\mathbf{A}_p^D$ is given by

$$\mathbf{A}_p^D = \frac{\mathbf{U}_s - \mathbf{U}_p}{\tau_p} . \tag{11.3}$$

Having defined the elementary back force on the fluid, we need to work out its statistical properties to obtain the source terms representing the net momentum and kinetic energy exchanges between the fluid and the particles which are added to the fluid velocity moment equations. From a physical standpoint, we can get a fair view of such terms by resorting to the hypothesis of local spatial homogeneity in a small volume $\delta\mathcal{V}^{(\mathbf{x})}$ around a point $\mathbf{x}$ containing $N_p^{(\mathbf{x})}$ discrete particles. Then, the momentum exchange $\delta\mathbf{M}_{p\rightarrow f}$ from these particles to the fluid flow is simply the sum of the $N_p^{(\mathbf{x})}$ elementary back forces. This gives

$$\delta\mathbf{M}_{p\rightarrow f} = -\sum_{i=1}^{N_p^{(\mathbf{x})}} m_p^{[i]} \left( \mathbf{A}_p^D \right)^{[i]} = -\rho_p \sum_{i=1}^{N_p^{\mathbf{x}}} \mathcal{V}_p^{[i]} \left( \mathbf{A}_p^D \right)^{[i]} , \tag{11.4}$$

with $m_p = \rho_p \mathcal{V}_p$ and $\mathcal{V}_p = \pi d_p^3/6$ the volume for spherical particles. This is rearranged with the particle volume fraction (since $\alpha_p \simeq \sum_{i=1}^{N_p^{(\mathbf{x})}} \mathcal{V}_p^{[i]}/\delta\mathcal{V}^{(\mathbf{x})}$) as

$$\delta\mathbf{M}_{p\rightarrow f} = -\rho_p \left( \sum_{i=1}^{N_p^{(\mathbf{x})}} \mathcal{V}_p^{[i]} \right) \frac{\displaystyle\sum_{i=1}^{N_p^{(\mathbf{x})}} m_p^{[i]} \left( \mathbf{A}_p^D \right)^{[i]}}{\displaystyle\sum_{i=1}^{N_p^{(\mathbf{x})}} m_p^{[i]}} \simeq -\rho_p \alpha_p \delta\mathcal{V}^{(\mathbf{x})} \langle \mathbf{A}_p^D \rangle , \tag{11.5}$$

where the estimation of $\langle \mathbf{A}_p^D \rangle$ corresponds to the one in Eq. (4.5) in Sect. 4.1.1. This gives the expression for the source term of momentum per unit fluid volume, $\delta\mathbf{M}_{p\rightarrow f}/\delta\mathcal{V}^{(\mathbf{x})}$, as $\alpha_p \rho_p \langle \mathbf{A}_p^D \rangle$ while, by similar steps, we can guess that the source term of kinetic energy is the average value of the work performed by $\mathbf{F}_{p\rightarrow f}$. Indeed, in the frame of this local homogeneous situation, all the particles contained in $\delta\mathcal{V}^{(\mathbf{x})}$ represent equivalent samples of the distribution at point $\mathbf{x}$, so that summing effects over them amounts to expressing Monte Carlo estimates of the averaged effects defined through proper integration against the MDF. Thus, although these expressions constitute an easier-to-understand approach, it must be remembered that the rigorous formulation of the fluid velocity moment equations requires the complete PDF formalism to be applied.

Such a derivation of the mean field equations for the fluid phase, as well as for the particle phase, is described in detail in [1, section 8] and in subsequent works [2, 3] following the methodology outlined in Sect. 4.1.1. Consequently, we limit ourselves to stating the main results and introducing useful relations for our present purpose.

The first important equation is the mean Navier-Stokes equation for the average fluid velocity $\langle \mathbf{U}_f \rangle$ which, for high Reynolds-number flows, reads

$$
\alpha_f \rho_f \left( \frac{\partial \langle U_{f,i} \rangle}{\partial t} + \langle U_{f,k} \rangle \frac{\partial \langle U_{f,i} \rangle}{\partial x_k} \right) + \frac{\partial \left( \alpha_f \rho_f \langle u_{f,i} \, u_{f,k} \rangle \right)}{\partial x_k} = - \alpha_f \frac{\partial \langle P_f \rangle}{\partial x_i}
$$

$$
- \alpha_p \rho_p \left\langle A_{p,i}^D \right\rangle , \qquad (11.6)
$$

where the last term on the right-hand side of Eq. (11.6) is the source term accounting for the mean momentum exchange between the fluid and particle phases. The terms on the left-hand side of Eq. (11.6) stem from the integration of derivatives along a fluid particle trajectory against the MDF of the fluid phase. Indeed, writing $D/Dt$ for such derivatives, i.e. $D/Dt = \partial/\partial t + U_{f,k} \partial/\partial x_k$, and with the definitions given in Sect. 4.1.1, we define an operator $\mathcal{L}(\langle \Phi \rangle)$ for any fluid-attached property $\Phi$, by

$$
\mathcal{L}(\langle \Phi \rangle) = \int \left( \frac{D\Phi}{Dt} \right) F_f^E(t, \mathbf{x}; \mathbf{z}_c) \, d\mathbf{z}_c , \qquad (11.7)
$$

using the same notation $\mathbf{z}_c$ as in Sect. 4.1.1 to denote the extra variables apart from the fluid particle position (in the following, we consider $\mathbf{z}_c = \mathbf{V}_f$ the fluid velocity value in sample space). From the definition of averages given in Eq. (4.3) applied to the fluid phase and the expression of $D/Dt$, this leads to

$$
\mathcal{L}(\langle \Phi \rangle) = \frac{\partial \left( \alpha_f \rho_f \langle \Phi \rangle \right)}{\partial t} + \frac{\partial \left( \alpha_f \rho_f \langle U_{f,k} \rangle \langle \Phi \rangle \right)}{\partial x_k} + \frac{\partial \left( \alpha_f \rho_f \langle u_{f,k} \, \Phi \rangle \right)}{\partial x_k} , \qquad (11.8)
$$

$$
= \alpha_f \rho_f \left( \frac{\partial \langle \Phi \rangle}{\partial t} + \langle U_{f,k} \rangle \frac{\partial \langle \Phi \rangle}{\partial x_k} \right) + \frac{\partial \left( \alpha_f \rho_f \langle u_{f,k} \, \Phi \rangle \right)}{\partial x_k} . \qquad (11.9)
$$

In Eqs. (11.8) and (11.9), we have used the fluid phase mean mass conservation equation while we can express Eq. (11.7) with the average of $D\Phi/Dt$, so that we get

$$
\alpha_f \rho_f \left\langle \frac{D\Phi}{Dt} \right\rangle = \alpha_f \rho_f \left( \frac{\partial \langle \Phi \rangle}{\partial t} + \langle U_{f,k} \rangle \frac{\partial \langle \Phi \rangle}{\partial x_k} \right) + \frac{\partial \left( \alpha_f \rho_f \langle u_{f,k} \, \Phi \rangle \right)}{\partial x_k} . \qquad (11.10)
$$

This formula turns out to be useful and allows the transport equations of the fluid velocity moments to be formulated in a tractable manner for later manipulations.

Note that the correlation $\langle u_{f,k} \Phi \rangle$ in the last term on the right-hand side of Eq. (11.10) can also be written as $\langle u_{f,k} \phi \rangle$, with $\phi = \Phi - \langle \Phi \rangle$.

For instance, the mean Navier-Stokes equation, Eq. (11.6), becomes

$$\mathcal{L}(\langle U_{f,i} \rangle) = \alpha_f \rho_f \left\langle \frac{D U_{f,i}}{Dt} \right\rangle = -\alpha_f \frac{\partial \langle P_f \rangle}{\partial x_i} - \alpha_p \rho_p \left\langle A_{p,i}^D \right\rangle . \tag{11.11}$$

The second important set of equations are the transport equations for the second-order fluctuating velocity moments of the fluid phase. When the source terms due to the action of particles on the fluid flow are included, the equations for $\langle u_{f,i} u_{f,j} \rangle$ have the following form (using Eq. (11.10) for each component, i.e. $\Phi = u_{f,i} u_{f,j}$)

$$\alpha_f \rho_f \left\langle \frac{D \left( u_{f,i} \, u_{f,j} \right)}{Dt} \right\rangle = -\alpha_f \rho_f \langle u_{f,i} \, u_{f,k} \rangle \frac{\partial \langle U_{f,j} \rangle}{\partial x_k} - \alpha_f \rho_f \langle u_{f,j} \, u_{f,k} \rangle \frac{\partial \langle U_{f,i} \rangle}{\partial x_k}$$

$$+ \alpha_f \rho_f \left( G_{ik} \langle u_{f,j} \, u_{f,k} \rangle + G_{jk} \langle u_{f,i} \, u_{f,k} \rangle \right) + \alpha_f \rho_f \langle B_{f,ik} \, B_{f,jk} \rangle$$

$$- \alpha_p \rho_p \left( \langle u_{s,i} \, A_{p,j}^D \rangle + \langle u_{s,j} \, A_{p,i}^D \rangle \right) . \tag{11.12}$$

In this equation, $\mathbf{G}$ and $\mathbf{B}_f$ are the matrices entering the drift and diffusion coefficients of the GLM selected for the description of the fluid phase. This is the same GLM used as the building block to derive the corresponding two-phase GLM for the discrete particles and which is, for the sake of providing a self-contained presentation, rewritten here as

$$DU_{s,i} = DU_{f,i} = -\frac{1}{\rho_f} \frac{\partial \langle P_f \rangle}{\partial x_i} \, dt + G_{ik} \left( U_{f,k} - \langle U_{f,k} \rangle \right) \, dt + B_{f,ik} \, dW_k . \tag{11.13}$$

For instance, if the expression of $\mathbf{G}$ is as in Eq. (6.19), $\mathbf{B}_f$ has the form given in Eq. (6.21). Note that the last two terms on the right-hand side of Eq. (11.12) corresponds to the gain or loss due to the fluctuating part of the work performed by the drag force. Since the drag forces are point forces in the present approach, due to the point-particle approximation, these works imply the velocity of the fluid seen. Anticipating on what is introduced below, these two terms are written assuming that there is no drift velocity, i.e. $U_{d,i} = \langle U_{s,i} - \langle U_{f,i} \rangle \rangle = 0$, which allows the instantaneous expression of $\mathbf{A}_p^D$ to be retained.

## 11.3   Current Formulations for the Velocity of the Fluid Seen

Taking into account two-way coupling is not limited to adding source terms representing the net exchange of momentum and kinetic energy between the two phases in the fluid mean-field equations, as in Eqs. (11.11) and (11.12). We need also to modify the model for the velocity of the fluid seen accordingly. As it transpires from its name, $U_s$ stands for the instantaneous fluid velocity at discrete particle locations and is thus impacted by particle back effects. In the frame of hybrid methods, where the fluid mean fields are calculated separately from the PDF model used to describe discrete particle dynamics, this aspect is sometimes overlooked. However, keeping the same form of the stochastic model for $U_s$ as in dilute flows results in inconsistent formulations. This is easily realized from situations where the statistics of the velocity of the fluid seen are similar to the fluid velocity ones. It is then clear that we need to have an extra term in the evolution equation of $U_s$ if we are to recover the same two-way coupling terms as on the right-hand side of Eqs. (11.11) and (11.12). Though this latter argument is more an a posteriori technical point, it nevertheless brings support to the former which is more physical in nature. For these reasons, it follows that we must devise an extra term, noted $S_{p \to s}$, so that

$$d\mathbf{U}_s = (d\mathbf{U}_s)_{\text{dilute}} + \mathbf{S}_{p \to s} \, dt. \tag{11.14}$$

From the above indications and from the more detailed analysis given in [1, section 8.4], we already know that we should have

$$\alpha_f \rho_f \langle S_{p \to s,i} \rangle = -\alpha_p \rho_p \langle A^D_{p,i} \rangle , \tag{11.15a}$$

$$\alpha_f \rho_f \langle u_{s,i} \, S_{p \to s,j} \rangle = -\alpha_p \rho_p \langle u_{s,i} \, A^D_{p,j} \rangle . \tag{11.15b}$$

Given that $A^D_p$ has a close expression in terms of the variables entering the particle state vector $\mathbf{Z}_p = (\mathbf{X}_p, \mathbf{U}_p, \mathbf{U}_s)$, cf. Eq. (11.3), the most obvious choice for $\mathbf{S}_{p \to s}$ is to assume that

$$\mathbf{S}_{p \to s} = -r_{\text{mp}} \mathbf{A}^D_p , \quad \text{with} \quad r_{\text{mp}} = \frac{\alpha_p \rho_p}{\alpha_f \rho_f} . \tag{11.16}$$

This is the formulation proposed in [1] and in subsequent works [4, 5]. If we retain this expression, we arrive at a complete form of a two-phase GLM including two-way coupling by adding the extra term defined in Eq. (11.16) to the two-phase GLM for dilute flows given in Eq. (8.3) in Sect. 8.1, which becomes

$$dU_{s,i} = -\frac{1}{\rho_f} \frac{\partial \langle P_f \rangle}{\partial x_i} \, dt + \langle U_{r,j} \rangle \frac{\partial \langle U_{f,i} \rangle}{\partial x_j} \, dt - r_{\text{mp}} A^D_{p,i} \, dt$$
$$+ G^*_{ij} \left( U_{s,j} - \langle U_{f,j} \rangle \right) dt + B_{s,ij} \, dW_j . \tag{11.17}$$

The simplicity of the closure proposed for $S_{p \to s}$ in Eq. (11.16) should not hide the fact that this remains a model, as discussed in [1, section 8.4]. This may even be a rather 'stiff model' in the sense that we use the very same variables of the particle state vector, namely $U_p$ and $U_s$, and simply weight the corresponding instantaneous value of $A^D_p$ by the probable local mass fractions of the particle and fluid phases. Other, somewhat 'smoother', models could be developed for $S_{p \to s}$, provided that the two requirements in Eqs. (11.15) are met. Thus, the model in

Eq. (11.16) is to be regarded as one possibility among a class of similar models for $S_{p \to s}$. It is nevertheless retained in the following since it appears as the reference formulation and allows to study the important tracer-particle limit in order to bring out conclusions that remain valid for such a class of models as to the behavior of such source terms.

## 11.4    The Tracer-Particle Limit of Exchange Terms

Studying the tracer-particle limit of the particle dynamical model provides valuable insights into the role played by the exchange terms representing particle back effects on the fluid and into the question of whether the reduced particle momentum equation can be deemed sufficient or the extended formulation is required. To that effect, we analyze the fluid limit in the following sense: we consider that the discrete particles become small isolated pockets of fluid but with a non-vanishing volume fraction. This corresponds to taking the limit $\tau_p \to 0$ but with $\alpha_p$ remaining constant. Given that we are interested in retrieving the single-phase description, we also consider that these fluid particles are well-mixed and that there is no drift velocity. Actually, it is as if we had started from a single-phase flow and decided to mark and track small areas of fluid to create an artificial two-phase flow system. Combining the fluid and particle descriptions should therefore yield the single-phase flow formulation.

### 11.4.1  Do We Recover the Single-Phase Moment Equations?

In the present statistical framework, answering this question amounts to checking that the single-phase models for the fluid velocity first- and second-order moment transport equations are retrieved. We start with the high Reynolds-number form of the mean momentum equation, Eq. (11.11), which, using the extended form of the particle momentum equation, becomes

$$\alpha_f \rho_f \left\langle \frac{DU_{f,i}}{Dt} \right\rangle = -\alpha_f \frac{\partial \langle P_f \rangle}{\partial x_i} + \alpha_p \rho_f \left\langle \frac{DU_{s,i}}{Dt} \right\rangle - \alpha_p \rho_p \left\langle \frac{dU_{p,i}}{dt} \right\rangle . \qquad (11.18)$$

Regardless of the choice of a fluid GLM, we have $\langle DU_{s,i}/Dt \rangle = -1/\rho_f \partial \langle P_f \rangle / \partial x_i$, which gives

$$\alpha_f \rho_f \left\langle \frac{DU_{f,i}}{Dt} \right\rangle + \alpha_p \rho_p \left\langle \frac{dU_{p,i}}{dt} \right\rangle = -\alpha_f \frac{\partial \langle P_f \rangle}{\partial x_i} - \alpha_p \frac{\partial \langle P_f \rangle}{\partial x_i} = -\frac{\partial \langle P_f \rangle}{\partial x_i} . \qquad (11.19)$$

As indicated above, we are considering the special case of the limit of small particles which become fluid-like elements and, in that sense, we can assume that $\rho_p \simeq \rho_f$, leading to

$$\alpha_f \left\langle \frac{DU_{f,i}}{Dt} \right\rangle + \alpha_p \left\langle \frac{dU_{p,i}}{dt} \right\rangle = -\frac{1}{\rho_f} \frac{\partial \langle P_f \rangle}{\partial x_i} . \tag{11.20}$$

We therefore retrieve the proper single-phase mean Navier-Stokes equation provided that we can show that $dU_p/dt$ tends towards $DU_s/Dt$ when $\tau_p \to 0$ (remember that, with present notations, we have in fact $DU_s/Dt = (DU_f/Dt)(t, \mathbf{X}_p(t)))$. Given that we are handling stochastic diffusion processes (for which acceleration is not defined), note that the mean value of the derivatives appearing in the previous equations are to be regarded as the first-order in time variations of the mean values of the increments, e.g. $\langle DU_s/Dt \rangle = \langle DU_s \rangle/Dt$.

## 11.4.2 The Fluid Limit of the Particle Momentum Equation

For small values of $\tau_p$, further light on the behavior of $dU_p/dt$ can be shed by manipulating the extended particle momentum equation, written as

$$\frac{d\mathbf{U}_p}{dt} = \frac{\mathbf{U}_s - \mathbf{U}_p}{\tau_p} + \mathbf{A}_s = \mathbf{A}_p^D + \mathbf{A}_s , \tag{11.21}$$

with $\mathbf{A}_s = D\mathbf{U}_s/Dt$ the acceleration of the fluid seen or the acceleration of the fluid at the discrete particle location (still using $\rho_p \simeq \rho_f$). For constant $\tau_p$, we can integrate this equation from an initial time $t_0$, which yields for $t \geq t_0$

$$\mathbf{U}_p(t) = \mathbf{U}_p(t_0) \exp(-(t - t_0)/\tau_p) + \frac{1}{\tau_p} \exp(-t/\tau_p) \int_{t_0}^{t} \mathbf{U}_s(u) \exp(u/\tau_p) du$$

$$+ \exp(-t/\tau_p) \int_{t_0}^{t} \mathbf{A}_s(u) \exp(u/\tau_p) du . \tag{11.22}$$

Integrating by parts the second term on the right-hand side of Eq. (11.22) leads to

$$\mathbf{U}_p(t) = \mathbf{U}_p(t_0) \exp(-(t - t_0)/\tau_p) + \mathbf{U}_s(t) \left[ 1 - \exp(-(t - t_0)/\tau_p) \right]$$

$$- \exp(-t/\tau_p) \int_{t_0}^{t} \left( \frac{d\mathbf{U}_s}{dt} \right)(u) \exp(u/\tau_p) du$$

$$+ \exp(-t/\tau_p) \int_{t_0}^{t} \mathbf{A}_s(u) \exp(u/\tau_p) du . \tag{11.23}$$

For long elapsed time after the initial condition (i.e. $(t - t_0)/\tau_p \gg 1$), this gives

$$\frac{\mathbf{U}_s - \mathbf{U}_p}{\tau_p} \simeq \frac{1}{\tau_p} \exp(-t/\tau_p) \int_{t_0}^{t} \left(\frac{d\mathbf{U}_s}{dt}\right)(u) \exp(u/\tau_p) du$$

$$- \frac{1}{\tau_p} \exp(-t/\tau_p) \int_{t_0}^{t} \mathbf{A}_s(u) \exp(u/\tau_p) du . \quad (11.24)$$

When $\tau_p$ is very small, $1/\tau_p \exp((u - t)/\tau_p)$ is a highly peaked function around $u = t$ which behaves as a Dirac function, and we obtain therefore that

$$\frac{\mathbf{U}_s - \mathbf{U}_p}{\tau_p} \simeq \frac{d\mathbf{U}_s}{dt} - \mathbf{A}_s , \quad \text{or} \quad \frac{d\mathbf{U}_p}{dt} \simeq \frac{d\mathbf{U}_s}{dt} . \quad (11.25)$$

Yet, this does not necessarily entail that $d\mathbf{U}_p/dt \simeq D\mathbf{U}_s/Dt$ in the small $\tau_p$ limit. Moreover, as already pointed out above, note that we have carried out these manipulations as if we are handling deterministic processes rather than stochastic diffusion ones, for which care must be taken though these results remain valid when written under incremental forms. With these caveats in mind, we are then led to consider whether $d\mathbf{U}_s$ tends towards $D\mathbf{U}_s$ when $\tau_p \to 0$.

### 11.4.3 Remaining Modeling Issues

At this stage, the developments presented in Sect. 8.2 reveal their usefulness. Indeed, if we refer to the alternative form of the complete particle dynamics model presented in Eqs. (8.26), we have seen that the two-phase GLM for $d\mathbf{U}_s$ contains a sub-model for $D\mathbf{U}_s$ to which it reverts when $\delta b$ goes to zero. Since $\delta b$ is a vanishing function in the limit of small $\tau_p$, that is $\delta b \to 0$ when $\tau_p \to 0$, we can therefore conclude that $d\mathbf{U}_p/dt \simeq D\mathbf{U}_s/Dt$ in the limit of vanishing particle inertia. This is true, however, with the form of the two-phase GLM handled throughout Sect. 8.2, that is in the dilute regime and in the absence of the additional source term, namely $r_{mp}\mathbf{A}_p^D$, which is now added to the Langevin model for $\mathbf{U}_s$. It remains to assess whether this conclusion is still valid when particle back effects to the fluid are accounted for in the modeled equation for $d\mathbf{U}_s$ and to monitor how the newly-added source term, $r_{mp}\mathbf{A}_p^D$, behaves in the limit of small $\tau_p$ since it involves the ratio of two vanishing quantities.

To investigate this question, we analyze a simplified version of the particle dynamical model by considering the situation of homogeneous stationary mean fields in a one-dimensional space setting (meaning that we handle scalars instead of vectors). This is convenient to concentrate on the key physical issues at stake and, furthermore, it allows analytical results to become tractable without loss of generality.

## 11.4.4 A Physical Case Study

In this simplified yet relevant case study, the particle state vector is limited to $Z_p = (U_p, U_s)$ (since $X_p$ does not play any role) and the mean field values, such as $\langle U_f \rangle$ and $\langle \epsilon_f \rangle$, are constant as well as the timescale $T_L$. Following the notation already introduced in Sect. 8.2, we write the timescale of the velocity of the fluid seen $T_L^*$ as $T_L = T_L^* (1 + \delta b)$ with $\delta b \geq 0$.

### 11.4.4.1 Limit Behavior with the Reduced Particle Momentum Equation

When we retain only the reduced form of the particle momentum equation, the SDEs describing the evolution of $(U_p, U_s)$ write

$$dU_p = \frac{U_s - U_p}{\tau_p} \, dt = A_p^D \, dt \, , \tag{11.26a}$$

$$dU_s = -\frac{U_s}{T_L}(1 + \delta b) \, dt - r_{mp} A_p^D \, dt + \sqrt{C_0 \langle \epsilon_f \rangle (1 + \delta b)} \, dW \, . \tag{11.26b}$$

In homogeneous situations and with constant $\tau_p$, $A_p^D$ is a Gaussian random variable (at each given time $t$ and a Gaussian process when considered in continuous time) fully characterized by its first and second moments. To work out the expressions of these two moments in the long-time limit when they reach their stationary value, it is convenient to manipulate directly the SDE satisfied by $A_p^D$. From its definition and the modeled evolution equations of $(U_p, U_s)$ in Eqs. (11.26), this equation reads

$$dA_p^D = -\left(\frac{1 + r_{mp}}{\tau_p}\right) A_p^D \, dt - \left(\frac{1 + \delta b}{\tau_p T_L}\right) U_s \, dt + \frac{1}{\tau_p} \sqrt{C_0 \langle \epsilon_f \rangle (1 + \delta b)} \, dW \, . \tag{11.27}$$

Since $\langle U_s \rangle$ is constant and can be taken as zero with a change of the coordinate system, applying the averaging operator to Eq. (11.27) yields that $\langle A_p^D \rangle$ is the solution of

$$d\langle A_p^D \rangle = -\left(\frac{1 + r_{mp}}{\tau_p}\right) d\langle A_p^D \rangle \, dt \, , \tag{11.28}$$

from which it follows that $\langle A_p^D \rangle(t) \to 0$ when $t \to +\infty$. To obtain the variance of $A_p^D$, we rely on stationary arguments for the various second-order moments, i.e. we write that $d\langle (A_p^D)^2 \rangle = d\langle U_s^2 \rangle = d\langle U_s A_p^D \rangle = 0$. Using Ito calculus, we obtain by multiplying Eq. (11.26b) with $U_s$ that

$$\langle U_s^2 \rangle = \frac{1}{2} C_0 \langle \epsilon_f \rangle T_L - \left(\frac{r_{mp} T_L}{1 + \delta b}\right) \langle U_s A_p^D \rangle \, , \tag{11.29}$$

and by multiplying Eq. (11.27) with $A_p^D$

$$\langle (A_p^D)^2 \rangle = -\left( \frac{1 + \delta b}{(1 + r_{mp}) T_L} \right) \langle U_s A_p^D \rangle + \frac{1}{2} C_0 \langle \epsilon_f \rangle \left( \frac{1 + \delta b}{(1 + r_{mp}) \tau_p} \right) , \qquad (11.30)$$

while the cross product of Eqs. (11.26b) and (11.27) with $A_p^D$ and $U_s$ gives

$$\left[ \frac{1 + r_{mp}}{\tau_p} + \frac{1 + \delta b}{T_L} \right] \langle U_s A_p^D \rangle = -\left( \frac{1 + \delta b}{T_L \tau_p} \right) \langle U_s^2 \rangle$$

$$- r_{mp} \langle (A_p^D)^2 \rangle + \frac{1}{\tau_p} C_0 \langle \epsilon_f \rangle (1 + \delta b) . \qquad (11.31)$$

Straightforward calculations show that the long-time value for the variance is

$$\langle (A_p^D)^2 \rangle = \frac{1}{2} C_0 \langle \epsilon_f \rangle \frac{1}{\tau_p} \frac{(1 + \delta b) T_L}{(1 + r_{mp}) T_L + (1 + \delta b) \tau_p} . \qquad (11.32)$$

For small values of $\tau_p$, the last factor on the right-hand side of Eq. (11.32) is of order one (given that $\delta b = \delta b(\tau_p) \ll 1$ and $0 \leq r_{mp} \leq 1$) and the variance of $A_p^D$ scales as

$$\langle (A_p^D)^2 \rangle \sim \frac{1}{2} C_0 \langle \epsilon_f \rangle \frac{1}{\tau_p} . \qquad (11.33)$$

This indicates, however, that $\langle (A_p^D)^2 \rangle$ grows unbounded as $\tau_p \to 0$ so that $A_p^D$ fluctuates wildly in this limit making it an ill-defined term. From Eq. (11.32), it appears also that this unsatisfactory behavior is not due to the particle back effect onto the fluid, since putting $r_{mp} = 0$ or keeping $r_{mp} \neq 0$ does not prevent the unbounded growth of $\langle (A_p^D)^2 \rangle$. Clearly, this points to a shortcoming of using the reduced form of the particle momentum equation.

### 11.4.4.2 Limit Behavior with the Extended Particle Momentum Equation

When the extended particle momentum equation is selected, the evolution equations for $(U_p, U_s)$ take now the following form

$$dU_p = \frac{U_s - U_p}{\tau_p} dt + DU_s = A_p^D dt + DU_s , \qquad (11.34a)$$

$$dU_s = DU_s - \frac{\delta b\, U_s}{T_L} dt - r_{mp} A_p^D dt + \sqrt{C_0\, \delta b\, \langle \epsilon_f \rangle}\, dW , \qquad (11.34b)$$

$$DU_s = -\frac{U_s}{T_L} dt + \sqrt{C_0 \langle \epsilon_f \rangle}\, dW' , \qquad (11.34c)$$

where $W$ and $W'$ are two independent Wiener processes. From this system of equations, the stochastic model for $A_p^D$ is easily derived from

$$dA_p^D = -\frac{1}{\tau_p} A_p^D dt + \frac{1}{\tau_p} (dU_s - DU_s) \tag{11.35}$$

and writes

$$dA_p^D = -\left(\frac{1+r_{mp}}{\tau_p}\right) A_p^D dt - \left(\frac{\delta b}{\tau_p T_L}\right) U_s dt + \frac{1}{\tau_p} \sqrt{C_0 \delta b \langle \epsilon_f \rangle} dW . \tag{11.36}$$

The resulting equation satisfied by the mean value indicates that we have again $\langle A_p^D \rangle \rightarrow 0$. The same methodology as with the reduced particle momentum equation can be applied to derive the expression of the variance $\langle (A_p^D)^2 \rangle$. After other tedious but still straightforward calculations, this yields the equilibrium value of the variance

$$\langle (A_p^D)^2 \rangle = \frac{1}{2} C_0 \langle \epsilon_f \rangle \frac{\delta b}{\tau_p} \frac{(1+r_{mp}) T_L + \tau_p}{(1+r_{mp})^2 T_L + (1+\delta b + r_{mp}) \tau_p} . \tag{11.37}$$

For small values of $\tau_p$, the last factor on the right-hand side of Eq. (11.37) is also of order one, whether $r_{mp} = 0$ or not, and the important resulting scaling is

$$\langle (A_p^D)^2 \rangle \sim \frac{1}{2} C_0 \langle \epsilon_f \rangle \frac{\delta b}{\tau_p} . \tag{11.38}$$

The essential difference comes from the non-dimensional small factor $\delta b = \delta b(\tau_p)$ which depends on how we model the timescale of the velocity of the fluid seen compared to the fluid one. When this results in $\delta b$ going to zero faster than $\tau_p$, then $\delta b/\tau_p \rightarrow 0$ as $\tau_p \rightarrow 0$. Note that this condition is met with the Csanady factors, cf. Eqs. (6.31), since we have $\delta b \simeq (\langle U_r \rangle)^2/\langle U_s^2 \rangle \sim \tau_p^2$ in the small $\tau_p$ limit. In that case, the mean value and the variance of $A_p^D$ vanish proving that $A_p^D \rightarrow 0$. If we remember that we are considering the limit behavior of the two-phase flow model when the discrete particle become fluid elements, the fact that the extra source term tends to zero is the expected outcome and can be regarded as translating a sound behavior of the modeled particle dynamical system.

## 11.4.5 The Limit Form of the Fluid Second-Order Moment Equation

Having validated that when the extended particle momentum equation is retained we have $dU_p/dt \rightarrow DU_s/Dt$ when $\tau_p \rightarrow 0$ even in the presence of exchange source terms, we can see from Eq. (11.20) that the correct form of the single-phase mean

Navier-Stokes equation is retrieved. To complement this result, we need to ascertain that the proper form of the second-order velocity moment equation is also recovered.

Using the extended particle momentum equation, cf. Eq. (11.21), to replace the drag acceleration force term $\mathbf{A}_p^D$, the transport equation for the fluid second-order velocity moment $\langle u_{f,i}\, u_{f,j}\rangle$, cf. Eq. (11.12), becomes

$$
\alpha_f \rho_f \left\langle \frac{\mathrm{D}\left(u_{f,i}\, u_{f,j}\right)}{\mathrm{D}t} \right\rangle + \alpha_p \rho_p \left\langle u_{f,i}\left(\frac{\mathrm{d}U_{p,j}}{\mathrm{d}t}\right)\right\rangle + \alpha_p \rho_p \left\langle u_{f,j}\left(\frac{\mathrm{d}U_{p,i}}{\mathrm{d}t}\right)\right\rangle =
$$

$$
-\alpha_f \rho_f \langle u_{f,i}\, u_{f,k}\rangle \frac{\partial \langle U_{f,j}\rangle}{\partial x_k} - \alpha_f \rho_f \langle u_{f,j}\, u_{f,k}\rangle \frac{\partial \langle U_{f,i}\rangle}{\partial x_k}
$$

$$
+\alpha_f \rho_f \left(G_{ik}\langle u_{f,j}\, u_{f,k}\rangle + G_{jk}\langle u_{f,i}\, u_{f,k}\rangle\right) + \alpha_f \rho_f \langle B_{f,ik} B_{f,jk}\rangle
$$

$$
+\alpha_p \rho_f \left\langle u_{f,i}\frac{\mathrm{D}U_{s,j}}{\mathrm{D}t}\right\rangle + \alpha_p \rho_f \left\langle u_{f,j}\frac{\mathrm{D}U_{s,i}}{\mathrm{D}t}\right\rangle . \tag{11.39}
$$

From Eq. (11.13) and Ito calculus, we have $\langle u_{f,j}\, \mathrm{D}U_{s,i}\rangle = G_{ik}\langle u_{f,j}\, u_{f,k}\rangle$ so that we can gather the terms involving the matrix $\mathbf{G}$ on the last two lines of Eq. (11.39). Using now that $\mathrm{d}U_p/\mathrm{d}t \to \mathrm{D}U_s/\mathrm{D}t$ when $\tau_p \to 0$ and taking $\rho_p \simeq \rho_f$ as already indicated above for the present analysis, we obtain

$$
\alpha_f \left\langle \frac{\mathrm{D}\left(u_{f,i}\, u_{f,j}\right)}{\mathrm{D}t} \right\rangle + \alpha_p \left\langle u_{f,i}\left(\frac{\mathrm{D}U_{f,j}}{\mathrm{D}t}\right)\right\rangle + \alpha_p \left\langle u_{f,j}\left(\frac{\mathrm{D}U_{f,i}}{\mathrm{D}t}\right)\right\rangle =
$$

$$
-\alpha_f \langle u_{f,i}\, u_{f,k}\rangle \frac{\partial \langle U_{f,j}\rangle}{\partial x_k} - \alpha_f \langle u_{f,j}\, u_{f,k}\rangle \frac{\partial \langle U_{f,i}\rangle}{\partial x_k}
$$

$$
+G_{ik}\langle u_{f,j}\, u_{f,k}\rangle + G_{jk}\langle u_{f,i}\, u_{f,k}\rangle + \alpha_f \langle B_{f,ik} B_{f,jk}\rangle . \tag{11.40}
$$

Since $\mathrm{D}/\mathrm{D}t$ stands for the instantaneous variation along the fluid particle trajectory, when we decompose $\mathbf{U}_f$ into its mean and fluctuating components, i.e. $\mathbf{U}_f = \langle \mathbf{U}_f\rangle + \mathbf{u}_f$, we can write that

$$
\left\langle u_{f,i}\left(\frac{\mathrm{D}U_{f,j}}{\mathrm{D}t}\right)\right\rangle = \left\langle u_{f,i}\left(\frac{\mathrm{D}\langle U_{f,j}\rangle}{\mathrm{D}t}\right)\right\rangle + \left\langle u_{f,i}\left(\frac{\mathrm{D}u_{f,j}}{\mathrm{D}t}\right)\right\rangle
$$

$$
= \langle u_{f,i}\, u_{f,k}\rangle \frac{\partial \langle U_{f,j}\rangle}{\partial x_k} + \left\langle u_{f,i}\left(\frac{\mathrm{D}u_{f,j}}{\mathrm{D}t}\right)\right\rangle . \tag{11.41}
$$

This relation is used for the last two terms on the left-hand side of Eq. (11.40), which is transformed into

$$
\alpha_f \left\langle \frac{\mathrm{D}\left(u_{f,i}\, u_{f,j}\right)}{\mathrm{D}t} \right\rangle + \alpha_p \left\langle u_{f,i}\left(\frac{\mathrm{D}u_{f,j}}{\mathrm{D}t}\right) + u_{f,j}\left(\frac{\mathrm{D}u_{f,i}}{\mathrm{D}t}\right)\right\rangle =
$$

$$- \langle u_{f,i}\, u_{f,k} \rangle \frac{\partial \langle U_{f,j} \rangle}{\partial x_k} - \langle u_{f,j}\, u_{f,k} \rangle \frac{\partial \langle U_{f,i} \rangle}{\partial x_k}$$

$$+ G_{ik} \langle u_{f,j}\, u_{f,k} \rangle + G_{jk} \langle u_{f,i}\, u_{f,k} \rangle + \alpha_f \langle B_{f,ik} B_{f,jk} \rangle . \tag{11.42}$$

At this point, it must be remembered that we are manipulating a stochastic diffusion process, cf. Eq. (11.13), so that derivatives should be handled through a careful application of Ito calculus. For $\left( u_{f,i}\, u_{f,j} \right)$, this is translated by the following formula

$$\langle D \left( u_{f,i}\, u_{f,j} \right) \rangle = \langle u_{f,i}\, Du_{f,j} \rangle + \langle u_{f,j}\, Du_{f,i} \rangle + \langle B_{f,ik} B_{f,jk} \rangle Dt , \tag{11.43}$$

which is rewritten as

$$\left\langle u_{f,i} \left( \frac{Du_{f,j}}{Dt} \right) + u_{f,j} \left( \frac{Du_{f,i}}{Dt} \right) \right\rangle = \left\langle \frac{D \left( u_{f,i}\, u_{f,j} \right)}{Dt} \right\rangle - \langle B_{f,ik} B_{f,jk} \rangle . \tag{11.44}$$

When injected in Eq. (11.42), we get therefore

$$\left\langle \frac{D \left( u_{f,i}\, u_{f,j} \right)}{Dt} \right\rangle = - \langle u_{f,i}\, u_{f,k} \rangle \frac{\partial \langle U_{f,j} \rangle}{\partial x_k} - \langle u_{f,j}\, u_{f,k} \rangle \frac{\partial \langle U_{f,i} \rangle}{\partial x_k}$$

$$+ G_{ik} \langle u_{f,j}\, u_{f,k} \rangle + G_{jk} \langle u_{f,i}\, u_{f,k} \rangle + \langle B_{f,ik} B_{f,jk} \rangle , \tag{11.45}$$

which is the correct form for the transport equation of the Reynolds-stress equation for a single-phase turbulent flow when a GLM is adopted.

## 11.5   The Resulting Form of the Particle Dynamical Model

Two conclusions can be drawn from these developments:

(i) The first one is that the extended form of the particle momentum equation is needed to capture the correct tracer-particle limit when particle inertia vanishes, regardless of whether we include the exchange source terms accounting for particle back effects on the fluid phase;

(ii) The second conclusion is that this correct behavior is ensured provided that the normalized difference between the timescale of the velocity of the fluid seen and the fluid (represented by the dimensionless factor $\delta b$ above) goes to zero faster than $\tau_p$ when particle inertia becomes negligible. Said differently, this means that we must have $\delta/\delta\tau_p \left( T_L/T_L^* \right) = 0$. In the formulation of a general two-phase GLM, presented in Sect. 8.1, this is translated by the requirement that the mapping operator embodied by the matrix $\mathbf{H}$, cf. Eqs. (8.3) and (8.4), must be such that we have $\delta H_{ij}/\delta\tau_p = 0$. This requirement provides therefore an additional condition to assess various proposals.

The main outcome is that we benefit now not only from guidelines but also from clear indications as to the general form of the discrete particle dynamical model. Indeed, taking up the notations introduced in Sect. 8.2 and used in the formulation for dilute flows, cf. Eqs. (8.26), an appropriate model for the particle state vector $\mathbf{Z}_\mathrm{p} = (\mathbf{X}_\mathrm{p}, \mathbf{U}_\mathrm{p}, \mathbf{U}_\mathrm{s})$ when two-way coupling is accounted for can be written as

$$dX_{\mathrm{p},i} = U_{\mathrm{p},i}\, dt \,, \tag{11.46a}$$

$$dU_{\mathrm{p},i} = \frac{U_{\mathrm{s},i} - U_{\mathrm{p},i}}{\tau_\mathrm{p}}\, dt + \frac{\rho_\mathrm{f}}{\rho_\mathrm{p}}\, DU_{\mathrm{s},i} \,, \tag{11.46b}$$

$$DU_{\mathrm{s},i} = -\frac{1}{\rho_\mathrm{f}}\frac{\partial \langle P_\mathrm{f} \rangle}{\partial x_i}\, dt$$
$$+ G_{ij}\left(U_{\mathrm{s},j} - \langle U_{\mathrm{f},j} \rangle\right) dt + \sqrt{\left(C_0\langle \epsilon_\mathrm{f} \rangle - \frac{2}{3}\mathrm{Tr}(\mathbf{G^a R_f})\right)}\, dW_i^{\mathrm{s-p}} \,, \tag{11.46c}$$

$$dU_{\mathrm{s},i} = DU_{\mathrm{s},i} + \langle U_{\mathrm{r},j} \rangle \frac{\partial \langle U_{\mathrm{f},i} \rangle}{\partial x_j} dt - \frac{\alpha_\mathrm{p}\rho_\mathrm{p}}{\alpha_\mathrm{f}\rho_\mathrm{f}}\left(\frac{U_{\mathrm{s},i} - U_{\mathrm{p},i}}{\tau_\mathrm{p}}\right) dt$$
$$+ \left(\delta b\, G_{ij} + ((\delta\mathbf{H})\mathbf{G})_{ij}\right)\left(U_{\mathrm{s},j} - \langle U_{\mathrm{f},j} \rangle\right) dt + B_{ij}^{\mathrm{t-p}}\, dW_j^{\mathrm{t-p}} \,. \tag{11.46d}$$

A few additional remarks can be made. First, although we have used the alternative formulation for $\mathbf{U}_\mathrm{s}$ in Eq. (11.46d) to bring out the correlation between $d\mathbf{U}_\mathrm{s}$ and $D\mathbf{U}_\mathrm{s}$, the two-phase GLM accounting for two-way coupling is the same as the one already expressed in Eq. (11.17). Second, in spite of its apparently exact expression, it is worth repeating that the added term in the equation of $d\mathbf{U}_\mathrm{s}$ when two-way coupling effects are present, namely $r_{\mathrm{mp}}\mathbf{A}_\mathrm{p}^\mathrm{D}$ (the third term on the right-hand side of Eq. (11.46d)), remains a model and could be replaced by another random process albeit with the same statistics. However, since we have essentially used these statistics to investigate its behavior in the tracer-particle limit, the conclusions that have been reached appear valid for the class of possible models for the source term $\mathbf{S}_{\mathrm{p}\to\mathrm{s}}$ introduced in Sect. 11.3. A third remark concerns the extended form of the particle momentum equation which is a SDE, as observed when plugging Eq. (11.46c) into Eq. (11.46b). It may be tempting to keep only the mean pressure-gradient term for $D\mathbf{U}_\mathrm{s}$ (thus keeping only the first term on the right-hand side of Eq. (11.46c)) when used in Eq. (11.46b). It follows from the developments in Sect. 11.4 that we still retrieve the mean Navier-Stokes equation in the tracer-particle limit but not the transport equation for the Reynolds-stress tensor components. We can only claim full consistency with the first two velocity moments of the fluid with the complete model as written above in Eqs. (11.46).

## 11.6   A Numerical Illustration

To illustrate a turbulent dispersed two-phase flow model belonging to the class of stochastic models developed in the previous chapters, we describe the numerical simulation of a turbulent air jet flowing along a vertical plane wall and laden with spherical glass particles whose mass fraction is high enough to modify the air jet.

### 11.6.1  A Wall-Jet Test Case

The flow situation is sketched in Fig. 11.2. It consists in an air jet laden with discrete particles flowing along a vertical wall and being mixed through an injector of width $b = 5$ mm with a co-current unladen stream. The width of the domain is 150 mm and its height is 350 mm. The air jet at the inlet section has a maximum velocity of 10 m s$^{-1}$ with a profile outlined in Fig. 11.2 and diminishing to 2 m s$^{-1}$ at $z = b$, which is the constant velocity of the co-current unladen stream. Based on the maximum inlet velocity and the width of the plane jet, the Reynolds number of the air jet is 3300. The discrete particles have a density of $\rho_p = 2590$ kg/m$^3$ and are poly-dispersed with a mean diameter of $d_p = 49.3\ \mu$m and a standard deviation of 4.85 $\mu$m. They are injected with a velocity profile close to the air one while the particle mass loading is around 1 which is sufficient to modulate the turbulence of the air jet. At the inlet section, data are provided for the first- and second-order velocity moments at 40 locations for the air jet and at 10 locations across the injector width for the particle phase. Measurements were taken at eight downstream locations corresponding to $x/b = 1, 5, 10, 15, 20, 30, 40$ and $50$, respectively. These measurements include the particle volume fraction, the mean vertical and horizontal particle velocities and their fluctuating components (i.e. their standard deviations) and also the air mean vertical velocities. Measurements are plotted later as a function of the non-dimensional wall-normal coordinate $z/b$. Although the measurements of the air velocities were made only at the last four downstream sections, they include data for the unladen and the laden situations, which is useful to assess particle back effects on the air flow. The flow is homogeneous in the spanwise

**Fig. 11.2** Representation of the particle-laden turbulent wall jet with two-way coupling between the air jet and the discrete particles

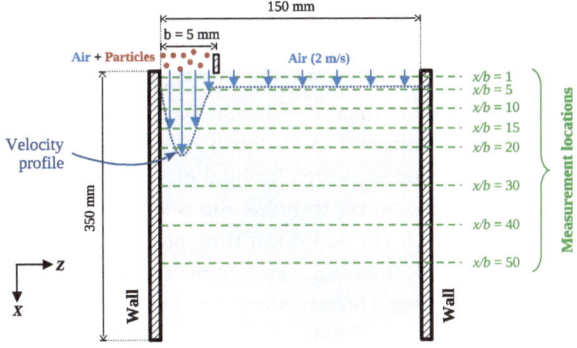

direction (here $y$) corresponding to a plane wall-jet where the flow is statistically two-dimensional.

### 11.6.1.1  A Word on the Turbulent Dispersed Two-Phase Flow Model

For the present simulations, we retain for the velocity of the fluid seen the two-phase SLM presented in Sect. 6.3.1 with the modification of the mean drift term described in Sect. 8.1.3, although this does not change results significantly since we are considering inertial particles rather than tracers. For similar reasons, we used the extended particle momentum equation given in Eq. (3.23) but where only the mean pressure-gradient term is considered. This corresponds to the formulation discussed above in Sect. 11.5 but, as we are not concerned with the fluid limit for this test case, the limitations analyzed there do not constitute a real issue (since the density ratio is very small in our case, the expression of the pressure-gradient term is secondary). To be consistent with the two-phase SLM, the corresponding second-moment closure is the Rotta model, given in Eq. (6.27) in Sect. 6.2.2. For the particle back effects on the fluid flow, the source terms described in Sect. 11.2 are implemented and the corresponding modification of the velocity of the fluid seen is as expressed in Eq. (11.17) in Sect. 11.3 above.

### 11.6.1.2  A Word on the Numerical Simulation

For the numerical computations, a mesh of the domain made up by 4646 hexahedral cells was built. This mesh is used to solve the fluid mean field transport equations in the moment solver with finite-volume techniques. The same mesh is also used to track and locate the discrete particles as well as to extract the statistics of interest by ensemble averaging over the particles located within each cell at a given time. In terms of mesh/Monte-Carlo methods, this corresponds to the classical nearest-grid-point approach [5, 6]. Other numerical details pertaining to the PDF solver, such as the exponential scheme for the time-integration of the SDEs, can be found in [4, 5]. The single-phase calculation of the air jet is first run with the moment solver and provides the results discussed below for the fluid phase when a one-way coupling is assumed. Then, discrete particles are injected according to the experimental values of the particle mass flow rate, which is obtained from the given value of the particle volume fraction and vertical velocity, and with instantaneous velocities sampled in a Gaussian distribution having the mean and correlation values given by the experimental data. Since this is a situation involving a two-way coupling between the fluid and particle phases, the moment and PDF solvers are run simultaneously following the description given in Sect. 7.1 and exchange information within each time step, as illustrated in Fig. 7.1. At the solid wall, wall-function type of boundary conditions are applied for the air flow while for particles the formulation proposed in Sect. 7.3.2 reduces to the classical elastic boundary conditions since particles have a high-enough inertia to cross the near-wall layer very quickly. For the fluid and particle coupled simulation, a time step of $1 \times 10^{-4}$ s was used ensuring that the CFL is smaller than one. To benefit from the statistically-stationary characteristic of this flow, once a steady state is reached, particle statistics are further averaged in time to reduce statistical noise.

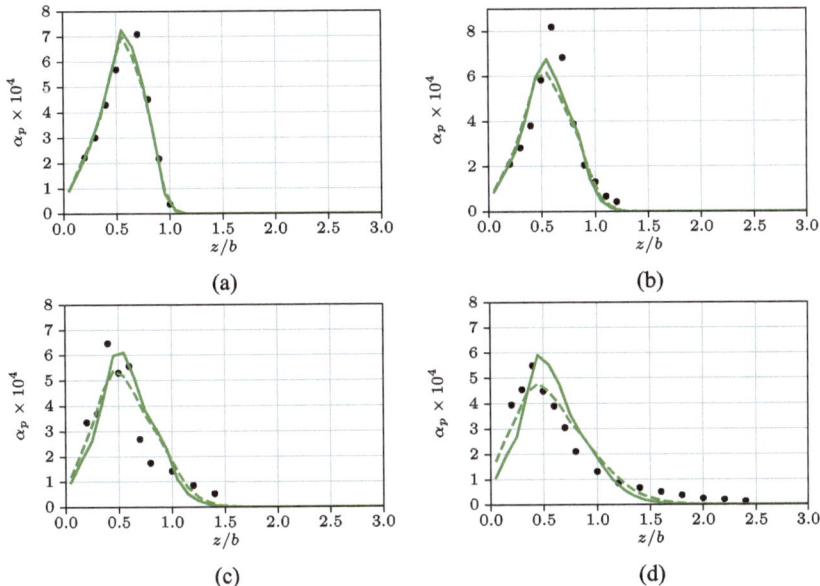

**Fig. 11.3** Profiles of the particle volume fraction at the first four downstream locations: comparison between experimental data (filled circle), simulations with one-way (dashed lines) and two-way coupling (continuous lines). (**a**) Plane $x/b = 1$. (**b**) Plane $x/b = 5$. (**c**) Plane $x/b = 10$. (**d**) Plane $x/b = 15$

### 11.6.1.3 Some Comments on the Numerical Results

The particle volume fraction is an important variable which is sensitive to the numerical estimation of the fluid mean pressure-gradient term entering the SDE of the velocity of the fluid seen. For this issue, we benefit from the analysis developed in Sect. 7.3 and have been careful to implement the same term used to ensure the respect of the fluid continuity equation. We present in Figs. 11.3 and 11.4 the complete evolution of the profiles of the particle volume fraction downstream of the inlet section with the last two results at $x/b = 40$ and $50$ being probably the most significant ones, as explained below. The sensitivity of the profiles of the particle volume fraction $\alpha_p$ can already be observed by comparing the results obtained when two-way coupling is taken into account and when only one-way coupling (i.e. no particle back effects on the fluid) is assumed. In both cases, the same numerical method to derive the mean pressure-gradient is applied but for the differences in the air mean velocity field. The discrepancy between the two profiles of $\alpha_p$ is not marked at the first sections (see Fig. 11.3) but become visible when $x/b \geq 30$ (see Fig. 11.4) and, especially, at the last two sections where the one-way-coupling simulation predicts an increase of the particle volume fraction in the vicinity of the wall which does not seem balanced by a counter-effect whereas the two-way-coupling simulation captures a much better profile compared to the experimental data.

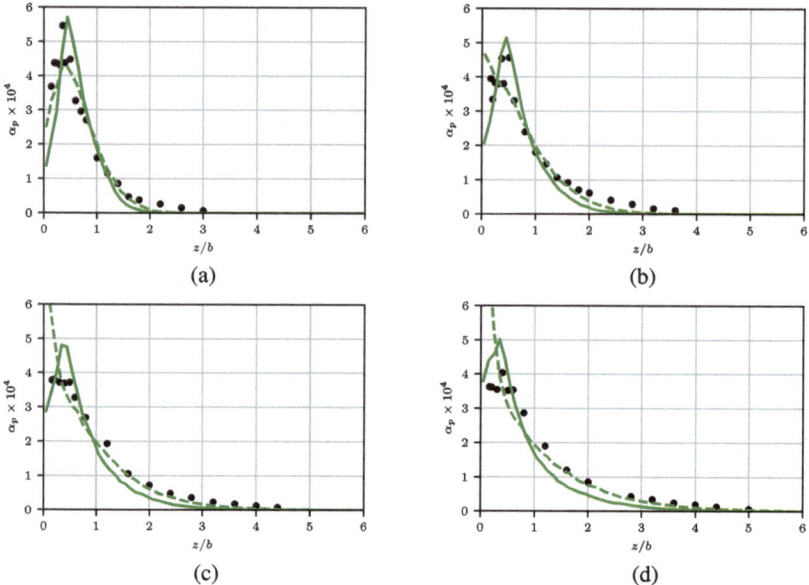

**Fig. 11.4** Profiles of the particle volume fraction at the last four downstream locations: comparison between experimental data (filled circle), simulations with one-way (dashed lines) and two-way coupling (continuous lines). (**a**) Plane $x/b = 20$. (**b**) Plane $x/b = 30$. (**c**) Plane $x/b = 40$. (**d**) Plane $x/b = 50$

The relevance of how the mean pressure-gradient in the Langevin model for the velocity of the fluid seen is evaluated is revealed by the profiles shown in Fig. 11.5 when another estimation (which does not respect the mass continuity equation at the numerical level) is used. It can be seen that these profiles of the particle volume fraction tend to deviate from the ones in Fig. 11.4 when $x/b \geq 30$ and even exhibit a hollow shape in the near-wall region.

The fact that these profiles manifest noticeable differences at the last three, even the last two, sections can be explained by the memory of the initial conditions, as we already analyzed in Sect. 7.4 for the case of particle dispersion in a turbulent round jet. Indeed, with the air and particle properties given above, the particle relaxation timescale is $\tau_p \simeq 2 \times 10^{-2}$ s from which, using a mean downward particle velocity of about 8 m s$^{-1}$ (taking a rough estimate over the injection or from the numerical results), we get that particles tend to loose the memory of their initial conditions when their residence time is greater than $\tau_p$, which translates into a distance from the injector of about $x/b \geq 30$. Consequently, profiles plotted at the last two sections, $x/b = 40$ and 50, reflect more how the effects of the air turbulence are modeled than how initial conditions are transported. This does not mean that the ability to transport initial conditions and to reproduce how they fade away as particles move downstream and as the turbulence of the air jet develops is to be overlooked since it expresses non-local effects and, in that sense, is a reminder of the

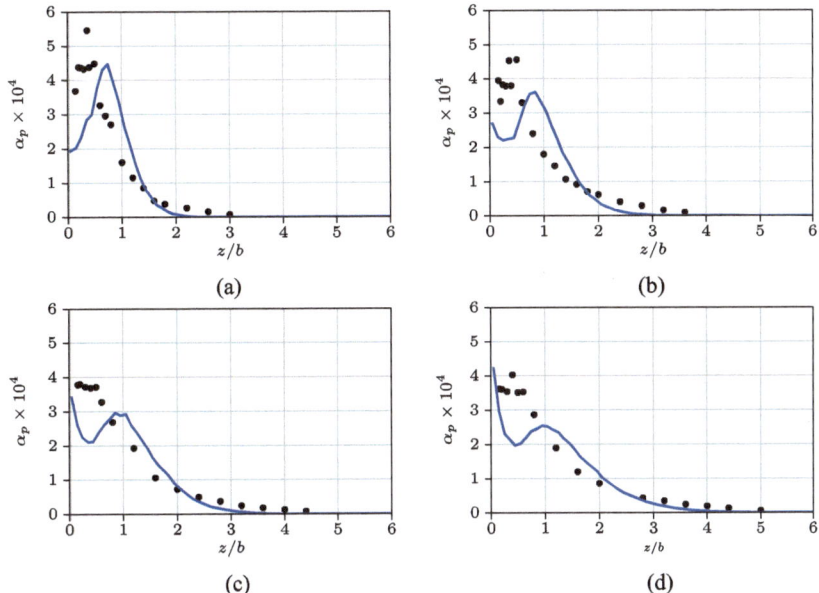

**Fig. 11.5** Profiles of the particle volume fraction at the last four downstream locations: comparison between experimental data (filled circle) and simulations with two-way coupling (lines) with a model that does not respect the mass continuity equation. (**a**) Plane $x/b = 20$. (**b**) Plane $x/b = 30$. (**c**) Plane $x/b = 40$. (**d**) Plane $x/b = 50$

discussions on local and non-closures detailed in Sect. 10.3. For our present concern, we nevertheless choose to focus on the last four, even the last two, sections for the different fluid and particle velocity statistics (i.e. first- and second-order moments).

The profiles of the particle and air mean vertical velocities are shown in Figs. 11.6 and 11.7, respectively. In both figures, two simulation results are plotted: one corresponding to the complete treatment of the air-and-particle system, that is including two-way coupling (represented by the continuous lines); and one corresponding to a one-way simulation in which the air jet is therefore treated as a single-phase flow and for which the source terms accounting for particle back effects on the fluid flow are switched off (represented by the dashed line). In Fig. 11.7, there are two sets of experimental data as measurements were also performed in the single-phase flow situation, allowing to shed light on both the single- and two-phase simulations. The one-way coupling situation is the starting point from which the source terms representing the exchanges of momentum and kinetic energy between the fluid and particle phases are switched on, and is therefore useful to bring out the impact of these source terms. In Fig. 11.7, it is seen that while the single-phase simulation tend to move closer to the experimental values at the last two sections, there are more noticeable discrepancies at the two preceding sections. It would be appreciable to carry out a similar analysis to the one used for particles to assess how initial conditions are affecting these plots but this turns out to be more uncertain

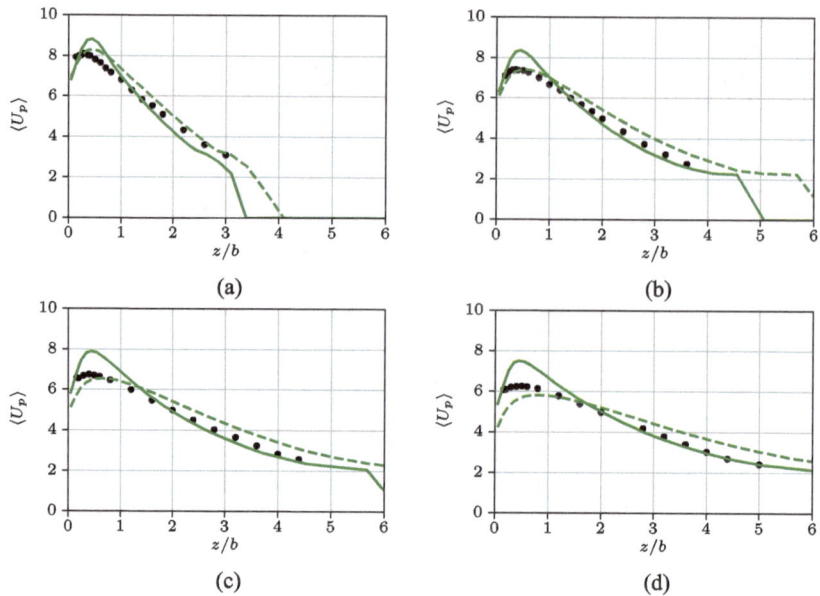

**Fig. 11.6** Profiles of the particle mean vertical velocity at the last four downstream locations: comparison between experimental data (filled circle), simulations with one-way (dashed lines) and two-way coupling (continuous lines). (**a**) Plane $x/b = 20$. (**b**) Plane $x/b = 30$. (**c**) Plane $x/b = 40$. (**d**) Plane $x/b = 50$

as the relevant timescale is $T_L \sim k_f/\langle \epsilon_f \rangle$ which is not available at the inlet section because $\langle \epsilon_f \rangle$ is unknown. We need also to account here for the relative simplicity of the turbulence model since the Rotta model does not include responses to the fluid mean gradients in its closure expression. For our purpose and in relation with the theme of the present chapter, we can observe that including particle back effects on the air jet has a significant impact on the mean velocities, for the particle as well as for the fluid phases. The curves shown in Figs. 11.6 and 11.7 reveal that the drop-off of the mean velocity profiles towards the constant velocity of $2 \text{ m s}^{-1}$ of the outer stream is well reproduced, although there is an overestimation of the peak velocities, both for particles and for the air flow. In this thin near-wall region which corresponds to the peak of the particle volume fraction, cf. Fig. 11.4, it appears that both the air and particle mean velocities diminish as we move downstream but with a timescale larger than the one deduced from the evolution of the experimental data. The comparison with the results obtained in the one-way-coupling situation suggests that, due to their mutual influence in the two-way-coupling case, they tend to lock each other in a slow decrease. On these aspects, further investigations are needed to determine the influence of the turbulence models and of the details of the numerical implementations in order to pinpoint more precisely what this simulation result manifests.

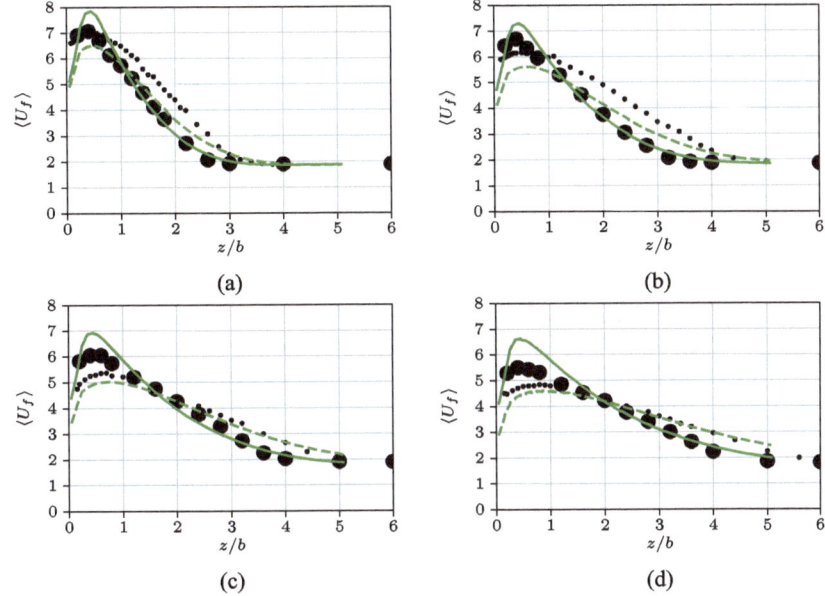

**Fig. 11.7** Profiles of the fluid mean vertical velocity at the last four downstream locations: comparison between experimental data (filled circle), simulations with one-way (dashed lines) or two-way coupling (continuous lines) and with a single-phase (black dot). (**a**) Plane $x/b = 20$. (**b**) Plane $x/b = 30$. (**c**) Plane $x/b = 40$. (**d**) Plane $x/b = 50$

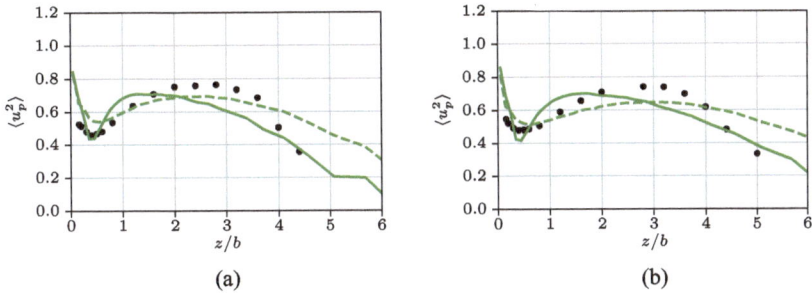

**Fig. 11.8** Profiles of the particle fluctuating vertical velocity at the last two downstream locations: comparison between experimental data (filled circle), simulations with one-way (dashed lines) and two-way coupling (continuous lines). (**a**) Plane $x/b = 40$. (**b**) Plane $x/b = 50$

These results are complemented by the profiles of the second-order particle velocity moments, which are shown in Figs. 11.8 and 11.9 for the particle velocity standard deviations in the vertical and horizontal directions, respectively. In spite of the simplicity of the turbulence model, i.e. the hybrid Rotta-model/SLM formulation, the anisotropy of the velocity components and their level are accurately reproduced. If we remember that for $z/b \geq 3 - 4$ the particle volume fraction is very small so that there are few particles in this outer region, it can also be said that

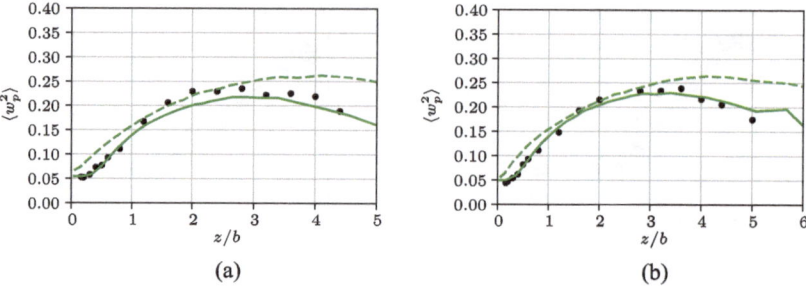

**Fig. 11.9** Profiles of the fluctuating horizontal velocity at the last two downstream locations: comparison between experimental data (filled circle), simulations with one-way (dashed lines) and two-way coupling (continuous lines). (**a**) Plane $x/b = 40$. (**b**) Plane $x/b = 50$

no marked or significant differences appear between the one- and two-way coupling simulations, contrary to what was observed for the results for the mean velocities.

---

## References

1. J.P. Minier, E. Peirano, Phys. Rep. **352**(1–3), 1 (2001).    https://doi.org/10.1016/S0370-1573(01)00011-4
2. E. Peirano, J.P. Minier, Phys. Rev. E **65**(4), 046301 (2002). https://doi.org/10.1103/PhysRevE.65.046301
3. J.P. Minier, Prog. Energy Combust. Sci. **50**, 1 (2015). https://doi.org/10.1016/j.pecs.2015.02.003
4. J.P. Minier, E. Peirano, S. Chibbaro, Phys. Fluids **16**(7), 2419 (2004). https://doi.org/10.1063/1.1718972
5. E. Peirano, S. Chibbaro, J. Pozorski, J.P. Minier, Prog. Energy Combust. Sci. **32**(3), 315 (2006). https://doi.org/10.1016/j.pecs.2005.07.002
6. R.W. Hockney, J.W. Eastwood, *Computer Simulation Using Particles* (CRC Press, 2021). https://doi.org/10.1201/9780367806934

# Statistical Modeling of Particle Collisions in Turbulent Flows

**Abstract**

Open any textbook on non-equilibrium statistical mechanics and you will find the Boltzmann equation as the archetypal description of particle-particle interactions. The Boltzmann equation is undoubtedly a major achievement and a well-established framework for short-range particle-particle interactions, especially for rarefied gases. It is therefore justified to refer to this framework when considering particle collisions in another context. The specific assumptions underlying the classical form of the Boltzmann equation are, however, often taken for granted. This can turn into a regrettable oversight since collisions of particles transported by random media challenge some aspects of the Boltzmann picture. With a view toward modeling particle collisions in turbulent flows, the purpose of this chapter is to revisit the key characteristics of the Boltzmann equation to bring out the limitations that need to be addressed.

**Chapter Content** After a reminder of the classical kinetic theory in Sect. 12.1, we discuss formulations in sample and physical spaces in Sect. 12.2. The limitations of present collision kernels for discrete particles in turbulent flows are brought out in Sect. 12.3 in relation with the molecular chaos assumption discussed in Sect. 12.4. We then discuss how stochastic jump-diffusion processes can be used in Sect. 12.5 before introducing numerical examples in Sect. 12.6.

## 12.1 A Reminder of Classical Kinetic Theory

Since there are several detailed accounts of the Boltzmann theory [1–4], we limit ourselves to recalling the main points of interest for the present discussion. In its final formulation, the Boltzmann approach is a statistical description developed in terms of the single-particle distribution function $f(t; \mathbf{y}_p, \mathbf{V}_p, d_p)$ in which only

© The Author(s) 2025
J.-P. Minier et al., *Understanding Turbulent Systems*,
Lecture Notes in Physics 1039, https://doi.org/10.1007/978-3-031-84466-9_12

the kinetic variables $(\mathbf{X}_p, \mathbf{U}_p)$ are retained as well as the particle diameter $d_p$ (or its volume $\mathcal{V}_p$). The physical picture consists in a process where particles undergo free flights and binary collisions. Note that the Boltzmann equation can easily account for the effects of an external and conservative force field (thus, deriving from a potential) so that particles do not necessarily follow straight lines in-between collisions but this does not change our discussion and we can still refer to free flights. In the Boltzmann-Grad limit where particle diameters are very small but with finite cross-section (i.e., in the limit when the number of particles $N_p \rightarrow +\infty$ and $d_p \rightarrow 0$, but with $N_p\, d_p^2$ remaining constant), collisions are regarded as instantaneous and point-wise elastic interactions so that mass, momentum and kinetic energy are conserved during collisions which appear as random redistributive events. In the following, binary collisions are assumed to occur between a particle with a state vector $(\mathbf{X}_p, \mathbf{U}_p, d_p)$ and a collisional partner whose variables are noted $(\mathbf{X}_{c,p}, \mathbf{U}_{c,p}, d_{c,p})$. The transformation rules are expressed writing the post-collisional velocities $\mathbf{U}_p^{ac}$ and $\mathbf{U}_{c,p}^{ac}$ in terms of the pre-collisional ones $\mathbf{U}_p^{bc}$ and $\mathbf{U}_{c,p}^{bc}$, which gives

$$\mathbf{U}_p^{ac} = \mathbf{U}_p^{bc} - \frac{2m_{c,p}}{m_p + m_{c,p}} \left( \left[ \mathbf{U}_p^{bc} - \mathbf{U}_{c,p}^{bc} \right] \cdot \mathbf{e} \right) \mathbf{e}\,, \tag{12.1a}$$

$$\mathbf{U}_{c,p}^{ac} = \mathbf{U}_{c,p}^{bc} + \frac{2m_p}{m_p + m_{c,p}} \left( \left[ \mathbf{U}_p^{bc} - \mathbf{U}_{c,p}^{bc} \right] \cdot \mathbf{e} \right) \mathbf{e}\,, \tag{12.1b}$$

with $\mathbf{e}$ a unit vector on the unit sphere $\mathcal{S}^2$ representing the orientation of the collision event (e.g., the lines connecting the center of two spheres in contact), and $m_p = \rho_p \pi d_p^3 / 6$ (resp. $m_{c,p} = \rho_p \pi d_{c,p}^3 / 6$) the mass of the target particle (resp. its collisional partner). These relations are easily inverted to provide the pre-collisional velocities $\mathbf{U}_p^{bc}$ and $\mathbf{U}_{c,p}^{bc}$ as a function of the post-collisional ones. With these notations and leaving out external force fields, the Boltzmann equation is expressed in sample space for the single-particle distribution function $f(t; \mathbf{y}_p, \mathbf{V}_p, \widehat{d}_p)$ as

$$\frac{\partial f}{\partial t} + V_{p,k} \frac{\partial f}{\partial y_{p,k}} = \frac{\partial \phi_{ext}(\mathbf{y}_p)}{\partial y_{p,k}} \frac{\partial f}{\partial V_{p,k}} + Q_{coll}(f)\,, \tag{12.2}$$

where $\widehat{d}_p$ is the sample space variable for $d_p$ and $-\partial \phi_{ext}(\mathbf{x})/\partial x_k$ an external force field (per unit mass) which is left out from now on for the sake of simplicity. In Eq. (12.2), $Q_{coll}(f)$ is a gain and loss term accounting for the collisional events in Eqs. (12.1) over all possible collisional partners which gives

$$Q_{coll}(f) = \int_{\mathbb{R}^3} d\mathbf{V}_{c,p} \int_{\mathcal{S}^2} \beta_{coll}(\mathbf{V}_p, \mathbf{V}_{c,p}, \widehat{\mathbf{e}})$$

$$\left[ f_2(t; \mathbf{y}_p, \mathbf{V}_p^{bc}, \widehat{d}_p, \mathbf{V}_{c,p}^{bc}, \widehat{d}_{c,p}) - f_2(t; \mathbf{y}_p, \mathbf{V}_p, \widehat{d}_p, \mathbf{V}_{c,p}, \widehat{d}_{c,p}) \right] d\widehat{\mathbf{e}} \tag{12.3}$$

where $\widehat{\mathbf{e}}$ is the sample space variable for $\mathbf{e}$. In Eq. (12.3), $\beta_{\text{coll}}(\mathbf{V}_{\text{p}}, \mathbf{V}_{\text{c,p}}, \mathbf{e})$ is the collision kernel measuring the likelihood of collisions and determined by the interaction potential between particles (thus related to the scattering problem). In the present case, it is sufficient to consider the classical hard-sphere model according to which the collision kernel takes the form

$$\beta_{\text{coll}}(\mathbf{U}_{\text{p}}, \mathbf{U}_{\text{c,p}}, \mathbf{e}) = \pi \left( \frac{d_{\text{p}}}{2} + \frac{d_{\text{c,p}}}{2} \right)^2 \left| (\mathbf{U}_{\text{p}} - \mathbf{U}_{\text{c,p}}) \cdot \mathbf{e} \right| . \tag{12.4}$$

Once a collisional kernel is chosen, it is seen that Eq. (12.3) represents the most general expression of the gain and loss term for binary collisions. A key point is that it involves the two-particle distribution function, noted $f_2$, written at the same location (reflecting the hypothesis of point-like collisions for simplicity). As such, the Boltzmann equation in Eqs. (12.2) and (12.3) is unclosed. It is clear that expressing the evolution equation in sample space for $f_2$ results in an open equation involving the three-particle distribution $f_3$, etc., so that we are faced with a BBGKY-like hierarchy. The closed form of the Boltzmann equation is obtained by resorting to the molecular chaos hypothesis according to which the velocities of colliding partners are uncorrelated, which means that we can write

$$f_2(t; \mathbf{y}_{\text{p}}, \mathbf{V}_{\text{p}}, \widehat{d}_{\text{p}}, \mathbf{V}_{\text{c,p}}, \widehat{d}_{\text{c,p}}) = f(t; \mathbf{y}_{\text{p}}, \mathbf{V}_{\text{p}}, \widehat{d}_{\text{p}}) f(t; \mathbf{y}_{\text{p}}, \mathbf{V}_{\text{c,p}}, \widehat{d}_{\text{c,p}}) . \tag{12.5}$$

This yields the classical expression of the Boltzmann equation, i.e. with a collision term that is a quadratic function of the single-particle distribution function $f$ and, for that reason, is often noted $Q_{\text{coll}}(f, f)$ in textbooks. From this brief reminder, it appears that, at least, three characteristics of the Boltzmann description of particle collisions are worth mentioning:

(a) It is developed in sample space retaining only the kinetic state vector $(\mathbf{X}_{\text{p}}, \mathbf{U}_{\text{p}})$;
(b) It involves collision kernels based on free flights between particle collisions;
(c) It relies on the molecular chaos assumption and assumes Markov behavior.

When considering particle collisions in random media with non-zero space and time correlations, each of these characteristics raises questions that need to be addressed. Furthermore, when dealing with mesoscopic particles such as colloids or discrete inertial particles, we are concerned not only with elastic collisions but also with inelastic ones (for instance, in granular matter) and even with sticky collisions (for example, in particle agglomeration or droplet coalescence). This is illustrated in Fig. 12.1 and it is interesting to discuss the points (a)–(c) above with respect to a wide range of outcomes for particle collisions.

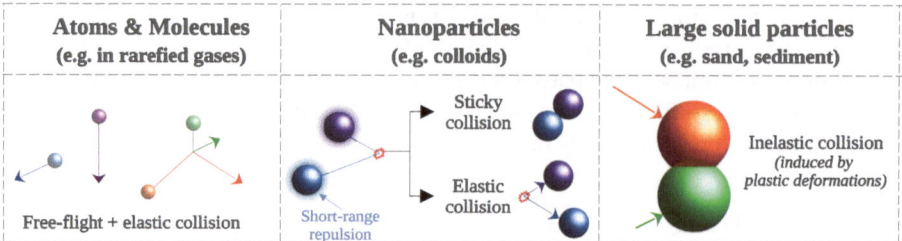

**Fig. 12.1** Sketch of the various outcomes of particle collisions depending on the type of forces involved between particles

## 12.2    Developments in Sample and Physical Spaces

The Boltzmann equation was originally developed in sample space and traditional presentations have retained this formulation. It is seen from Eqs. (12.2)–(12.5) that the description of the physical process is reduced to a single function which is the solution of a PDE written in a six-dimensional space involving a non-linear collisional term, which makes solving it (in a strong sense) a formidable task. Yet, it has great merit in that it constitutes the main pathway for the derivation of balance equations for continuum fluid mechanics in the hydrodynamical limit through, for instance, the Chapman-Enskog or the Grad approaches. As such, things may look similar for the statistical description of turbulent particle-laden flows with collisions. There are, however, differences that suggest to adopt another point of view.

First, we have seen that the kinetic description is ill-based for particle transport in non-fully resolved turbulent flows and that the particle state vector must include, at least, the velocity of the fluid seen $\mathbf{Z}_p = (\mathbf{X}_p, \mathbf{U}_p, \mathbf{U}_s)$. We are then handling a distribution function (or a PDF) that has, at the minimum, ten independent variables if we include the particle diameter $d_p$ or its volume $\mathcal{V}_p$. Using the general form in Eqs. (6.9) to model particle transport in the absence of collisions and retaining only the reduced particle momentum equation for the sake of keeping simple notations without loss of generality, the evolution equation in sample space for $p(t; \mathbf{y}_p, \mathbf{V}_p, \mathbf{V}_s, \upsilon_p)$, where $\upsilon_p$ is the sample space variable corresponding to $\mathcal{V}_p$, becomes

$$
\frac{\partial p}{\partial t} + V_{p,k} \frac{\partial p}{\partial y_{p,k}} = -\frac{\partial}{\partial V_{p,k}} \left[ \left( \frac{V_{s,k} - V_{p,k}}{\tau_p} \right) p \right]
$$

$$
- \frac{\partial \left[ A_{s,k} p \right]}{\partial V_{s,k}} + \frac{1}{2} \frac{\partial \left[ \left( B_s B_s^{\perp} \right)_{kl} p \right]}{\partial V_{s,k} \partial V_{s,l}} + Q_{\text{coll}}(p) , \qquad (12.6)
$$

which makes strong solutions even more difficult to obtain. Second, PDF equations, such as the Boltzmann-Fokker-Planck one in Eq. (12.6), are useful to derive the corresponding transport equations in physical space for statistics of interest. At this stage, there is, however, a crucial difference with the steps leading from the Boltzmann equation to the hydrodynamical ones. In turbulent two-phase flow modeling, we are not interested in reducing the description to a set of PDEs for a small number of one-point moments because these equations are unclosed and, as indicated in Sect. 10.4, there is little hope to rely on a physically-meaningful turbulent-viscosity concept to help us. In fact, we are led to solving (in a weak sense) the PDF equation by developing stochastic particle systems that act as Monte Carlo solutions. It follows that, just as the Langevin equations are used to model and simulate particle transport, we need to devise stochastic particle systems in physical space that play the same role for the collision term.

Although more limited than the one on the operator formalism, there is now a large-enough literature dedicated to such stochastic particle systems, often mathematically-oriented [5–7], as well as more synthetic accounts [8] to which interested readers are referred to. A nice presentation can be found in [9]. As indicated in these works, stochastic particle systems are essentially built on jump processes (cf. Sect. 4.3) and can be represented by the generic algorithm for $N_p$ particles contained in a small volume $\mathcal{V}$ and described with a reduced particle state vector $\mathbf{Z}_p^{(N)} = (\mathbf{U}_p^{[1]}, \ldots, \mathbf{U}_p^{[N]})$ limited to particle velocities when a locally homogeneous hypothesis is made inside the volume $\mathcal{V}$. The $N_p$ particles interact through the following generalized Poisson process

1. The system waits in the same state during a time which is an exponentially-distributed random variable whose parameter is

$$\lambda_c(\mathbf{Z}_p^{(N)}) = \frac{1}{2\,N_p\,\mathcal{V}} \sum_{\substack{k,l=1 \\ k \neq l}}^{N_p} \int_{\mathcal{S}^2} \beta_{\mathrm{coll}}(\mathbf{U}_p^{[k]}, \mathbf{U}_p^{[l]}, \mathbf{e})\, \mathrm{d}\mathbf{e}\,. \tag{12.7}$$

2. At the time of a jump, two particles, labeled $i$ and $j$, are chosen in the volume $\mathcal{V}$ and a direction $\mathbf{e}$ on the unit sphere $\mathcal{S}^2$ is sampled with the probability

$$p_c(i, j, \mathbf{e}) = \frac{\beta_{\mathrm{coll}}(\mathbf{U}_p^{[i]}, \mathbf{U}_p^{[j]}, \mathbf{e})}{2\,N_p\,\mathcal{V}\,\lambda_c(\mathbf{Z}_p^{(N)})} = \frac{\beta_{\mathrm{coll}}(\mathbf{U}_p^{[i]}, \mathbf{U}_p^{[j]}, \mathbf{e})}{\sum_{\substack{k,l=1 \\ k \neq l}}^{N_p} \int_{\mathcal{S}^2} \beta_{\mathrm{coll}}(\mathbf{U}_p^{([k]}, \mathbf{U}_p^{[l]}, \mathbf{e})\, \mathrm{d}\mathbf{e}}\,.$$

$$\tag{12.8}$$

3. For the selected particle pair, velocities are updated according to Eqs. (12.1).

This presentation follows one approach to simulate jump processes by generating the random interval times between jumps. As indicated in Sect. 4.2.3, another approach consists in using fixed time intervals (or time steps) during which the random numbers of events that take place are sampled in a Poisson distribution whose parameter is determined from the local value of the collision kernel or from the mean free path. Sometimes just the mean numbers of events are considered when they are very large. In the latter formulation, this corresponds to the collision

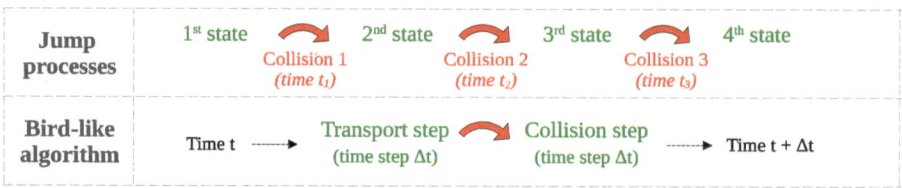

**Fig. 12.2** Representation of the two different ways to account for jumps in a generalized Poisson process

step of the well-known direct simulation Monte Carlo (DSMC) method which relies on a time-splitting algorithm where the transport and collision steps are treated sequentially (first, transport without collision; then, collisions without transport) and where the above stochastic particle model is applied in each cell [10, 11] (see also Fig. 12.2).

Stochastic particle systems are flexible in the sense that they allow more general collision outcomes to be treated. Since we simulate explicitly a number of collisions between selected pairs of particles, it is indeed straightforward to introduce criteria (based on the values of particle-attached variables or properties) to decide whether each collision is elastic, inelastic or sticky. The formulation of these stochastic particle systems is discussed in several works [8] and the corresponding expressions in terms of operators in sample space are well described in [7]. To illustrate this point with a specific example, it is interesting to consider the case of colloidal particles interacting with a wall surface where they can deposit or bounce off. Similarly to the colloid-colloid interactions outlined in Sect. 2.2.3, when transported by a fluid flow to the immediate vicinity of a bounding wall surface, colloids are subject to short-ranged potential forces [12] (see also Fig. 12.3): an attractive one (e.g., due to van der Waals forces) and a repulsive one (e.g., due to electrostatic double-layer forces between similar particles). The key observation is that these forces are extremely short-range (typically from a few Å to a few nanometers). This implies that they act over scales that are a few orders of magnitude below typical hydrodynamical ones.

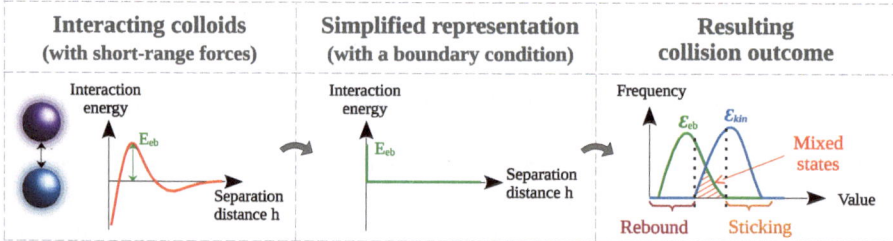

**Fig. 12.3** Sketch of the potential interaction between a colloid and a wall, its simplified representation as a boundary condition, and its resulting effect on the collision outcome (deposition or rebound)

For that reason, the exact potential is replaced by a step function that corresponds to the maximum of the potential, called the energy barrier $E_{eb}$. This step function acts as a boundary condition for colliding particles [13]: deposition occurs when the kinetic energy is higher than the energy barrier; otherwise, an elastic rebound occurs. Since the value of the energy barrier is sensitive to the presence of small protuberances on the particle surface (called roughness), it is actually a random variable whose distribution $p(\mathcal{E}_{eb})$ needs to be generated, with $\mathcal{E}_{eb}$ the sample space variable corresponding to $E_{eb}$. In short, each particle-particle interaction involves a double randomness coming from the particle kinetic energy and the value of the local energy barrier, cf. Fig. 12.3. However, as developed in [8, section 6.1], the formulation of the same boundary condition in sample space becomes quite intricate with a random combination of absorbing and reflecting conditions which goes far beyond the standard ones usually considered for the PDF equation [14–16].

Several remarks can be made:

(i) From the descriptions already given in Sect. 2.2.3 and the one above, it is clear that the same processes are at play between pairs of colloids in a solution and between colloids and a wall surface. In a way, we are already considering two-particle interactions but with one 'particle' being fixed (i.e., the surface wall). This allows to introduce the notions we wish to discuss below while still retaining the one-particle PDF framework;

(ii) It also appears that the modeling approach to colloids deposition follows in the footsteps of DSMC-like methods since transport and particle-wall interactions are treated sequentially, with one step where particles are transported without short-range interactions until they are detected as having hit a wall, and a second step where the outcome of the interaction is determined without transport;

(iii) In the frame of one-particle models, the occurrence of a particle-wall impact can be obtained from the resolved particle dynamics through detection algorithms (no collision kernels are needed).

We can elaborate on the last remark, which corresponds to the question: how do we detect a particle-wall interaction? At the moment, there are two situations where solutions can be worked out (cf. Fig. 12.4). The first one is when particle inertia is high enough so that, within a small time interval (one time step of numerical simulations), the trajectories of these particles are straight lines. Then, the detection criterion consists simply in checking whether the predicted location of a particle at the end of each time interval is still within the domain or lies outside, in which case this implies that an interaction has occurred during the time interval (simple geometrical considerations give the position where the interaction took place and, if necessary, an estimation of the time at which it happened). We can refer to this situation as the 'local ballistic regime' (by local, it is meant here local in time since we are considering particle evolution within a small time increment). In the immediate vicinity of a wall, more precisely in the so-called viscous sublayer [17], turbulent fluctuations are strongly dampened by viscous effects so that randomness

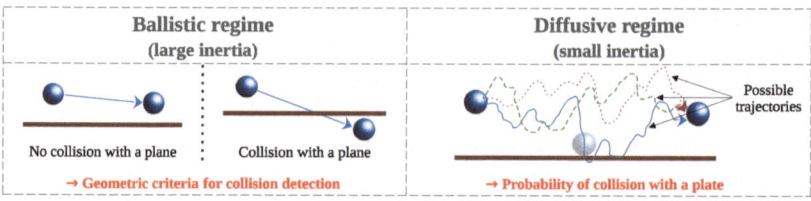

**Fig. 12.4** Detection of a particle-wall interaction during a small time interval. In the ballistic regime, the position of the end point of the particle trajectory is enough to detect an interaction. In the diffusive regime, there is a non-zero probability to hit the wall even for start and end points within the domain

due to turbulence is lessened. By reducing the observation time intervals (or, conversely, the numerical time steps), it can be assumed that the local ballistic regime is always reached. This is not so, however, for particles small enough to be sensitive to Brownian fluctuations. In fact, the second situation where we can come up with a tractable detection criterion for particle-wall interactions is when Brownian effects dominate completely so that we can, locally, neglect fluid velocity and consider that particle motion is governed by Brownian noise and the Stokes-Einstein diffusion coefficient, i.e., $d\mathbf{X}_p \simeq \sqrt{2\mathcal{D}}\,d\mathbf{W}$. In that case, we are in the 'local diffusive regime' which is the counterpart of the ballistic one. As illustrated in the right panel of Fig. 12.4, even if the predicted location of a particle at the end of a time interval is still within the domain (of course, if that position is outside, an interaction has occurred just as in the ballistic regime), there is now a non-zero probability that this particle has touched the wall during the time interval. With the assumption mentioned for the diffusive regime, we can, however, use the distance between a particle and the wall which is a Bessel process. This corresponds to the notion of Brownian (or Bessel) bridges and we can derive an analytical expression for the probability that the distance becomes smaller than the particle radius $R_p$ knowing the initial $a$ and final $b$ distances

$$P_a^b(R_p) = \mathcal{P}\left[\min_{t<s<t+\Delta t} (|\mathbf{X}_p(s)|) \le R_p \,\Big|\, |\mathbf{X}_p(t)| = a, |\mathbf{X}_p(t + \Delta t)| = b\right].$$
(12.9)

From the mathematical theory of diffusive bridges [18–20], we obtain

$$P_a^b(R_p) = \min\left(1\,;\, \frac{\exp\left(\dfrac{R_p(a + b - R_p)}{\mathcal{D}^2\Delta t}\right) - 1}{\exp\left(\dfrac{ab}{\mathcal{D}^2\Delta t}\right) - 1}\right).$$
(12.10)

Apart from these two asymptotic regimes, the mathematical derivation appears intractable (to the best of the authors' knowledge, no analytical formula has been obtained), which means that for particles described by a Langevin-type of model, no detection criterion is available. This is a serious limitation since a wide range of particles belong precisely to that intermediate category.

## 12.3   Limitations of Present Collision Kernels

As mentioned, the processes involved in particle-particle interactions within a flow are the same as the ones between a particle and a wall but for the fact that they involve relative particle motions instead of the dynamics of individual particles. Their treatments depend on the information available.

If we have access to the two-particle PDF, we can deduce the relative motion between a pair of particles as well as their relative kinetic energy upon collision to evaluate if the energy barrier between these particles is overcome or not. Note that this criterion is relevant for colloidal particles. For larger or inertial particles, we can use other criteria to determine whether the collision results in a plastic deformation, or involves whereby kinetic energy is dissipated into heat inside both particles, or a simple elastic rebound. We are, in fact, in the same situation as the one for a particle interacting with a wall surface and we can use detection algorithms (see Fig. 12.5). For particles in the ballistic regime, this leads to cylinders of collisions or similar deterministic methods based on particle free flights between collisions [21]. In the diffusive regime, similar reasoning based on diffusive bridges can be applied [19]. Nevertheless, for particles in the intermediate regime, we are faced with the limitations pointed out at the end of Sect. 12.2.

If we do not have access to the two-particle PDF, detection algorithms cannot be used and we must rely on pre-determined collision kernels or collision frequencies. Given the amount of work on particle-particle interactions, it might be believed that this is a well-addressed problem. It turns out that this is not the case. The classical collision kernel coming from the usual formulation of kinetic theory is given in Eq. (12.4). This is, of course, valid but only insofar as we have particles traveling in straight lines (or deterministic ones), which is the case for atoms or molecules in a void or in an external and conservative force field as in the original Boltzmann picture. For discrete particles in a fluid flow, this remains acceptable but only for

**Fig. 12.5** Different particle regimes used to detect a collision between a pair of particles within a small time interval. Current expressions of the detection algorithms are based on cylinder of collisions for the ballistic regime and on diffusive bridges for the diffusive regime but nothing is available in the intermediate regime

'bullet-like' particles in the ballistic regime for which the effects of the underlying fluid can be ignored during the time interval considered. At first sight, the situation seems also well-established in the diffusive regime, i.e., for very small particles governed by Brownian motion. Indeed, another well-known result is the Brownian collision kernel which, for two classes of particles labeled $i$ and $j$ corresponding to diameters $d_\mathrm{p}^{[i]}$ and $d_\mathrm{p}^{[j]}$ respectively, is expressed by the following formula

$$\beta_\mathrm{coll}^{[i-j]} = \frac{2k_\mathrm{B}\Theta_\mathrm{f}}{3\rho_\mathrm{f}\nu_\mathrm{f}} \frac{\left(d_\mathrm{p}^{[i]} + d_\mathrm{p}^{[j]}\right)^2}{d_\mathrm{p}^{[i]}d_\mathrm{p}^{[j]}} . \tag{12.11}$$

The original derivation, which goes back to the original contribution of Smoluchowski, can be found in several textbooks [22] but rests upon the use of a special boundary condition. The above formula is indeed obtained by considering the interaction of Brownian particles with a 'target particle' at the surface of which a perfect-sink condition is applied. In other words, every time a particle touches the surface of the target one, it is removed and replaced far from it (to ensure a constant influx of particles toward the target one). In more mathematical terms, this killing condition amounts to enforcing that the first-passage time (the first time when a particle hits the target one) is a stopping-time for the process governing the particle trajectory. It is important to realize that, in the traditional approach where the Brownian collision kernel is deduced from the particle concentration profile around the target particle, the resulting expression depends explicitly on the choice of this boundary condition. If we consider, for instance, that particles are reflected at the surface of the target one, it is known from the mathematical studies of such processes (Brownian or Bessel reflected stochastic diffusion processes) that there is a near-one probability that this particle hits again the boundary condition (i.e., the surface of the target particle) an infinite number of times in any small time interval, in which case we would get that the Brownian collision kernel is infinite! It appears therefore that the expression in Eq. (12.11) is still subject to some uncertainty as to its exact validity and that additional studies are needed to reconcile proper mathematical treatment and physical pictures to obtain non-zero but finite predictions of Brownian collision kernels with more support than presently.

For the wide range of particles whose behavior is intermediate between the diffusive and ballistic regimes, the situation is even more puzzling but, quite surprisingly, the formulation of collision kernels accounting for the influence of fluid flows has received scarce attention. Yet, we cannot do without the presence of an underlying fluid flow since particles in that range are correlated with fluid patterns. For example, two particles which seem on a collision path can avoid one another due to lubrication forces between them as they come closer. The presence of some typical fluid flow structures can have also a marked influence on the resulting collision kernels. For instance, particles caught within a convergent zone interact more frequently than they would have done without the fluid bringing them together. In a mirror situation, particles in a divergent zone, such as a vortex, can be dragged

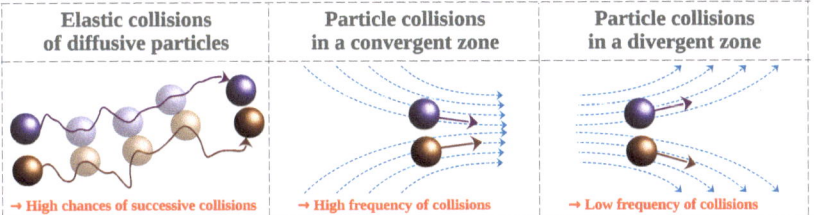

**Fig. 12.6** Representation of fluid-related issues for collision kernels. For Brownian particles, elastic interactions without artificially removing one collision partner is predicted to yield an infinite number of successive collisions. For non-Brownian particles, the influence of the underlying fluid flow can prevent or induce collisions at rates different than what criteria based only on particle velocities predict

away and prevented from interacting even though they seem on a course to do so had we just considered their ballistic regime. These different qualitative situations are pictured in Fig. 12.6.

Note that we are not referring to particle preferential concentration, which is related to local particle accumulation or depletion in some typical fluid structures (cf. Sect. 6.1.3), but to the fluid-induced changes of particle encounter rates. Though preferential concentration modifies the number of interactions in some zones due to an increase or decrease of the local particle concentration, we are here concerned with the probability of such interactions. On the other hand, it must be noted that the collision probability itself can vary significantly depending on the instantaneous flow structures in which particles are trapped and on the intermittent nature of turbulent flows [23, 24], so that even in a statistically homogeneous flow, collisions can take place at different rates in different zones of the flow.

## 12.4   The Molecular Chaos Assumption

For discrete particles in the intermediate regime between the ballistic and diffusive ones, the presence of the fluid flow has far-reaching consequences that go beyond questioning present closures of collision kernels: it challenges the molecular chaos hypothesis. Indeed, the velocities of particles with inertia neither very large nor negligible are correlated with the fluid velocity seen. Since they collide, or interact at distances small compared to hydrodynamical scales, they are at the same location and therefore see the same fluid velocity. Given the turbulent flow non-zero time and length correlations and that these fluid velocities are the driving forces for particle motion, it follows that the velocities of two colliding particles must be correlated. There is no dodging the fact that the hypothesis of 'molecular chaos', through which colliding particle velocities are assumed independent (cf. Eq. (12.5)), is not satisfied.

To the authors' knowledge, this issue does not seem to be well-recognized, apart from some interesting considerations in [22] for agglomeration problems in the context of population balance methods which rely on the same assumption. Yet, the failure of the molecular chaos hypothesis implies that modifications must be brought to current formulations in physical as well as in sample spaces. In physical space, for example, we cannot apply anymore the Bird-like algorithm, at least as it is formulated in Eqs. (12.7)–(12.8) with Eqs. (12.1). Indeed, even if we select a given collision kernel or collision frequency, the molecular chaos assumption is present in the treatment of the $N_p$-PDF as the product of the $N_p$ single-particle ones

$$p^{(N)}(t; \mathbf{V}_p^{[1]}, \ldots, \mathbf{V}_p^{[N_p]}) = \prod_{i=1}^{N_p} p^{(1)}(t; \mathbf{V}_p^{[i]}) , \qquad (12.12)$$

as manifested by the number of particle pairs taken to be $N_p \times N_p$ (note that particles are sampled independently to make up a pair). We need to generate 'true pairs' in the sense that their velocities must be correctly correlated. In sample space, this is clearly translated by the fact that in the Boltzmann equation, cf. Eq. (12.2), we must retain the gain and loss term as in Eq. (12.3) (without resorting to Eq. (12.5)).

What is needed to close the system is to build a model for the two-particle PDF. Recognizing the role played by the velocity of the fluid seen, it is proposed to do so by devising a model for the joint state vectors $\mathbf{Z}_p = (\mathbf{X}_p, \mathbf{U}_p, \mathbf{U}_s)$ of two discrete particles labeled $i$ and $j$, that is for the joint PDF

$$p = p(t; \mathbf{y}_p^{[i]}, \mathbf{V}_p^{[i]}, \mathbf{V}_s^{[i]}, \upsilon_p^{[i]}; \mathbf{y}_p^{[j]}, \mathbf{V}_p^{[j]}, \mathbf{V}_s^{[j]}, \upsilon_p^{[j]}) . \qquad (12.13)$$

Since the two velocities of the fluid seen, $\mathbf{U}_s^{[i]}$ and $\mathbf{U}_s^{[j]}$, are driving the two-particle motion, this indicates that the actual challenge is to develop a two-fluid-particle-PDF model that would therefore contain time and length information. At the moment, the development of such models for general non-homogeneous turbulent flows remains an unexplored domain.

## 12.5 Jump-Diffusion Processes as Modeling Tools

In classical kinetic theory, the driving picture is one in which particles (i.e., atoms and molecules) undergo free flights and collisions. In contrast, the overriding image in turbulent dispersed two-phase flow is one where particles (i.e., small discrete elements) evolve through random motions and collisions. This is well captured by diffusion-jump stochastic processes and most of the physical aspects discussed previously can be cast into this framework provided that we retain the extended particle state vector $\mathbf{Z}_p = (\mathbf{X}_p, \mathbf{U}_p, \mathbf{U}_s)$.

With respect to the issue of particle collisions, the formulation depends on the level of description at which PDF models are developed. If we can write a two-particle PDF model in turbulent flows, we can build (at least, in principle!) a detection algorithm for binary collisions to generate the probability of a collision event and its corresponding characteristics (angle of collision, outcome of the collision, etc.). On the other hand, if we stay at the level of one-particle PDF models, we cannot do without assuming a collision kernel, or a collision frequency, and also conditional statistical information (conditional angle of collision and velocity of the collision partner). Provided that this information is available, the evolution equations for the particle state vector have the form

$$d\mathbf{X}_p = \mathbf{U}_p dt \,, \tag{12.14a}$$

$$d\mathbf{U}_p = \frac{\mathbf{U}_s - \mathbf{U}_p}{\tau_p} \, dt + \sqrt{\frac{2k_B \Theta_f}{m_p \tau_p}} \, d\mathbf{W}_p + \mathbf{C} \, dN_t \,, \tag{12.14b}$$

$$d\mathbf{U}_s = \mathbf{A}_s dt + \mathbf{B}_s \cdot d\mathbf{W}_s \,, \tag{12.14c}$$

where we have retained only the drag force in the particle momentum equation, Eq. (12.14b), but have accounted for Brownian motion as in Eqs. (9.5)–(9.6), with $\mathbf{W}_p$ and $\mathbf{W}_s$ two independent vectorial Wiener processes. The last term on the right-hand side of Eq. (12.14b) involves a non-homogeneous Poisson process $N_t$, which is a scalar process applied to the velocity components of that particle. This Poisson process accounts for the number of collisions along a particle trajectory, since $dN_t$ is non-zero only at the times when a collision takes place, and is defined by its parameter which is equal to the local value of the collision frequency $\lambda_c(\mathbf{X}_p)$. The process $\mathbf{C}$ governs the amplitude of the different jumps of the particle velocity due to collision events and is a vectorial process since it affects the particle velocity components differently. It is seen that we are still following the main lines of the Bird-like algorithm given in Eqs. (12.7)–(12.8) and the transformation rules in Eqs. (12.1) albeit with noticeable changes. The first step is similar and is simulated by the increments of the non-homogeneous Poisson process $N_t$ (hence the choice of its frequency $\lambda_c(\mathbf{X}_p)$, which plays the same role as in Eq. (12.7)). The vectorial process $\mathbf{C}$ represents the second step where we select a collision partner, with the difference compared to the original algorithm that the velocity of this partner particle must be sampled in the conditional velocity PDF (conditioned on the pre-collision velocity $\mathbf{U}_p$ or, more precisely, on the two velocities of the fluid seen by the two colliding particles having close values), before applying the transformation rules in Eqs. (12.1). This is illustrated in Fig. 12.7.

To connect with previous developments, it is illustrative to retain the same collision kernel, i.e. $\beta_{coll}(\mathbf{U}_p, \mathbf{U}_{c,p}, \mathbf{e})$, and to work out the resulting conditional probability, noted $W(t; \mathbf{U}_p^{ac} \,|\, \mathbf{U}_p)$, for a particle velocity to jump from $\mathbf{U}_p$ to $\mathbf{U}_p^{ac}$ at time $t$ due to a collision with a collisional partner. In sample space, this gives the following expression

$$W(t; \mathbf{V}_p^{ac} \,|\, \mathbf{V}_p) \, d\mathbf{V}_p^{ac}$$

$$= \left( \int_{S^2} \lambda_c(\mathbf{X}_p) \left( \frac{\beta_{coll}(\mathbf{V}_p, \mathbf{V}_{c,p}, \widehat{\mathbf{e}})}{\lambda_c(\mathbf{X}_p)} \right) \widetilde{p}(t; \mathbf{V}_{c,p} \,|\, \mathbf{V}_p) \, d\widehat{\mathbf{e}} \right) d\mathbf{V}_{c,p} \,, \tag{12.15}$$

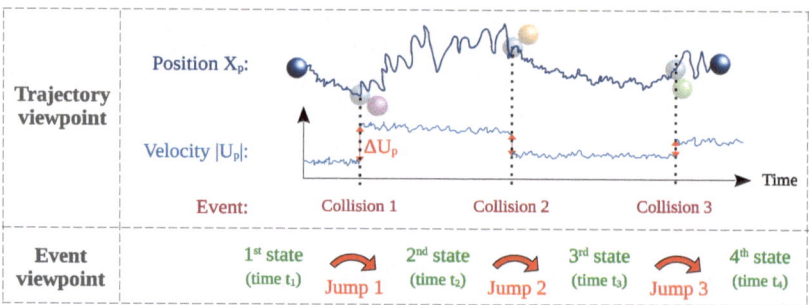

**Fig. 12.7** Sketch of a diffusion-jump process on a particle position and velocity. The velocity jumps at random times due to collisions with other particles and follows a diffusion process in between. The position trajectory remains continuous but exhibits different fluctuations due to velocity jumps

where the term in the integral corresponds to the probability that a jump takes place multiplied by the probability that a specific event governed by the choice of the collision angle $\mathbf{e}$ occurs. Indeed, for two chosen values of the particle velocity before and after the jump, $\mathbf{U}_p$ and $\mathbf{U}_p^{ac}$ respectively, and a chosen value of the angle $\mathbf{e}$, the transformation rules in Eqs. (12.1) show that the velocity of the collision partner $\mathbf{U}_{c,p}$ is fully determined $\mathbf{U}_{c,p} = \mathbf{U}_{c,p}(\mathbf{U}_p^{ac}, \mathbf{U}_p, \mathbf{e})$. Note that the conditional velocity PDF is noted $\widetilde{p}$ to indicate that this is an assumed PDF since we are in the frame of single-particle PDF where this information is not accessible. With this understanding of the (pre)-collisional velocity of the collision partner, the integrand in Eq. (12.15) is a function of $\mathbf{e}$ and the integral is thus over all possible values of $\mathbf{e}$ on the unit sphere $\mathcal{S}^2$. When integrating $W(t; \mathbf{V}_p^{ac} \mid \mathbf{V}_p)$ over all possible velocity jumps, we obtain

$$\int_{\mathbb{R}^3} W(\mathbf{V}_p^{ac} \mid \mathbf{V}_p)\, d\mathbf{V}_p^{ac} = \int_{\mathbb{R}^3} \int_{\mathcal{S}^2} \beta_{coll}(\mathbf{V}_p, \mathbf{V}_{c,p}, \widehat{\mathbf{e}})\, \widetilde{p}(t; \mathbf{V}_{c,p} \mid \mathbf{V}_p)\, d\widehat{\mathbf{e}}\, d\mathbf{V}_{c,p}\,.$$

(12.16)

If we consider, for the sake of simplicity, the loss term in the Chapman-Kolmogorov equation due to all possible jumps, we have simply to multiply by the probability of the particle velocity $p(t; \mathbf{V}_p)$

$$I_{loss}(t; \mathbf{V}_p) = -\int_{\mathbb{R}^3} W(\mathbf{V}_p^{ac} \mid \mathbf{V}_p)\, p(t; \mathbf{V}_p)\, d\mathbf{V}_p^{ac}\,,$$

(12.17)

which gives

$$I_{loss}(t; \mathbf{V}_p) = \int_{\mathbb{R}^3} \int_{\mathcal{S}^2} \beta_{coll}(\mathbf{V}_p, \mathbf{V}_{c,p}, \widehat{\mathbf{e}})\, \widetilde{p}(t; \mathbf{V}_{c,p} \mid \mathbf{V}_p)\, p(t; \mathbf{V}_p)\, d\widehat{\mathbf{e}}\, d\mathbf{V}_{c,p}\,.$$

(12.18)

It is straightforward to show that similar results are obtained with the gain term. Should the assumed conditional PDF $\tilde{p}$ be equal to the true conditional one, we would retrieve $p(t; \mathbf{V}_p, \mathbf{V}_{c,p})$ in the previous expressions, which would indicate that we have the Fokker-Planck-Boltzmann equation, cf. Eq. (12.6), with the correct form of the collision terms, as in Eq. (12.3). Actually, such a statement would be overreaching but the expressions of the collision terms as in Eq. (12.18) show that we are nevertheless consistent with a 'linearized version' of the Boltzmann collision integral. In the frame of single-particle PDF models, it seems difficult to achieve better results. Nevertheless, these formulations confirm that diffusion-jump stochastic processes are useful tools to capture the physics of turbulent dispersed two-phase flows with collisions.

## 12.6 Numerical Illustrations

To illustrate the issues related to particle collisions in turbulent flows, we briefly describe three simple test cases: (a) the first one demonstrates how fine-scale simulations can be applied to extract detailed information on aggregates (also called flocs) in homogeneous isotropic turbulence; (b) the second one displays how Monte-Carlo approaches can be used to simulate the aggregation of nanoparticles in suspension; (c) the third one shows what can be learned from models based on diffusive bridges when Brownian effects dominates.

### 12.6.1 Flocculation of Cohesive Sediments in Homogeneous Isotropic Turbulence

The case of particles suspended in homogeneous isotropic turbulence (HIT) is an interesting situation to consider since all fluid statistics are uniform. Yet, behind this apparent simplicity, the agglomeration of discrete particles can display intriguing features, like an abrupt growth of aggregates [24]. This is partly related to the intense turbulent fluctuations that occur in HIT (which can speed-up the formation of large flocs due to multiple collision between different objects as well as lead to the fragmentation of already formed flocs) and to the preferential concentration of inertial particles (which do not remain uniformly distributed, as revealed in Fig. 6.3).

#### 12.6.1.1 A Word on the Simulation Method
The flocculation (or agglomeration) of large inertial sediments in HIT can be explored using fully-resolved numerical simulations. For instance, one can rely on the so-called DNS-DEM method [25] (DEM meaning Discrete Element Method). This consists in resolving explicitly all the turbulent scales in a flow using the

Navier-Stokes equation for an incompressible and constant-property Newtonian fluid [17]:

$$\frac{\partial U_{f,k}}{\partial x_k} = 0 \,, \tag{12.19a}$$

$$\frac{\partial U_{f,i}}{\partial t} + U_{f,k}\frac{\partial U_{f,i}}{\partial x_k} = -\frac{1}{\rho_f}\frac{\partial P_f}{\partial x_i} + \nu_f\frac{\partial^2 U_{f,i}}{\partial x_k \partial x_k} + (\text{forcing}) \,. \tag{12.19b}$$

where the forcing term represents an energy contribution needed to compensate the turbulent kinetic energy dissipation rate and maintain a steady state for the isotropic turbulent flow. Meanwhile, the dynamics of each individual particle is computed by solving the equations for the particle position $\mathbf{X}_p$ and velocity $\mathbf{U}_p$:

$$\frac{d\mathbf{X}_p}{dt} = \mathbf{U}_p \,, \tag{12.20a}$$

$$m_p\frac{d\mathbf{U}_p}{dt} = \mathbf{F}_{f\to p} + \mathbf{F}_{p\to p} \,. \tag{12.20b}$$

where the drag, pressure-gradient and buoyancy terms are included in the hydro-dynamic forces $\mathbf{F}_{p\to p}$, while short-range double-layer interactions and adhesive contributions are included in the inter-particle forces $\mathbf{F}_{p\to p}$ [25].

The simulations were performed considering particles with a size of 20 $\mu$m, which are representative of clay-based flocculi. The shear rate for the homogeneous isotropic turbulence is taken equal to 350 s$^{-1}$. The particle stickiness was varied to explore how cohesive forces between particles affect the evolution of floc sizes.

### 12.6.1.2  Some Comments on the Numerical Results

Monitoring the evolution of floc sizes is key to understand how fast flocculation takes place. For that purpose, different parameters can be used to evaluate the size of an aggregate: the amount of primary particles forming each individual floc $n_f$, the floc size (usually along its major axis length) or the fractal dimension (which relates the amount of particles to the floc size through a power law). For the sake of simplicity, we focus here on the amount of particles forming each floc and on the floc size.

Numerical results show that larger flocs are formed when cohesive forces between particles are increased, as revealed in Fig. 12.8 by comparing panels (a–c) for mildly sticky objects to panels (d–e) for highly sticky objects. In addition, the combined visualization of the particle positions and the nearby turbulent coherent structures (shown in panels a and d) allows to identify the regions where aggregation/breakup are more frequent. This analysis confirms that intense coherent structures facilitate the growth of aggregates (by more frequent collisions) and inducing more frequent breakups (due to the higher shear exerted on large aggregates).

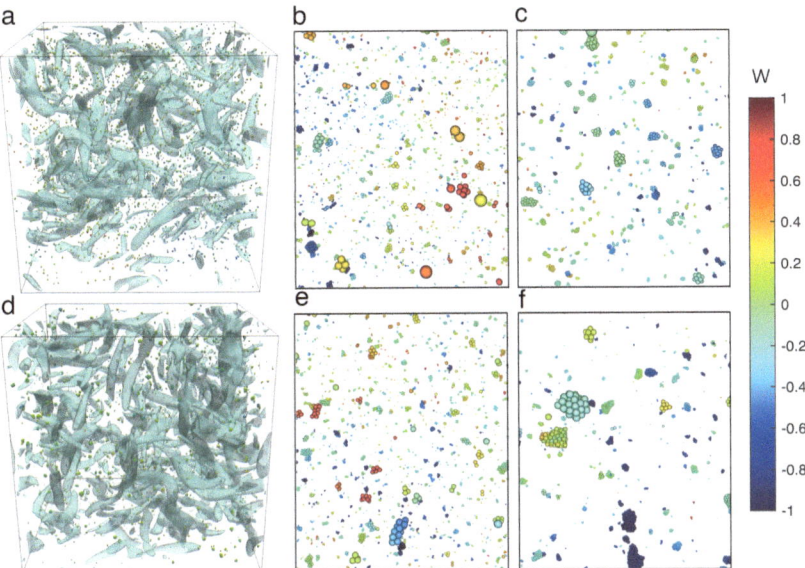

**Fig. 12.8** Snapshot of simulations for mildly sticky particles (panels **a–c**) and highly sticky particles (panels **d–f**). (**a** and **d**) Turbulent coherent structures (revealed using the iso-surfaces of vorticity) and the flocs, (**b** and **e**) Zoomed in view of the flocs at early stage, and (**c** and **f**) Zoomed in view of the flocs at late stage. Particles are colored by the vertical component of the particle velocity $W$. Reprinted with permission from [25]. © 2022 American Geophysical Union. All rights reserved

Figure 12.9 shows the time evolution of the floc size distribution for both cases of mildly and highly sticky objects. It appears that a steady-state is reached after some time (the time to reach this steady-state is shorter for more sticky objects). This steady-state corresponds to the stage at which aggregation is balanced by breakup. In addition, the floc size distribution depends on the ratio between the turbulent shear and the floc cohesive strength.

These results illustrate what we can learn from such simulations. Since they provide information on the collision and breakup rates, they can be used to estimate such kernels in complex situations that are difficult to investigate experimentally.

## 12.6.2  A Monte Carlo Method for Nanoparticle Aggregation

Going back to statistical models, Monte-Carlo approaches can be applied to obtain information on the morphology of aggregates. Instead of solving all the scales of the fluid and particle dynamics as in the DNS-DEM method described in Sect. 12.6.1, the idea here is to simplify the dynamics of particles using random motion. Such approaches prove very useful to estimate the morphology of aggregates formed by nanoparticles undergoing primarily Brownian motion, as in soot formation [26, 27].

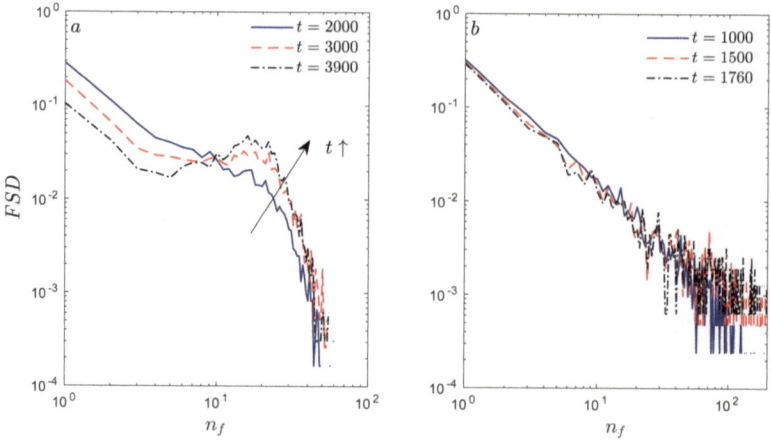

**Fig. 12.9** Time evolution of the floc size distribution (FSD) for the case of mildly sticky particles (**a**) and highly sticky particles (**b**). The floc size is represented by the number of primary particles ($n_f$) in the floc. Reprinted with permission from [25]. © 2022 American Geophysical Union. All rights reserved

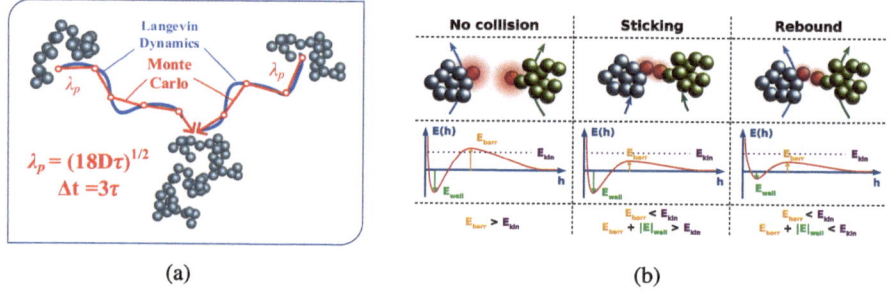

**Fig. 12.10** Illustrations of some features occurring in molecular clouds in space. (**a**) Illustration of the algorithm of the Monte-Carlo Aggregate Code (MCAC). Reprinted with permission from [28]. © 2020 Elsevier Inc. All rights reserved. (**b**) Illustration of how the particle-particle interactions are accounted for in MCAC for collision outcomes. Reprinted with permission from [26]. © 2021 Elsevier Inc. All rights reserved

### 12.6.2.1  A Word on the Simulation Method

The principle of the Monte-Carlo Aggregation Code (MCAC) is to mimic the Brownian motion of nanoparticles. To that effect, the Monte-Carlo approach described in [28] consists in iteratively selecting a particle at a given time (with some probability) and then moving this particle in a random orientation for a given persistent distance unless a collision with neighbors is detected. When a collision is detected (using simple geometric criteria), the two particles stick together exactly at the first point of contact as shown in Fig. 12.10a. Particles within the same aggregate are then considered to be moving concomitantly. This process is repeated sequentially until a user-defined criteria is met (e.g. a given elapsed time).

As depicted in Fig. 12.10b, the MCAC model has been recently improved by differentiating between three collision outcomes [26]: the two colliding objects can (a) repel each other, (b) stick to each other or (c) rebound away from each other after impact. The collision outcome actually depends on the ratio between the relative kinetic energy upon collision and the inter-particle potential energy. The latter is computed assuming a Lennard-Jones type of potential (that includes an attractive contribution related to van der Waals forces and an short-range repulsive term related to the non-overlap of molecular clouds) together with electrostatic interactions between primary particles.

The main interest of this type of Monte-Carlo method is that each collision between particles can be detected and treated. This means that one can extract information on the collision rates and on the morphology of aggregates formed by particles undergoing purely diffusive motion, without resorting to more complete Langevin models for particle dynamics (such as the ones described in Sect. 6.3.1).

### 12.6.2.2  Some Comments on the Numerical Results

The numerical results were obtained with particles having an initial random distribution in a cubic box without overlaps and by applying periodic boundary conditions.

The evolution of the aggregate size over time can be tracked numerically by computing the volume-equivalent radius, which corresponds to the radius of a sphere with a volume equivalent to that occupied by the aggregate. The results obtained for the geometric mean of this volume-equivalent radius are displayed in Fig. 12.11. Several conclusions can be drawn. First, the aggregate size increases monotonically over time. Second, the increase in aggregate size is faster for young (nascent) soot than for mature soot. This can be related to the lower bulk density of young soot. Third, it appears that electrostatic repulsion has no effect on the size of the aggregates obtained. This observation implies that, for soot nanoparticles, it is possible to neglect the effect of charges on the agglomeration process. In general, one usually decomposes the agglomeration kernel as the product of a collision

**Fig. 12.11** Time evolution of the geometric mean volume-equivalent radius obtained from a MCAC simulation of nascent and mature soot. Reprinted with permission from [26]. © 2021 Elsevier Inc. All rights reserved

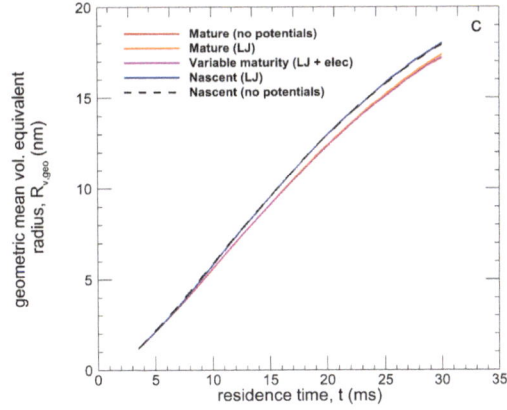

rate $\beta_{coll}$ (measuring how frequent collisions are) and a collision efficiency $\beta_{eff}$ (measuring how frequent collisions lead to agglomeration), i.e., $\beta_{agglo} = \beta_{coll}\,\beta_{eff}$. But in the present case, these results indicate that the Brownian collision kernel (such as the one given by Eq.(12.11)) does not need to be modified to account for the role of electrostatic charges in this case, i.e. $\beta_{agglo} = \beta_{coll}$.

### 12.6.3  Collision Probabilities of Particles Undergoing Brownian Motion

As mentioned before in Sect. 12.2, one can rely on the notion of Brownian (or Bessel) bridges to derive an analytical expression for the collision probability between particles undergoing purely diffusive motion. This case can be representative of nanoparticles in a fluid at rest, where the fluid velocity is negligible while particle dynamics is governed solely by Brownian effects.

#### 12.6.3.1  A Word on the Simulation Method
When Brownian effects dominate, the particle dynamics simplifies to the over-damped Langevin equation

$$d\mathbf{X}_p \simeq \sqrt{2\mathcal{D}}\,d\mathbf{W} \tag{12.21}$$

where $\mathcal{D}$ is the Stokes-Einstein diffusion coefficient. By solving Eq. (12.21) for each particle in suspension, one has access to the distance between each pair of particles at the beginning and end of each time step. Hence, relying on the notion of Brownian (or Bessel) bridges, the collision probability between a pair of particles during a given time $P_a^b(R_p)$ is given by Eq. (12.10), with $a$ (resp. $b$) the initial (resp. final) distance between the two particles [20]. Once this collision probability is evaluated, the collision between the pair of particles can be randomly generated using the following Monte Carlo method: a collision occurs only if the value of a random number—sampled in a uniform distribution in [0, 1]—is greater than $P_a^b(R_p)$.

The notion of Brownian bridges not only provides information on the probability of collision but also on the time of collision [19, 20]. In fact, an analytical solution gives the distribution of the first hit time, i.e. the time at which the first collision occurs. As shown in Fig. 12.12, once a first collision time is generated, the position of both particles at the collision time can be generated accordingly. The dynamics in the time remaining after a collision event then depends on the collision outcome: (1) if an elastic collision occurs, the motion of only one partner is generated again (the position of the other partner being fully determined from the center of mass of the pair of particles, which remains unaffected by the collision event); (2) if agglomeration occurs, the dynamics of the aggregate is given by Eq. (12.21) and the size of the aggregate is updated.

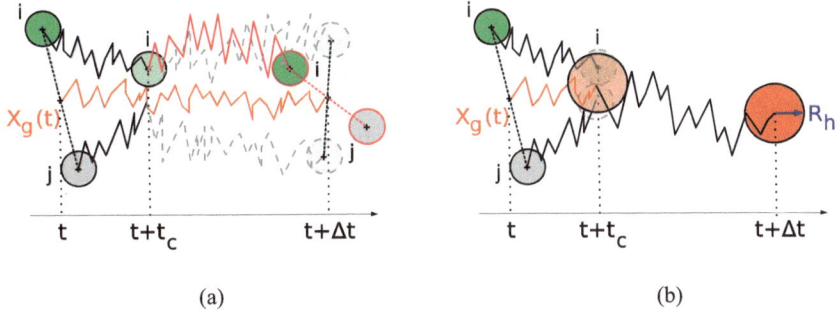

**Fig. 12.12** Sketch showing how a stochastic model can be used to detect and treat the collision between a pair of particles in the diffusive regime. Reprinted with permission from [29]. © 2013 American Chemical Society. All rights reserved. (**a**) Case of an elastic collision. (**b**) Case of a sticky collision

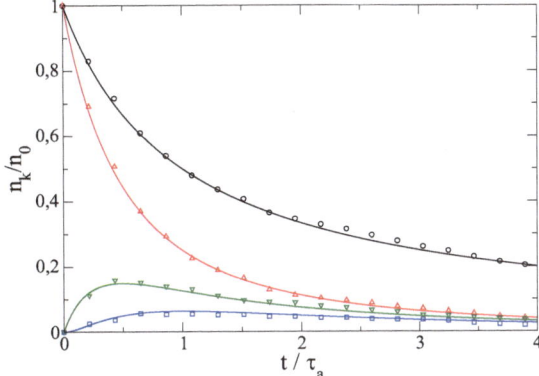

**Fig. 12.13** Evolution of the relative number of primary particle $n_1$ (red open triangle), doublets $n_2$ (green inverted triangle), triplets $n_3$ (blue open square) and total number of objects $n_T$ (black open circle) as a function of the dimensionless time $t/\tau_a$: comparison between theoretical values (lines) and simulations with the probabilistic model (symbols). Reprinted with permission from [29]. © 2013 American Chemical Society. All rights reserved

### 12.6.3.2 Some Comments on the Numerical Results

The results described in the following have been obtained for inertialess particles suspended in air at room temperature and in the case of sticky collisions, i.e. assuming that each collision event leads to the formation of an aggregate.

Figure 12.13 shows the evolution of the number of primary particles, the number of doublets (aggregates formed by two primary particles), the number of triplets (aggregates formed by three particles) and the total number of particles/aggregates in the domain. These concentrations are plotted as a function of time scaled by the half-life of the agglomeration process $\tau_a = 3\mu_f/(4k_B\Theta_f n_0)$ (with $\mu_f$ the fluid dynamic viscosity, $\Theta_f$ its temperature, $k_B$ the Boltzmann constant and $n_0$ the initial number of particles). This half-life of agglomeration corresponds to the time after which the total number of particles equals half of the initial number of primary objects $n_0$. The results obtained with the probabilistic model are in very good agreement with the theoretical values, even when starting from a relatively limited number of particles initially in the domain (in this case $n_0 = 1000$). These results confirm the interest of relying on such probabilistic approaches in the context of particle collision/agglomeration/breakup. Nevertheless, as mentioned before in Sect. 12.2, no detection criterion is available in the more generic case with a Langevin-type model for the particle dynamics.

## 12.6.4 Current Issues and Open Questions

These three examples serve to outline various formulations for particle-particle interactions in turbulent flows. They serve also to illustrate the current limitations and, in a way, the wall of complexity we are facing in terms of statistical approaches. Indeed, we seem to have only two well-developed possibilities which are at two opposite ends of the range of complexity. On the one hand, we can track and calculate every degrees of freedom, as in the first example in Sect. 12.6.1. Note that, in that case, very detailed expressions of the inter-particles forces $\mathbf{F}_{p \to p}$ can be included. On the other hand, once we consider reduced statistical descriptions, we are limited to Brownian dynamics, as in the two examples in Sects. 12.6.2 and 12.6.3. For these applications, elaborate formulations are being proposed, showing the inventiveness stimulated by these problems. However, in these two examples, we have reduced the problem to only Brownian particles. As mentioned in Sect. 12.6.3, even for Brownian noise, we are not able, at the moment, to work out similar formulas for the probability of the collisional events with a Langevin model, as with Eqs. (9.5) and Eq. (9.10) in Sect. 9.3. As soon as we wish to go beyond the over-damped Langevin limit, developments seem intractable. And this is already true for Brownian noise which, as discussed in Sect. 9.2, is actually a much easier situation compared to turbulent noise since we can apply new ideas for one-particle or $N$-particle formulations indifferently contrary to present one-particle PDF models for turbulent flows.

Clearly, we are in dire need of new ideas, if we are to hope for breakthroughs.

## References

1. F. Reif, *Fundamentals of Statistical and Thermal Physics* (Waveland Press, 2009). https://waveland.com/browse.php?t=520
2. R.L. Liboff, *Kinetic Theory: Classical, Quantum, and Relativistic Descriptions* (Springer Science & Business Media, 2003). https://doi.org/10.1007/b97467
3. R. Balescu, *Statistical Dynamics: Matter Out of Equilibrium* (Published by Imperial College Press and Distributed by World Scientific Publishing Co., 1997). https://doi.org/10.1142/p036
4. J. Keizer, *Statistical Thermodynamics of Nonequilibrium Processes* (Springer Science & Business Media, 2012). https://doi.org/10.1007/978-1-4612-1054-2

5. W. Wagner, in *Monte Carlo and Quasi-Monte Carlo Methods 2002: Proceedings of a Conference held at the National University of Singapore, Republic of Singapore, November 25–28, 2002* (Springer, 2004), pp. 129–153. https://doi.org/10.1007/978-3-642-18743-8_7
6. A. Eibeck, W. Wagner, Ann. Appl. Probab. **13**(3), 845 (2003). https://doi.org/10.1214/aoap/1060202829
7. N. Fournier, S. Mischler, J. Math. Pures Appl. **84**(9), 1173 (2005). https://doi.org/10.1016/j.matpur.2005.04.003
8. J.P. Minier, Prog. Energy Combust. Sci. **50**, 1 (2015). https://doi.org/10.1016/j.pecs.2015.02.003
9. W. Wagner, Phys. Fluids **23**(3) (2011). https://doi.org/10.1063/1.3558866
10. G. Bird, Phys. Fluids **6**, 1518 (1963). https://doi.org/10.1063/1.1710976
11. G.A. Bird, *Molecular Gas Dynamics and the Direct Simulation of Gas Flows* (Oxford University Press, 1994). https://doi.org/10.1093/oso/9780198561958.001.0001
12. J.N. Israelachvili, *Intermolecular and Surface Forces* (Academic Press, 2011). https://doi.org/10.1016/C2011-0-05119-0
13. C. Henry, J.P. Minier, G. Lefèvre, Adv. Colloid Interface Sci. **185**, 34 (2012). https://doi.org/10.1016/j.cis.2012.10.001
14. C. Gardiner, *Stochastic Methods*, 4th edn. (Springer, Berlin, 2009). https://link.springer.com/book/9783540707127
15. H.C. Öttinger, *Stochastic Processes in Polymeric Fluids: Tools and Examples for Developing Simulation Algorithms* (Springer Science & Business Media, 2012). https://doi.org/10.1007/978-3-642-58290-5
16. H. Risken, *Fokker-Planck Equation* (Springer, 1996). https://doi.org/10.1007/978-3-642-61544-3
17. S. Pope, *Turbulent Flows* (Cambridge University Press, 2000). https://doi.org/10.1017/CBO9780511840531
18. A.N. Borodin, P. Salminen, *Handbook of Brownian Motion-Facts and Formulae* (Springer Science & Business Media, 2015). https://doi.org/10.1007/978-3-0348-8163-0
19. C. Henry, J.P. Minier, M. Mohaupt, C. Profeta, J. Pozorski, A. Tanière, Int. J. Multiphase Flow **61**, 94 (2014). https://doi.org/10.1016/j.ijmultiphaseflow.2014.01.007
20. M. Mohaupt, J.P. Minier, A. Tanière, Int. J. Multiphase Flow **37**(7), 746 (2011). https://doi.org/10.1016/j.ijmultiphaseflow.2011.02.003
21. M. Chen, K. Kontomaris, J. McLaughlin, Int. J. Multiphase Flow **24**(7), 1079 (1999). https://doi.org/10.1016/S0301-9322(98)00007-X
22. D. Ramkrishna, *Population Balances: Theory and Applications to Particulate Systems in Engineering* (Academic Press, 2000). https://doi.org/10.1016/B978-0-12-576970-9.X5000-0
23. J. Bec, H. Homann, S.S. Ray, Phys. Rev. Lett. **112**(18), 184501 (2014). https://doi.org/10.1103/PhysRevLett.112.184501
24. J. Bec, S.S. Ray, E.W. Saw, H. Homann, Phys. Rev. E **93**(3), 031102 (2016). https://doi.org/10.1103/PhysRevE.93.031102
25. M. Yu, X. Yu, S. Balachandar, Water Resources Res. **58**(6), e2021WR030402 (2022). https://doi.org/10.1029/2021WR030402
26. J. Morán, C. Henry, A. Poux, J. Yon, Carbon **182**, 837 (2021). https://doi.org/10.1016/j.carbon.2021.06.085
27. J. Morán, J. Yon, A. Poux, F. Corbin, F.X. Ouf, A. Siméon, J. Colloid Interface Sci. **575**, 274 (2020). https://doi.org/10.1016/j.jcis.2020.04.085
28. J. Morán, J. Yon, A. Poux, J. Colloid Interface Sci. **569**, 184 (2020). https://doi.org/10.1016/j.jcis.2020.02.039
29. C. Henry, J.P. Minier, J. Pozorski, G. Lefèvre, Langmuir **29**(45), 13694 (2013). https://doi.org/10.1021/la403615w

# Perspectives on the Roads Ahead

# 13

**Abstract**

In this conclusion, we summarize the key points as well as the main results obtained in the preceding chapters. We also suggest directions for future developments and revisit the nature of the invitation which underlies this book.

**Chapter Content** After summarizing results in Sect. 13.1 and proposing ideas in Sect. 13.2, the underlying purpose is outlined in Sect. 13.3.

## 13.1 Going Back on the Developments

The purpose of this book was to bring out the issues in statistical physics that appear when modeling discrete particle dynamics in non-fully resolved turbulent flows. As analyzed for the fluid-particle and particle-particle interactions, one key point is that the non-zero time and space correlations of turbulent flows challenge the framework of kinetic theory. To overcome this shortcoming and respect the Markovian approach, models based on sets of particle-attached variables that include fluid flow characteristics sampled along particle trajectories, such as the velocity of the fluid seen, need to be developed. A second key point is that we have mostly followed the trajectory point of view, as a modeling tool, to revisit how to account for noise and to present developments on fast-variable elimination techniques for the derivation of the diffusive limit.

There is a special benefit in adopting the trajectory standpoint. Traditionally, PDFs are used to derive a limited number of mean fields (the hydrodynamical level) which are the quantities to be solved. This is not the case for turbulent dispersed two-phase flows since the set of mean-field equations corresponding to the first and second-order moments extracted from the PDF models involves a large number of coupled partial differential equations whose solution becomes quickly

© The Author(s) 2025

J.-P. Minier et al., *Understanding Turbulent Systems*,

Lecture Notes in Physics 1039, https://doi.org/10.1007/978-3-031-84466-9_13

intractable [1,2]. This implies that we are interested not only in modeling but also in simulating the PDFs. To that end, formulating PDF models in terms of trajectories is useful since this leads directly to dynamical Monte Carlo methods [3,4] and, thus, to practical numerical implementations. In the course of the discussion, we have chosen to focus the analyses more on the statistical content of various modeling proposals than on their predictive capacities. The latter is an important aspect but a different one from the actual objective of the present work, even though it is already useful to limit the class of numerical formulations to be considered to those that are theoretically consistent. Nevertheless, numerical models raise specific issues that require a monograph of their own to be properly addressed.

## 13.2   The Possible Roads Ahead

What are the possible modeling roads ahead? Using the feedback provided by one-particle PDF models, a number of directions can be suggested:

(1)  While the development of particle transport models is more advanced, setting up a new framework for particle collisions which takes into account fluid-induced correlations between particle velocities appears as an important issue;

(2)  Some of the present difficulties to obtain closed expressions would be over-come by two-particle PDF models or, at least, formulations containing spatial information around discrete particle positions. This is a direction where new ideas would have a strong impact on the ability of PDF models to capture more properties of turbulent dispersed two-phase flows with less approximation;

(3)  A related theme of research is the construction of unified PDF models that include both the fluid and the discrete-particle phases into one stochastic system (containing two sets of stochastic particles, one for the fluid-phase description and one for the dispersed-phase one). This was first developed in [1] but these propositions need to be taken up and tested. At the moment, PDF models exist separately for the fluid phase and for the particle phase but rarely for the complete fluid-and-particle system;

(4)  Finally, though we are operating 'above the hydrodynamical level of description', the expression of such unified two-phase PDF models in terms of stochastic particles opens the door to connections with the particle-based stochastic methods used 'below the hydrodynamical level of description', i.e. MD, dissipative particle dynamics (DPD), smoothed dissipative particle dynamics (SDPD) [5]. Devising coarse-graining methods to drive us from one scale, or one level of description, to a higher one (or a reduced level of description) is indeed challenging even if the overriding image is one when we 'lump' particles together to form a new particle for the next level. There is, however, an additional difficulty in that we need to go from a N-particle PDF model in which particle-particle interactions are directly accounted for (as in MD, DPD, SDPD) to a one-particle PDF model where particle-particle interactions are handled through mean-field approaches. Yet, formulations in

terms of particle stochastic systems allows such possibilities to be considered and such developments to be attempted. In that sense, it is hoped that turbulent dispersed two-phase flow modeling will find its place among all these well-recognized domains of physics.

## 13.3 What Was the Invitation About?

At the beginning of this book, we referred to a multi-fold invitation. An invitation to, first, discover the richness of the subject and, second, to follow in the footsteps of the developments discussed from Chaps. 4 to 12. Supporting these steps lies a deeper invitation, which is to rekindle the waning art of modeling. Present-time research seems to be characterized by an emphasis put on high-performance computing, big data, artificial intelligence, etc., and to rely heavily on direct numerical simulations made possible by reducing our ambitions to much-simplified flow situations. These are indeed useful tools to help us sort out intricate issues, navigate through large result files, or to access to variables difficult to measure by real experimental systems such as the dissipation of turbulent kinetic energy in the immediate vicinity of a solid wall. They nevertheless belong to the category of tools whose interest is to provide means to an end which is to live up to the challenges raised by Nature. Acting at a human scale, turbulence and its range of interesting phenomena, such as turbulent dispersed two-phase flows, has the invaluable quality to remind us of our limited grasp on some of Nature's phenomena. In that respect, it is believed that devising models remains the best approach if we are to improve this understanding. And that this is still a human task, a pen and paper endeavor guided by ideas. The example of particle collisions in turbulent flows suggests that we may not need complicated and big-data calculations. We need ideas. It is also believed that keeping the modeling path thriving is beneficial. Most of time, our models do not work. Many times, they even fail miserably. Yet, by doing so, they prod us to delve further and, who knows, we could end up learning something about natural systems and, perhaps, also about ourselves. After all, it may very well be that there is no risk in walking along that road.

## References

1. J.P. Minier, E. Peirano, Phys. Rep. **352**(1–3), 1 (2001). https://doi.org/10.1016/S0370-1573(01)00011-4
2. J.P. Minier, S. Chibbaro, S.B. Pope, Phys. Fluids **26**(11), 113303 (2014). https://doi.org/10.1063/1.4901315
3. S.B. Pope, Prog. Energy Combust. Sci. **11**(2), 119 (1985). https://doi.org/10.1016/0360-1285(85)90002-4
4. S.B. Pope, Annu. Rev. Fluid Mech. **26**(1), 23 (1994). https://doi.org/10.1146/annurev.fl.26.010194.000323
5. H.J.C. Berendsen, *Simulating the Physical World: Hierarchical Modeling from Quantum Mechanics to Fluid Dynamics* (Cambridge University Press, 2007). https://doi.org/10.1017/CBO9780511815348